# Ecophysiology of Desert Reptiles

**Plate I** (Frontispiece) Male *Amphibolurus nuchalis* at Shark Bay exhibiting the 'elevated' high-temperature avoidance behaviour pattern which occurs at air temperatures in excess of 35°C. (Air temperature = 41.3°C.) (See Chapter 2.)

# Ecophysiology of Desert Reptiles

S. D. BRADSHAW

Department of Zoology
University of Western Australia
Perth, Western Australia

ACADEMIC PRESS
(Harcourt Brace Jovanovich, Publishers)
Sydney Orlando San Diego New York Austin
London Montreal Tokyo Toronto

*Pour Jill, qui n'a vu que le début . . .*

ACADEMIC PRESS AUSTRALIA
Centrecourt, 25–27 Paul Street North
North Ryde, N.S.W. 2113

*United States Edition published by*
ACADEMIC PRESS INC.
Orlando, Florida 32887

*United Kingdom Edition published by*
ACADEMIC PRESS, INC. (LONDON) LTD.
24/28 Oval Road, London NW1 7DX



Printed in Australia

*National Library of Australia Cataloguing-in-Publication Data*

---

Bradshaw, S. D. (Sidney Donald), date.
Ecophysiology of desert reptiles.

Bibliography.
Includes index.
ISBN 0 12 124575 6.
ISBN 0 12 124576 4 (pbk).

1. Reptiles — Physiology. 2. Reptiles — Ecology.
3. Desert fauna — Physiology. I. Title.

597.9'041

---

*Library of Congress Catalog Card Number:* 86–47558

# Contents

# Préface

C'est un grand honneur que m'a fait le Professeur Don Bradshaw en me demandant d'écrire la préface d'un ouvrage aussi remarquable que celui-ci et je le dois certainement beaucoup plus à l'amitié qui nous unit depuis tant d'années—amitié née à l'ombre d'une démarche physiologique de même philosophie—qu'à ma connaissance très restreinte en herpétologie. Peut-être, en effet, l'intérêt que nous portons depuis longtemps l'un et l'autre aux tentatives d'explication, par de nouvelles disciplines physiologiques que nous avons vu s'élaborer, soit de l'existence de la vie dans les milieux apparemment hostiles, soit des capacités de passage à un stade donné du cycle biologique d'un milieu donné dans un autre très différent (certaines migrations) a-t-il déterminé son choix?

Quoi qu'il en soit, il m'est agréable de reconnaître la valeur exceptionnelle de cet ouvrage due, sans doute, pour une large part, au fait que le Professeur Bradshaw fut dès sa jeunesse un observateur sagace et critique, dans son pays, d'une nature comblée de richesses naturelles originales encore imparfaitement connues; il constata que la vie ne s'y déroule pas toujours, comme l'enseignaient certains ouvrages venus d'autres continents et il fut stimulé par l'idée que l'étude de certains écosystèmes australiens allait lui permettre de faire progresser sensiblement l'état de nos connaissances.

Les Reptiles et, spécialement les Squamates, l'ont séduit de prime abord sans doute par leur merveilleux passé, qui n'est probablement pas sans rapport avec certains caractères du présent, mais aussi par les multiples caractères actuels—qu'il contribua d'ailleurs à mettre largement en évidence, ou à mieux faire connaître—d'animaux qui apparaissent comme préparés à ce climat hostile qu'est le désert (taux métabolique très bas, types d'excrétion économisant l'eau, résistance exceptionnelle à des perturbations métaboliques extraordinaires, hétérothermie, héliothermie, divers comportements salvateurs dans des situations extrêmes, etc . . .). Sa thèse, réalisée dans un laps de temps exceptionnel (deux années) et soutenue il y a exactement 20 ans, soulignait déjà de facon éclatante certaines de ces particularités communes à la majorité des Reptiles, désertiques ou non. L'auteur ne manque pas cependant de souligner certains types d'organes qui peuvent apparaître comme des adaptations, non seulement par leur présence mais par l'intensité de leur fonctionnement.

Toutefois, Bradshaw eut rapidement pleine conscience qu'il n'y a pas un désert, mais des déserts et qu'il importait d'élargir le champ de ses observations et expérimentations par des recherches sur les Reptiles de deux autres continents comprenant des zones désertiques: les Etats-Unis, d'une part, l'Afrique avec le rude Sahara d'autre part. Comme base expérimentale de recherche dans ce dernier écosystème, l'auteur choisit l'Europe et il nous apporte donc, dans ce livre, toutes les richesses que peut rassembler sur quatre continents, un naturaliste de cette clairvoyance, un physiologiste de cette habileté et de cette rigueur expérimentale et un philosophe de cette qualité avec qui les échanges de pensée sont toujours féconds.

En effet, indépendamment de sa valeur exceptionnelle pour la connaissance de la physiologie des Reptiles au sein des écosystèmes désertiques et en dehors de ceux-ci, cet ouvrage présente un intérêt exceptionnel pour les recherches écophysiologiques en général au sein desquelles l'auteur distingue l'"*environmental physiology*"*, *la physiologie écologique* et *l'écophysiologie*: la première considérant essentiellement la physiologie des organismes en fonction des facteurs physiques ou chimiques des biotopes naturels ou expérimentaux, la seconde, en fonction de tous les facteurs physiques, chimiques, biologiques des écosystèmes naturels ou artificiels, l'écophysiologie enfin qui, étendant les investigations précédentes aux populations souligne leur polymorphisme génétique, les rapports de celui-ci avec les phénomènes dits d'adaptation, ceux-ci discutés de facon si originale et judicieuse par l'auteur.

C'est un ouvrage qui allie excellemment l'observation la plus pertinente dans la nature, la récolte de données acquises sur le terrain par des méthodes

---

* Je conserve ici l'expression anglaise, car la traduction proposée par mes collègues: "physiologie de l'environnement" ne me paraît pas correcte puisqu'il s'agit évidemment de physiologie des organismes vivants, dans ses rapports avec l'environnement.

adaptées à ce type d'investigation très spéciale, l'expérimentation rigoureuse au laboratoire et l'interprétation sagace, souvent très originale, des résultats apportés par l'une et l'autre. Et, Bradshaw va bien au-delà. Suivant les conseils de Claude Bernard, il monte au sommet de l'arbre le plus élevé de la forêt dont il exploite les richesses et il nous apporte, après avoir passé au crible d'un esprit critique très aigu les idées jusqu'alors recues, des vues originales très neuves sur l'évolution, l'adaptation, la sélection naturelle, c'est-à-dire sur les domaines de la biologie aux horizons les plus fascinants. J'ai cité Claude Bernard et comment ne pas rappeler qu'en définissant la vie comme un conflit—ou une collaboration—entre l'organisme vivant et son milieu, celui-ci donna le coup d'envoi officiel à cet ensemble de disciplines que j'ai évoquées ci-dessus. Sans doute, l'approche de Claude Bernard se situait-elle proche de l''environmental physiology', car la notion d'écosystème dans toute sa complexité n'existait pas alors, la génétique était au berceau et ses interrelations avec l'environnement étaient des plus incertaines. Ce sont donc surtout les facteurs physiques et chimiques du milieu que considérait l'illustre physiologiste, mais l'idée était désormais admise qu'on ne pouvait comprendre la physiologie naturelle d'un être vivant sans connaître les conditions de son milieu et la fécondité de cette notion apparaît pleinement aujourd'hui dans l'épanouissement des disciplines que je viens d'évoquer.

Le présent ouvrage marque une étape capitale de l'évolution de cette notion, plus complexe qu'elle n'apparaissait au 19ème siècle et l'auteur éclaire si bien, de ses vues personnelles et originales, cette complexité, que celle-ci s'estompe à la lumière de ses réflexions. J'ai déjà dit combien m'apparaissent précieuses les définitions qu'il donne de trois titres de discipline qui étaient jusqu'ici plus ou moins confusément employés pour désigner des démarches voisines mais cependant qui apparaissent à Bradshaw, à certains égards, distinctes. Il est probable que cette distinction ne sera pas sans soulever quelques réserves, certains biologistes par exemple considérant l'environnement comme aussi complexe que l'écosystème. Cependant, il est évident que notre langage scientifique a moins besoin de synonymes que de mots qui expriment des idées voisines mais nettement distinctes et bien précises et je crois les définitions de Bradshaw judicieuses et utiles.

Je voudrais maintenant souligner par deux exemples plus ponctuels l'originalité des réflexions de l'auteur.

Celui-ci fait remarquer que, lorsqu'on transporte des populations sauvages au laboratoire, un certain pourcentage des individus les constituant succombe; il en résulte que la population sur laquelle l'homme de laboratoire expérimente n'est pas identique à la population sauvage et c'est l'une des sources des différences qui peuvent apparaître au cours de l'étude comparée d'une population en captivité et d'une autre, d'origine identique, laissée en condition naturelle ou placée dans des conditions semi-naturelles.

D'autre part, Bradshaw réhabilite fort opportunément les valeurs obtenues dans nos expériences et que nos calculs statistiques nous incitent à éliminer comme aberrants, c'est-à-dire qui sont en dehors de l'intervalle moyenne plus ou moins 2 écarts types. Cette manière de faire qui peut se justifier quand nous travaillons sur des animaux de laboratoire, est très dangereuse chez les populations d'animaux sauvages car ces sujets aux données "aberrantes" peuvent être les plus intéressants en tant que représentant des possibilités évolutives nouvelles ou d'ultimes témoins d'une étape antérieure de l'évolution.

Cet ouvrage est un modéle d'observation, d'expérimentation et de réflexion écophysiologique et sa portée est d'autant plus grande qu'en une époque où l'homme cherche de plus en plus à maîtriser la nature pas ses cultures, ses élevages, ses manipulations variées qu'il applique à maintes espèces, il est indispensable qu'il connaisse bien l'écophysiologie de celles-ci à l'état sauvage et dans les conditions de captivité qu'il leur impose pour essayer de savoir d'où elles viennent et où elles vont, et éviter certaines erreurs dont nous connaissons tous des exemples. Il n'instruit pas seulement le lecteur sur les divers aspects de la physiologie des Reptiles, il lui apporte des éléments indispensables pour nourrir une réflexion personnelle sur les plus vastes et les plus importants problèmes de la biologie moderne et c'est pourquoi il sera lu avec le plus grand profit par tous les biologistes curieux aussi bien de science fondamentale que des applications de la science, par tous les chercheurs comme par tous les étudiants qui trouveront là un modèle de réflexion et d'action pour les progrès de notre connaissance du monde vivant.

Professeur Maurice Fontaine,
*Membre de l'Académie des Sciences, Paris.*

# Preface

I was honoured to be asked by Professor Don Bradshaw to write the Preface to his book, an invitation which I owe more to our long friendship and common philosophical approach—which grew apace with our mutual interests in physiology—than to my own slender knowledge of herpetology. Perhaps, in fact, his choice was determined by the interest that we have shared for many years in attempting to explain—with the help of new physiological disciplines that we have seen emerge—on the one hand, the existence of living systems in apparently hostile environments, and, on the other, the capacities needed to ensure the passage from one milieu to another very different one at particular stages of the life cycle.

Whatever the reason, it is a pleasure for me to attest to the value of this work which is doubtless due in large part to Professor Bradshaw's early experiences in a country unique for the richness of its fauna, much of it still unstudied, and to his early realisation that ecological principles enunciated in other continents and consecrated in texts, may not apply universally and that a close study of Australian ecosystems could considerably advance the state of our knowledge.

Reptiles, and especially squamates, must first have attracted him because of their spectacular past, which is probably not unrelated to certain present characteristics, but also by their diversity—which he has helped to document and clarify. He shows that these are animals which appear preadapted to the hostile desert climate by virtue of their low rate of metabolism, their mode of excretion which economises water, their exceptional resistance to metabolic perturbations, their heterothermy and heliothermy and by a variety of other behavioural responses which ensure their survival in extreme situations. In his thesis, defended exactly twenty years ago and completed in the space of only two years, he already underlined these features which are common to all reptiles, whether desertic or not. He also identifies certain organs which may be considered as adaptations, not just by their presence, but because of their intense activity.

Although we may speak of 'the desert' because of certain features in common, deserts are in reality remarkably diverse and Bradshaw early on realised the need to expand the scope of his research to include reptiles occupying desertic zones in other continents, particularly North America and North Africa with the Sahara. As a comparative basis for his studies primarily on the Saharan ecosystem he chose Europe, so that in the book we find an impressive wealth of observations carried out on four continents.

However, over and above the obvious value of the book for the detailed information it gives on the physiology of reptiles in arid ecosystems, the work will be of great interest to those concerned with the relationship between physiology and the environment and Bradshaw distinguishes between the methodology and philosophy of *environmental physiology*, *physiological ecology* and *ecophysiology*. He considers the first to be essentially the study of the physiological responses of organisms to variations in simple physical and chemical factors of natural or experimental biotopes; the second, the response to a complex of factors—physical, chemical and biological—which characterise natural ecosystems and which experimenters attempt to simulate in laboratory reconstructions; the third, ecophysiology, differs by its emphasis on the response of populations, as well as of individuals, living in their natural habitat and underlines the importance of genetic polymorphism in the process of adaptation which is discussed in some detail.

This is a work which excels in combining pertinent observations derived from nature with data acquired in the field, often using new methods and techniques devised especially for this sort of work, and then juxtaposing these with results derived from rigorous laboratory experimentation such that all sources contribute equally to the interpretation. Bradshaw has obviously taken note of Claude Bernard's advice to researchers—to climb to the top of the highest tree in the forest—and from this vantage point, and after rigorously sieving conventional interpretations, he provides an original commentary on the processes of evolution, adaptation and natural selection: that is, the frontiers of biology which remain the most fascinating. In citing Claude Bernard it is impossible not to recall that in defining life as a conflict—or a collaboration—

between the organism and its environment, he effectively stimulated the development of the three disciplines we have just discussed. Almost certainly Claude Bernard's approach would have been closest to that of 'environmental physiology' since the concept of the ecosystem with all its complexity did not exist in his time, and the science of genetics was still in its cradle so that little if anything was known of its interrelationships with the environment. It was thus primarily the physical and chemical factors of the environment which were taken into consideration by this eminent physiologist but the idea was nevertheless sown that one could not comprehend the natural physiology of a living being without knowledge of its external environment, and the fecundity of this insight is reflected today in the flourishing of these same disciplines.

This book marks an important stage in the evolution of this concept, far more complex than it seemed in the 19th century, and the author does much to clarify this complexity which we glimpse through the light of his own experiences. I have already drawn attention to the value of the three definitions proposed by Bradshaw to distinguish between discrete disciplines, hitherto considered as more-or-less interchangeable. Doubtless these definitions will not meet with immediate acceptance, especially from those biologists who consider the environment to be as complex as the ecosystem itself. It is evident however that the scientific language has less need of synonyms than it does of words to distinguish concepts that are very closely related and I believe that these definitions will prove to be of value.

I should now like to highlight, with two concrete examples, the originality of the author's reflexions. He points out that, when wild animals are transported to the laboratory, a certain percentage of them invariably succumbs. The result is that the population used for experimentation in the laboratory differs genetically from that in the field, and this is one of the major sources of disparity which can emerge when two apparently-identical populations of a species are studied, one in the laboratory and the other in the field.

Bradshaw reminds us too that our rigorous statistical training leads us inevitably to consider as 'aberrant' any values which lie outside the 'norm' of two standard deviations from the mean. Such an approach may be justified when dealing uniquely with laboratory populations, of restricted genetic diversity, but is highly dangerous when applied to natural populations in the field where apparently aberrant forms may represent possibilities for the future evolution of the group or attest to past evolutionary stages.

The book is a model of observation, experimentation and ecophysiological reflexion and its relevance is all the greater in an era characterised by man's relentless efforts to tame nature through cultivation, domestication and the manipulation of countless species. It reminds us that an understanding of the ecophysiology of these same species in their natural state, as well as in the artificial conditions imposed upon them by captivity, is indispensable if we are ever to comprehend from whence they have come and what will be their future, and thereby avoid repeating just those sorts of errors which we all know so well. The book not only informs the reader at the level of reptilian physiology but also raises some of the most important questions in modern biology and this is why it will be read with profit by biologists interested both in fundamental aspects of science, as well as in its applications, and by researchers and students who will find in it a model for reflexion and action to advance our knowledge of the living world.

Professor Maurice Fontaine
*Membre de l'Académie des Sciences, Paris.*

# Acknowledgements

I acknowledge the help of many people in writing this book and in assisting me with much of the research on which it is based. Professors Bert Main and Harry Waring introduced me to the study of ecophysiology and I have spent many subsequent years discussing its various facets with Dr François Lachiver of the Muséum d'Histoire Naturelle in Paris, whose many insights have proven invaluable. The early chapters of the book were written in Paris, whilst on study leave from the University of Western Australia, and I am grateful to Professor Yvain Fontaine for providing the congenial conditions so necessary for writing, and for the courteous welcome to his laboratory. My other colleagues in Paris, Drs Michel Lemire, Claude Grenot and Roland Vernet, have all assisted me by their infectious enthusiasm for problems ecophysiological; I have relied heavily on the unrivalled knowledge and experience of Dr Hubert Saint Girons, particularly for the sections concerning reproduction.

In Australia, numerous colleagues and students have assisted me, both in the laboratory and in the field, including Brian Clay, Don Edward, Hylke Bakker, Greg Rice, Keith Morris, Terry Miller, Kent Williams, Ian McDonald, Jack Cannon, Brian Fergusson, Stephen Ambrose and my research officers, Julie McAllister and Felicity Owen. I am particularly grateful to Dr M. S. Johnson for his helpful critique on the nature of biological adaptation. Mr and Mrs J. Lee Steere gave access to Hamelin Pool Station at Shark Bay and Mr and Mrs D. Southall provided hospitality on untold occasions. Recent work at Port Hedland, in the Pilbara region of Western Australia, has been facilitated through the courtesy of Mr and Mrs J. Stickney of Mallina Station. All the work involving the use of $^{18}0$ to measure material and energy budgets of free-ranging reptiles has been done in collaboration with Dr Ken Nagy of the University of California at Los Angeles, and using the facilities of the Laboratory of Biomedical and Environmental Sciences.

My work on reptilian ecophysiology has been generously supported by the Australian Research Grants Scheme (Grants D71/17826, D1-77/15153 and D1-83/15122) and also, indirectly, by the University of Western Australia through the maintenance of items of equipment. My thanks also go to my secretary, Mrs Betty Curtis, whose secretarial capacities are only surpassed by her charm and equanimity when confronted with nigh-impossible requests. The figures were all drawn by Adrian Henderson whose artistic sensibility has added much to the illustrations.

I also wish to acknowledge for permission to reproduce the following Tables and Figures: Academic Press (Tables I, XIV, XVIII, XXVI, XXIX, XL, XLIII, XLV, XLVI, and Figs. 1.4, 2.1, 2.2, 2.28, 2.29, 2.30, 2.31, 2.32, 2.33, 2.39, 2.40, 2.41, 2.42, 3.1, 3.10, 3.11, 3.15, 3.18, 3.19, 3.20, 4.1, 4.5, 4.6), Alfred Benzon Symposium XI, Munksgaard (Table XXXII), American Midland Naturalist, University of Notre Dame (Table XXIII), American Physiological Society (Fig. 2.10), American Society of Ichthyologists and Herpetologists (Table XLIV, and Figs. 2.8, 2.9, 5.16), Australian Academy of Science, Canberra (Figs. 3.2, 3.3, 3.4, 3.8, 3.9), Beech Hill Publishing Co. (Fig. 4.7), Cambridge University Press (Table V and Figs 1.2, 1.8), Dowden, Hutchinson and Ross, Inc. (Fig. 5.19), Ecological Society of America (Tables XLI, XLVIII, XLIX, and Fig. 1.15), Elsevier Science Publishers (Tables LIV, LV, LVI, and Figs. 1.10, 5.3), Gustav Fischer Verlag, Stuttgart (Table VI), Harvard University Press, Massachusetts (Table XLII), Junk, den Haage (Tables VIII, IX, X, XI, and Figs. 1.12, 1.13, 4.4), Pergamon Press (Tables XXI, XXII, XXIV, XXV, XXVI, and Figs. 2.6, 2.7, 2.14, 2.18, 2.19, 2.20, 2.22, 2.37, 2.38, 4.3). The Physiological Society, Cambridge University Press (Table XX), Plenum Publishing Corporation (Figs. 3.17, 3.21), Professor Théodore Monod (Table IV, and Figs. 1.1, 1.6), Scientific American, Inc. (Fig. 3.13), CSIRO Editorial and Publications Service (Fig. 2.17), Springer-Verlag, Heidelberg (Tables XV, XXVII, XXVIII, LIX, and Figs. 2.11, 2.12, 2.13, 2.21, 2.23, 2.24, 2.25, 2.26, 3.22, 3.23, 5.13, 5.14, 5.20), Society for the Study of Reproduction (Fig. 3.14), UNESCO Press (Figs. 1.5, 1.7, 1.9), University of California Press (Fig. 3.5), University of Oklahoma (Figs. 1.14, 1.16), The Zoological Society of London (Table XVI, and Figs. 2.3, 2.4, 2.16, 4.2), Miscellaneous Publications, Museum of Zoology, University of Michigan

(Table L), Comptes-Rendus de la Société de Biogéographie (Fig. 1.11), Bulletin de la Société Zoologique de France (Fig. 3.6).

Permission has been sought for publication approval from the University of Chicago Press (American Naturalist and Physiological Zoology (Tables II, XIII, XXXVIII, LI and Figs. 2.15, 5.17, 5.18), and from the Herpetologists League Inc. (Herpetologica) (Tables XXXVII, LII, LIII, and Fig. 3.7).

# Introduction

The terms 'environmental physiology', 'physiological ecology', and 'ecophysiology' are employed almost interchangeably to denote studies which are held nevertheless to differ fundamentally in their aims from those of general and comparative physiology. Comparative physiology as a discrete discipline grew in part in response to the need to extend physiological observations beyond humans and their familiar animals to lower vertebrates and invertebrates, and we owe much to pioneers such as Claude Bernard, August Krogh, Homer Smith and Knudt Schmidt-Nielsen. Central to this science, however, is the notion of 'phylogeny recapitulated through comparative anatomy' and the 'comparative method' relies for its success on the judicious choice of varied material for the revelation of general principles. Much of what we know of the mechanism of nervous transmission, for example, derives from studies on the giant axons of the squid. This material was chosen because of the difficulties experienced when working with myelinated nerves. The assumption implicit in these studies, however, is that the basic mechanism of transmission is the same in myelinated and in giant-unmyelinated nerves and that one can generalise from one to the other. Animals are thus chosen as objects for study, not for their own intrinsic merit, but because they present aspects of organisation which are not readily investigated in humans and laboratory animals. Other examples such as the highly developed renal medullae of desert rodents spring readily to mind. This approach forcibly takes no account of the environment from which the animal was derived, nor of the evolutionary pressures which have favoured such a development, but views them instead as serendipitous concatenations to be exploited by the alert experimentalist.

'Environmental physiology' gained considerable impetus after World War II because of technological advances in the realm of biometeorology and seeks to investigate the way in which physiological parameters vary as a function of the physical environment. Typically, parameters such as body temperature, rate of metabolism, heart rate, circulation and oxygen transport are measured and compared at a variety of different environmental temperatures, partial pressures of oxygen, barometric pressures, light intensities etc., all of which vary in a predictable manner between the major biotopes of the earth (mountain, desert, polar, submarine etc.). In one important sense, however, the aims of the environmental physiologists are not 'fundamentally' different from those of general and comparative physiologists: both are searching for generalisations with which to explain the apparent diversity of the animal kingdom, and both consider that the factors which have given rise to this diversity over time are without relevance to such a study. To quote Folk (1974), environmental physiology . . . 'is primarily concerned with the physiological responses of healthy mammals to natural changes or extremes of the physical environment'.

'Physiological ecology' on the other hand admits the need to study the animal in its natural environment, rather than in laboratory reconstructions which are thought to mimic this, and, according to Gans & Pough (1982), is a science which differs philosophically in its approach to the study of organismic adaptation and diversity from that of general and comparative physiology. Other than stressing a need for a 'functional' rather than a 'phylogenetic' approach (i.e. a reptile is a reptile and not an imperfect mammal) these authors do not clarify how physiological ecology differs fundamentally from the classical approach. From the examples they give, one may deduce that 'physiological ecology' is the study of physiological processes in an appropriate ecological context. Field studies are thus seen as being essential for the definition of the problem and for the correct framing of the questions to be addressed '. . . . it (physiological ecology) can be distinguished from the older disciplines, not so much by the measurements it makes, as by the context in which it phrases its questions' (Gans & Pough, 1982). Field work is thus an important part of the study but it is implied that the final analysis should proceed in the laboratory where experimental conditions can be controlled and manipulated at will (see also Barker-Jørgensen, 1983).

'Ecophysiology' goes further than this and attempts to integrate individual responses (which are rarely uniform) within the context of a single population or species. Although animals are usually studied singly in the laboratory (which facilitates the elucidation of significant mechanisms) they rarely exist alone in nature and occur in populations which are more or less isolated in space and time. Many populations are polymorphic in nature and much of this variety and its underlying genetic variability is lost when animals are transferred to the laboratory and exposed to markedly different conditions of climate and nutrition. This inherent variation found within natural populations is held to be an essential component of the adaptive responses of organisms to environmental pressures and one of the major foci of ecophysiological studies. Variation is intrinsically difficult to study (our minds function by association through similarity) and this is immediately evident when one considers normal laboratory procedures. Animals are studied in groups which are thought to be large enough to encompass the individual variability found within a given population, and responses are then averaged and compared using statistical tests which embody a typological concept. Typically, animals which behave 'aberrantly' (i.e. they fall more than two standard deviations away from the 'norm') are excluded from the analysis rather than being made the subject of further analysis to determine the reasons for their discrete behaviour. In classical physiology of course this process of elimination of variability goes even further by relying on stocks of laboratory species which have been selected over long periods of time and display greatly reduced genetic variability. Clearly this is a necessary adjunct to the study of mechanism where the aim is to produce a stable and reproducible preparation, but this begs the question of whether this variation is equally a nuisance to the animal in the pursuit of its normal existence and what rôle, if any, such variation plays at the level of the population.

Ecophysiology thus insists on the primacy of field over laboratory-derived data as the only means by which one may establish the 'realised niche' of the animal (Main, 1983). The rôle of the laboratory should be to facilitate the testing of hypotheses framed in the light of field data and to permit a full description of the physiological capacities of the animal, many of which may never be realised in the field. The Australian dingo for example, which was introduced by the Australian Aborigines some 30 000 years ago, can withstand an ambient temperature of −70°C, a temperature to which no individual would ever be exposed in its lifetime (Shield, 1972). Knowledge of this 'fundamental niche' (in the wide sense of Hutchinson, 1957) is essential if one is to predict the likely outcome of environmental changes which involve removal of competition, short-term perturbations, catastrophes etc., many of which are likely to be associated with changes in population size.

Opinions differ on the extent to which it is possible to carry out scientific experiments, rather than simply make observations, in the field. The 'scientific method' adhered to by most practitioners is that described most aptly by Popper (1959) in which predictions from hypotheses are verified or falsified ideally under controlled conditions where all variables, other than the one of interest, are held constant. In reality of course this is rarely the case, and laboratory experiments are usually chimeras of the real world and the ideal experimental situation. In a properly controlled experiment, however, where due attention has been given to the selection of the experimental animals, support conditions, feeding régime etc., one may have some confidence in the generality of the results and, in the case of doubt, the experiment can be repeated *ad infinitum*. The situation is clearly somewhat different in the field, where nothing is under the control of the experimenter, except perhaps the disposition of the animals. It is reasonable to argue therefore that one can only observe events under these conditions and note correlations which may result in the formulation of testable hypotheses. Others (e.g. Cody & Diamond, 1975) would argue that experimentation is possible in the field and, moreover, is more realistically undertaken there, rather than in artificial laboratory reconstructions where the responses of animals are unpredictable and often aberrant. The difficulty with 'experiments' in the field is that it is easy to imagine the outcome being influenced by some environmental factor which has not been included in the design (e.g. a sudden change in barometric pressure) and which may not even be monitored by the experimenter. Changing light intensities over a period of, say, a month 'could' influence two species in a different way and profoundly modify the results of a long-term 'experiment' set up to investigate the effects of, say, crowding on rate of growth. Such is less likely to be the case in the event of field experiments carried out with a single species and within a relatively restricted portion of the available biotope. But even in the case of carefully planned single-species 'experiments' where, for example, a portion of the population is given access to supplementary feed (e.g. Hart, 1980) it is always possible to argue that the results were influenced by some substance in the artificial diet (e.g. a phytoestrogen or a preservative) which was not present in the diet of the nonsupplemented group of animals. Considerable caution is thus needed in interpreting the results of such studies and the laboratory must remain, for philosophical reasons if for no other, the environment of choice for the rigorous elimination of superfluous hypotheses.

There are a number of reasons why research on reptilian ecophysiology has undergone such a spectacular development in recent years and why reptiles present such a favourable material for studies of this sort. Primarily this is because of the marked disparity of behaviour when reptiles are observed in the laboratory and in their natural environment. Reptiles were long thought to be 'cold-blooded' because of their manifest inability to elevate their body temperature above ambient levels by physiological means when isolated from sources of radiant energy, and it was only as a result of the field studies of Cowles & Bogert in 1944 that their astonishing capacities for behavioural thermoregulation were appreciated. In the late 1960s and early 1970s an enormous effort was expended on field studies of temperature regulation in a great variety of species of reptiles (see Avery, 1982 for a review) and it was some time before the necessity for adjunct laboratory studies became evident as the analyses moved from the descriptive to the analytical phase. This dependence of reptiles on favourable configurations of the external environment, in the absence of which their true physiological capacities are not deployed, contrasts markedly with the situation in other amniotes (birds and mammals) and explains in large part why reptiles were long considered to be 'primitive' and to lack many of the basic physiological mechanisms which are normally considered essential for vertebrate homeostasis.

Many comparative physiologists in the 1960s had concluded, for example, on the basis of laboratory studies where body temperature was not controlled, that reptiles lacked an adrenocorticotrophic hormone (ACTH) in the pituitary gland (e.g. Macchi, 1967) and that their kidneys did not respond homeostatically to variations in electrolyte status of the body (e.g. Schmidt-Nielsen & Skadhauge, 1967). Subsequent studies in the laboratory have amply confirmed the existence of these fundamental aspects of vertebrate organisation in reptiles, but these are only manifested in laboratory situations if one respects the particular thermal preferenda of the species concerned.

This obvious dependence of reptiles on environmental sources and sinks of heat for the precise regulation of their body temperature would also appear to be one of the main reasons why this group of vertebrates has been so 'successful' in arid and desert environments where high temperatures *per se* are often inimical to the survival of birds and mammals. Such 'success' in terms of numerical abundance and taxonomic diversity implies the existence of specific adaptive mechanisms promoting individual survival and population persistence and reptilian ecophysiologists have devoted considerable time and energy to the search for these adaptations. The concept of biological adaptation is not easy to define however (see Main & Baker, 1981) and Peters (1976) has argued cogently that many of the familiar constructs employed by biologists to buttress their arguments, such as the competitive-exclusion principle, succession, and even natural selection, are in reality no more than tautologies. Population persistence in arid environments obviously necessitates that individuals survive for a period of time sufficient to secure their reproduction and the eventual recruitment of their young to the population; such survival again obviously depends upon the felicitous exploitation of limited resources for growth and maturation coupled with the capacity to endure extended periods of environmental stress. Whether such characteristics can be considered as 'adaptations', however, as is usually the case, is a moot point because a definition of adaptation in terms of survival would be clearly a tautology (i.e. 'adaptations' facilitate survival, therefore those which survive are forcibly the 'best-adapted', see also Williams, 1970). It would seem possible on the other hand to define 'specific adaptations' in terms of *qualities or capacities possessed by some, but not all, members of a given taxon which enable them to exploit a particular niche which is in consequence unavailable to those other members of the group*. These 'adaptations' may or may not lead in the long term to the differential survival of the group which possesses them (not all adaptations are necessarily 'successful' when judged in terms of persistence) and one may thus avoid the *a posteriori* definition of adaptation in terms of survival.

If we set aside for a moment then the assumption which underlies much reptilian ecophysiology, that reptiles are specifically adapted to the desert milieu, and ask instead in what way desert-living species show modifications of their behaviour and physiology, not shared by non-desertic species, which suit them peculiarly to the desert environment, then some progress might be possible. Gould & Lewontin (1979) have recently criticised what they call the 'adaptationist program' and its 'panglossian'* excesses. They take evolutionary biologists to task for their readiness and eagerness to interpret every 'trait' as a specific adaptation to a particular environmental constraint and thus raise what are secondary epiphenomena to the level of primary selection. They point out that the 'adaptationist program' has held sway, particularly in the English-speaking countries, during the last 40 years and the 'new synthesis' (Huxley, 1942) elevated natural selection to such a central rôle that adaptation was seen as the inevitable outcome of its operation and the sole legitimate explanation of diversity in organic form and function. Gould & Lewontin (1982) signal a timely need for a reappraisal of our ideas concerning the nature of adaptation in biology and the rôle of modalities other than simple natural selection in procuring evolutionary change (see also Jacob, 1977). They categorise much of the thinking in this area as essentially 'non-scientific' and Pielou (1981) echoes much of their concern in discussing the extent to which modelling in ecology has become an apparently acceptable substitute for the real discipline.

The aim of this book is to examine the proposition that the conspicuous presence of members of the Class Reptilia in the deserts of the world is a direct consequence of identifiable adaptations in their morphology, behaviour and physiology. Conventional evolutionary theory would argue that repeated periods of exposure to environmental stress should lead, ultimately, to a shaping of the genotype in desert animals through natural selection such that there is a gradual accumulation of behavioural and physiological responses appropriate for individual survival and thus commensurate with long-term species persistence. Ecophysiological studies have now been carried out with reptiles living in deserts on three continents (Australia, North America and Africa) and abundant data exist for the prosecution of such an investigation.

---

* Le Docteur Pangloss is a character in Voltaire's *Candide* who always maintains that everything is made for the best possible purpose: "Car tout étant fait pour une fin, tout eft néceffairement pour la meilleure fin". (*Candide ou l'optimise*, edition de 1759, traduit par le Dr Ralph).

An attempt will first be made to search for commonality amongst the problems with which desert reptiles must contend, and the extent to which desert reptiles are capable of maintaining physiological homeostasis in the field will be examined. Reproductive strategies of desert reptiles will be compared, particulartly in relation to their population dynamics, and recent studies on life-history 'tactics' of reptiles will be reviewed. Finally, an attempt will be made to identify the causes and monitor the incidence of mortality in desert reptiles in the field through case studies of the few species that have been studied intensively from this point of view.

The rigorous application of the principles outlined above does not result in the positive identification of many such specific adaptations with which to account for the spectacular success of reptiles as a group in desert and arid environments. Far from terminating the analysis, however, this conclusion leads to the formulation of a set of more fascinating questions. Why has there not been the evolution of adaptative traits in this group of animals which would considerably enhance the quality of their adaptive responses in the face of environmental exigencies (for example, why are there no functional nasal salt glands in lizards of the genus *Amphibolurus* who manifestly, have need of them?) and how can one account for the fact that reptiles obviously tolerate conditions which would not be supported by other vertebrates? Slobodkin & Rapoport (1974) have recently suggested that if environmental perturbations and stress merely kill members of a population, without eliciting any other responses, then selection will favour high reproductive rates, but not modifications of the behaviour and physiology of the animals which would enable future generations to contend with such stress. Obviously such questions can only be answered by ecophysiological studies which attempt to integrate the whole biology of the animal rather than study in isolation a single facet. Ecophysiology is thus of prime relevance to evolutionary theory and desert reptiles provide an ideal test case.

# 1
# The Desert Environment

## I. GENERALITIES

Deserts are regions characterised essentially by a deficit of precipitation relative to water loss by evaporation and may be typically hot or cold. They are usually classified into five major types depending upon their location (Logan, 1968): (a) *subtropical* (e.g. Sahara, Kalahari, Sonora, Chihuahua and Australian), (b) *cool coastal* (e.g. Namib, Chile and Peru), (c) *rain shadow* (e.g. Patagonia and to some extent the Mojave), (d) *Continental–interior* (e.g. Great Basin, Central Asian) and (e) *polar*. This classification is somewhat arbitrary and some deserts share features of more than one of the above groupings; the Mojave desert in North America for example displays aspects of both a rain-shadow and a continental-interior desert, as is the case with the Simpson desert in Australia. What are usually considered to be the major deserts of the world however are all subtropical in location (see Fig. 1.1). Two broad belts of aridity circle the globe in the vicinity of the Tropics of Cancer and Capricorn (roughly 20°–30° N and S) and these deserts are the end-result of common climatic factors experienced in these latitudes.

High-pressure anticyclones are a semi-permanent feature of the subtropics and cold high-altitude air descends towards the surface. During its descent the temperature of the air increases greatly due to adiabatic heating (10°C/1000m)

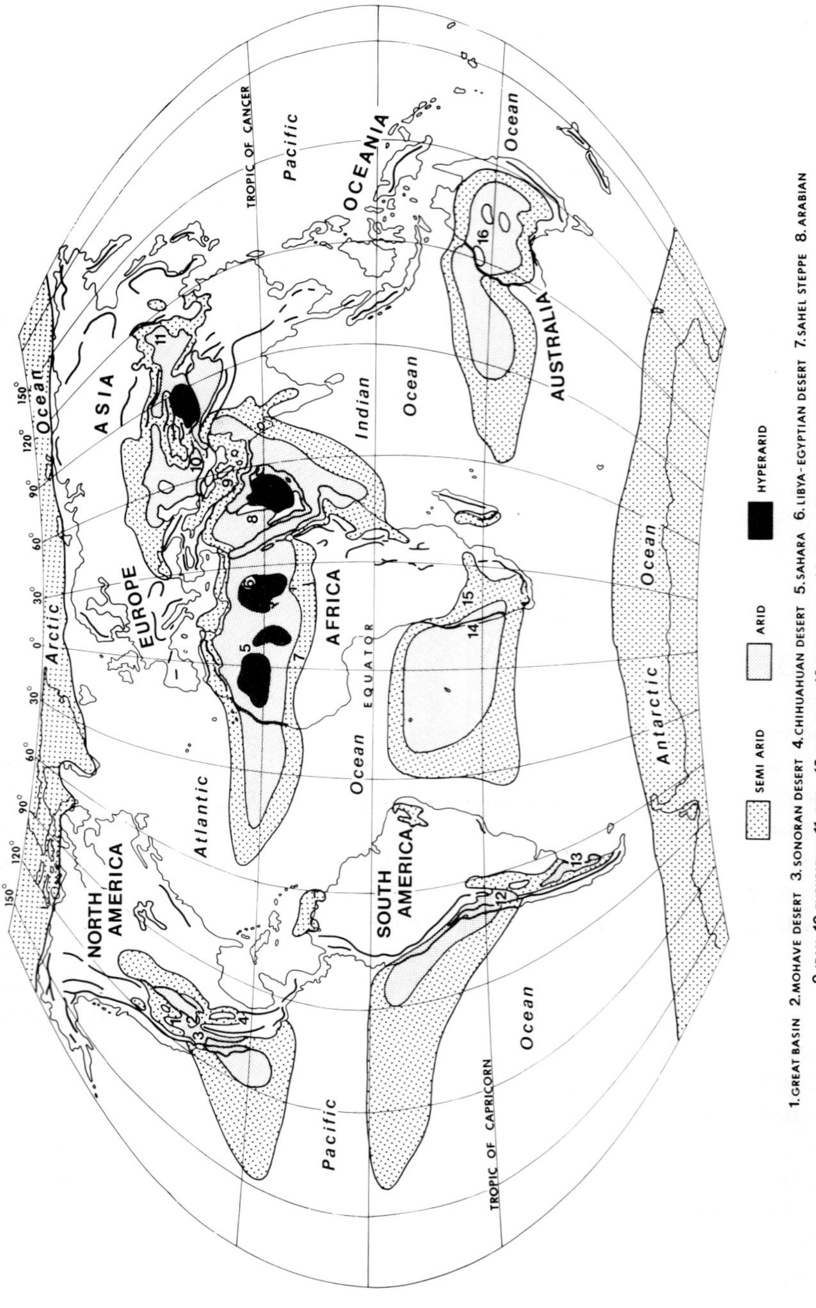
150°
120°
90°
60°
30°
0°
30°
60°
90°
120°
150°
Arctic
Ocean
NORTH AMERICA
EUROPE
ASIA
AFRICA
OCEANIA
AUSTRALIA
SOUTH AMERICA
Atlantic
Ocean
Pacific
Ocean
Pacific
Ocean
Indian
Ocean
Antarctic
Ocean
TROPIC OF CANCER
EQUATOR
TROPIC OF CAPRICORN
SEMI ARID
ARID
HYPERARID
1. GREAT BASIN 2. MOHAVE DESERT 3. SONORAN DESERT 4. CHIHUAHUAN DESERT 5. SAHARA 6. LIBYA-EGYPTIAN DESERT 7. SAHEL STEPPE 8. ARABIAN
9. IRAN 10. TURKESTAN 11. GOBI 12. ATACAMA 13. PATAGONIA 14. NAMIB 15. KALAHARI 16. AUSTRALIAN DESERT

and by the time it reaches the ground it is hot and dry with virtually no precipitation potential. This process occurs equally over land and sea and the subtropical deserts so formed cover equally vast areas of the ocean where their effects are of course less evident. These meteorological phenomena influence the local climate in a predictable way. The clarity of the dry air permits high penetrance of solar radiation which is in consequence little filtered, and the air is heated primarily by secondary long-wave radiation from heated surfaces. The lack of cloud cover at night facilitates heat loss by radiation from the surface and night temperatures are normally low in subtropical deserts. High diurnal temperature ranges are thus a feature of these deserts but, as Monod (1973) points out, they rarely exceed 55°C and can be greater in non-desert regions. Lyon for example in central France had a peak temperature range of 63.2°C between 1921 and 1940 and that for Paris was 53.3°C.

Humidity and precipitation are low throughout the year and the rate of evaporation is extremely high. Because the air is almost never fully-saturated with water vapour, dew, frost and fog are rare phenomena in the desert. Rain usually results from a moist air mass which has entered the desert region from elsewhere. Rainfall is thus always low but, in addition, its occurrence is highly variable and thus unpredictable. When the variance of the log of the mean annual precipitation from a number of biotopes is plotted against the annual precipitation for example (Fig. 1.2) it is evident that deserts are characterised by low and unpredictable falls of rain. Rainfall data from Reggan in the central Algerian Sahara from 1959–1965 are given in Table I and the hyperaridity of the site is evident. The total rainfall over this period of seven years amounted to only 53.3 mm with an annual mean of 7.6 mm. Low (1978) using Colwell's (1974) measures of Predictability (P), Constancy (C) and Contingency (M) has compared some of the world's deserts and it is of interest that the Australian Desert, represented by Alice Springs, has the lowest probability of rainfall, the second-lowest figure for constancy and the third-lowest figure for contingency (see Table II).

Persistent wind is another feature of subtropical deserts which should not be ignored. The descending cold air masses are displaced towards equatorial low pressure regions and are deflected by the Coriolis Force to become the Trade Winds. Dry desert winds enhance considerably the evaporation of water which exacerbates problems of water balance in both plants and animals. These winds are also important weathering agents shaping the topography, particularly that of dunes (Dubief, 1952) and 'dust storms' in the Sahara are proverbial. The 'sirocco' of the Sahara blows generally towards the Mediterranean basin and similar winds are found in other deserts (e.g. the 'chinook' in the deserts of North America).

---

< **Fig. 1.1** Distribution of the world's major deserts showing their concentration in subtropical latitudes and extensive westward oceanic extensions (adapted from Monod, 1973).

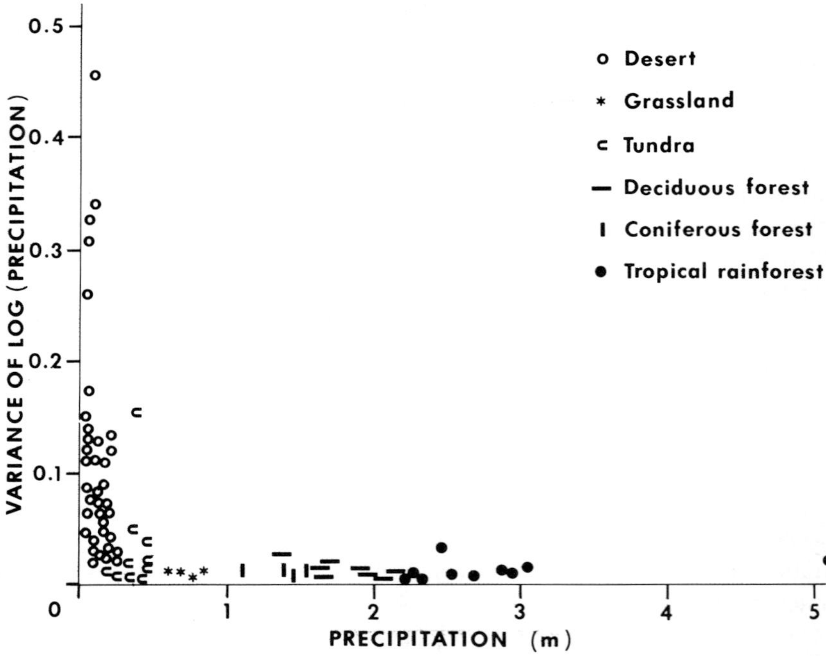


**Fig. 1.2** A plot of the *variance* of the mean annual precipitation against the mean annual precipitation itself for various habitat types, including desert (from MacMahon, 1981).

## II. MEASURES OF INTENSITY

Various methods for assessing and comparing the aridity of particular biotopes have been developed by geographers over the past 50 years and most of these are based on the interactions between temperature and moisture within a given environment. Emberger's Pluviometric Quotients $Q_1$ and $Q_2$ (Emberger, 1930, 1933) are still widely used by bioclimatologists in Mediterranean countries, especially as modified by Sauvage (1963) in the form of 'climagrams'.

$Q_1 = M - m$ where M = mean max. temperature (°K) of the hottest month

m = mean min. temperature (°K) of the coolest month

p = mean annual precipitation (mm)

$$Q_2 = \frac{2000P}{M^2 - m^2}$$

**Table I:** Precipitation at Reggan (Algeria) from February 1959 to January 1965* (from Grenot, C. J., 1974).

| Year | Jan. | Feb. | Mar. | Apr. | May | June | July | Aug. | Sep. | Oct. | Nov. | Dec. | Total |
|---|---|---|---|---|---|---|---|---|---|---|---|---|---|
| 1959 | | 0.1 | t | | t | t | | | 6.8 | t | t | | 6.9 |
| 1960 | | | 0.1 | | t | 2.7 | 0.2 | | t | t | t | t | 3.0 |
| 1961 | | | | | t | t | | | | t | | | t |
| 1962 | | 5.1 | t | t | t | 0.6 | | t | t | t | 6.9 | t | 12.6 |
| 1963 | t | t | t | | | t | ? | 0.3 | | 2.1 | t | | 2.4 |
| 1964 | | | | t | | | | t | | | t | 28.4 | 28.4 |
| 1965 | 33.7 | | | | | | | | | | | | |

* Rain in mm, t = traces.

**Table II:** Predictability (P), constancy (C) and contingency (M) of rainfall in 11 deserts (from Low, 1978).

| Desert type | Years Data | P | C | M |
|---|---|---|---|---|
| Subtropical: | | | | |
| Alice Springs, Australia | 86 | .111 | .069 | .042 |
| Baghdad, Iraq | 29 | .571 | .170 | .401 |
| El Centro, Calif., U.S.A. | 28 | .465 | .328 | .136 |
| Aoulef, Algeria | 29 | .815 | .755 | .060 |
| Cool coastal: | | | | |
| O'Okiep, South Africa | 21 | .259 | .041 | .218 |
| Arica, Chile | 20 | 1.000 | 1.000 | .000 |
| Onslow, N.C., U.S.A. | 74 | .379 | .199 | .180 |
| Rain shadow: | | | | |
| San Juan, Argentina | 84 | .257 | .202 | .055 |
| Phoenix, Ariz., U.S.A. | 64 | .184 | .037 | .147 |
| Polar: | | | | |
| Verkhoyansk, U.S.S.R. | 34 | .838 | .778 | .060 |
| Fort Good Hope, Canada | 51 | .904 | .868 | .036 |

**Source:** Clayton 1944 (1800–1940); U.S. Department of Commerce 1941–1974 (post–1940).
**Note:** Deserts were chosen for number of years of available data (at least 20). Slatyer's (1962) concept of effective rainfall was used to describe three states: (1) rain (R) $< 0.4$ evapotranspiration (E) (i.e., not enough for an initial effective rainfall to initiate grass growth); (2) $0.4E < R < 1.2E$ (enough to initiate grass growth but not to sustain it for more than 1 wk); and (3) $R > 1.2E$ (enough for both initial effective and effective carryover rain). The Australian subtropical desert (Alice Springs) has the least predictable rainfall of the subtropical deserts for which data were available.

**Table III** Bioclimatic Zones within the Mediterranean climate and associated measures of intensity of aridity (adapted from Quézel & Barbero, 1982; Le Houérou, 1982).

| Bioclimatic Zone | P (mm) | $Q_2$ | X | $Q_e$ | P/ETP | I | Thorn-thwaite | Meigs | Comments |
|---|---|---|---|---|---|---|---|---|---|
| Pre-humid | >1200 | >150 | | | | | | | |
| Humid | 800–1200 | 110–150 | | 0.6 | | | | | |
| Sub-humid | 600–800 | 70–110 | | 0.45 | | | | | Certain types of tropical savannah, maquis and chaparral. Agriculture is normal use. Interannual rainfall variability <25%. |
| Semi-arid | 400–600 | 45–70 | 200–300 | 0.25 | 0.2–0.5 | 4.5 | −20–−40 | S | Steppe with some savannahs & tropical scrob. Maquis, garigue & chaparral. mallee & mulga in Australia. Primarily grazing but rainfed agriculture possible. Interannual rainfall variability 25–50%. |
| Arid | 100–400 | 10–45 | 300–350 | 0.1 | 0.03–0.2 | 0.3–4 | −40–−57 | A | Vegetation scattered and usually woody, succulent, thorny or leafless shrubs. Light grazing possible. Interannual rainfall variability 50–100%. |
| Hyper-arid | < 100 | < 10 | >360 | <0.1 | <0.03 | <0.3 | >−57 | E | Low and irregular rainfall which may fall in any season. Little perennial vegetation; agriculture and grazing impossible. Interannual variability of rainfall may reach 100%. |

P = Annual precipitation (mm)
$Q_2$ = Pluviometric Quotient of Emberger (1933)
X = Xerothermic Index of Bagnouls & Gaussen (1952)
$Q_e$ = ETR/ETP, Evapotranspiration Quotient (see Calvet, 1982)
P/ETP = UNESCO Standard (MAB, 1979)
I = Index of Aridity of Capot-Rey (1951)

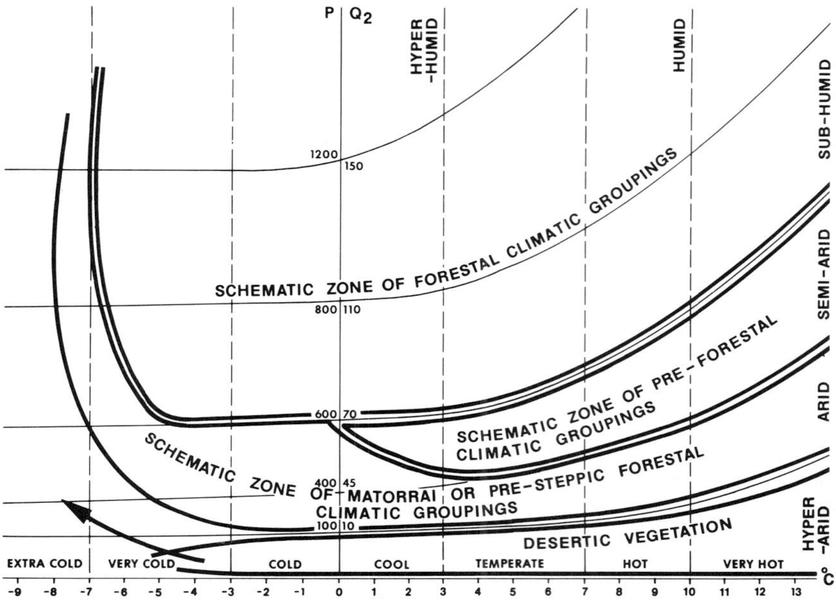


**Fig. 1.3** 'Climagram' developed by Sauvage, and generated by plotting Emberger's Pluviometric Quotient ($Q_2$) and annual rainfall (P) against the mean minimum temperature of the coldest month of the year (from Quézel & Berbero, 1982).

Sauvage's 'climagrams' plot $Q_2$ as an ordinate against 'm' the mean minimal temperature of the coldest month of the year, and enable a number of 'bioclimatic stages' to be defined. Normally six stages of the Mediterranean climate are recognised (see Table III) and a typical 'climagram' is shown in Fig. 1.3. Three stages of this classification are of interest to the desert biologist: the semi-arid, arid and hyper-arid regions. Emberger (1954) defined the Mediterranean climate as follows:

*'Le climat mediterranéen est un climat extratropical á photopériodisme saisonnier et quotidien á pluviosité concentrée sur les saisons froides ou relativement froides de l'année, l'éte, saison la plus chaude, étant sec'* which describes adequately the arid zones of much of Australia and the western United States. Hyper-arid regions do not occur to any appreciable extent in these continents as they do in the Sahara, Iran, and the Takla-Makan desert of central Asia. Purists such as Theodore Monod (Monod, 1973) argue therefore that these regions are not 'true' deserts in which, for example, the concept of average rainfall (P) has no meaning. This question is purely one of semantics however as the hyper-arid environments of the Sahara clearly form one extreme of a continuous spectrum of Mediterranean climates as defined by Emberger and differ in degree rather than in kind.

Emberger's Pluviometric Quotients supplanted essentially the simple Aridity Index of Martonne (1927) and Dubief's Coefficient (Dubief, 1950), although the Index of Aridity devised by Capot-Rey (1951) is still employed. In this,

$$I = \frac{100P/E + 12p/e}{2}$$

where E and P are total evaporation and rainfall for the year and e and p correspond to the wettest month.

The notion of potential and actual rates of evapotranspiration were first introduced by Thornthwaite (1948), the ratio between these two values being defined as the Evapotranspiration Coefficient ($Q_e$). Meigs (1952) used Thornthwaite's system to develop a world map of 'homoclimates' which was established as an international reference by UNESCO. Recent research has added to and refined these various measures of aridity but a single measure has yet to be adopted by all. The latest UNESCO map of the world distribution of arid regions (MAB, 1979) uses the ratio P/ETP where P stands for annual precipitation in millimetres and ETP is the mean annual potential evapotranspiration as an index of aridity. This ratio P/ETP was also adopted by the FAO (Riquier and Rossetti, 1976) and thus has the advantage of wide usage. Using this definition the arid zone can be defined as a region bounded by:

$$0.03 < P/ETP < 0.20$$

and the hyper-arid zone: $$P/ETP < 0.03$$

Koeppen & Geiger (1936) considered that the boundary between humid and semi-arid environments occurred when potential evaporation equalled annual precipitation (i.e. $P/ETP = 1.0$) and the boundary between the arid and semi-arid zones was placed arbitrarily where evaporation was twice as high as precipitation (i.e. $P/ETP = 0.5$). In the 1979 UNESCO map this latter boundary now becomes 0.2 (see Table III). ETP is now calculated using the formulae of Penman or Turc (1961) as Thornthwaite's method has been found to under-estimate potential evapotranspiration rates in Mediterranean climates (see Calvet, 1982).

$$ETP = 0.4\,t/15t\,(G+50)\,(1+[50-U]/70) \quad \text{Turc (1961)}$$

where ETP = potential evapotranspiration (mm)
t = mean monthly temperature (°C)
G = solar radiation (cal/cm$^2$/day)
U = relative humidity (%)

Turc (1951) similarly gives a formula from which actual evapotranspiration rates (ETR) can be calculated.

$$ETR = \frac{P}{0.9 + p^2/L^2}$$

where $L = 300 + 25t + 0.005t$
P and p = annual and monthly precipitation respectively
t = annual mean monthly temperature.

A number of methods also exist with which the length and severity of the dry period may be assessed in any given place. Aubreville (1949), considered any month with less than 30 mm rainfall to be 'dry' and Bagnouls & Gaussen (1953) established a 'Xerothermic Index' in which..

$$X = K((N-n) - [b+r]/2)$$

where N = number of dry days per month
n, b and r = days of rain, fog and dew per month
K = coefficient of relative humidity.

Bagnouls & Gaussen (1957) subsequently developed the method of 'ombrothermic diagrams' in which P in millimetres is graphed on a scale equal to 2t in degrees Celsius which illustrates graphically the dry period of the year (see e.g. Fig. 1.5). Dry periods of the year are defined as those where $P < 2t$ and these appear graphically as periods when temperature (and hence evapotranspiration) exceeds rainfall. These 'ombrothermic diagrams' of Bagnouls & Gaussen are also known as Gaussen-Walter diagrams and are employed widely in comparing Mediterranean climates (see Di Castri, 1981).

Table III compares some of these measures of aridity, particularly in relation to the semi-arid, the arid and the hyper-arid regions.

## III. CLIMATIC COMPARISONS

Durand-Dastes (1968) has offered a useful classification of world deserts in which a number of generic groupings are possible:

A. *Hot Deserts*

a. Subtropical

Sahara to Indus
Kalahari
Australian
Baja California, Sonora, Chihuahua

b. Coastal
   Namib
   Atacama-Peru

B. *Cold Deserts*
   a. Central Asian
      Takla-Makan
      Gobi
      Pei-Shan
      Ala-Shan
      Tibetan plateau
   b. Middle East
      Black Sea
      Aralo-Caspian
      Turkestan
   c. North American
      Great Basin
      Great Plains
      Mojave
   d. South American
      Altiplanos andins
      Patagonia

These deserts differ somewhat in their degree of aridity (as well as being classified basically into 'hot' or 'cold' deserts) and the world-wide distribution of the desert bioclimatic zones and their respective total areas are shown in Table IV. By far the greatest amount of biological data available is that for the deserts of North America, Australia and North Africa, although work has now commenced in the Kalahari and Namib deserts. Of the continents, Australia has the highest percentage of its total area as desert, (80%, whereas the overall figure for the African continent is of the order of 60%) but none of this is classified as hyper-arid whereas 93% of North Africa (Morocco, Algeria, Tunisia, Libya and Egypt) falls into this category. The amount of desert classified as hyper-arid is also very small in North America, the majority being semi-arid.

Subtropical deserts (e.g. Sahara, Australian, Sonora, Chihuahua) are usually bounded on their equatorward margin by semi-arid steppes and grasslands (savannas) with a typically summer rainfall régime. These are transitional zones and the southern Sahara ends where moist Trade Winds from the southern hemisphere manage to penetrate from the Atlantic Ocean

**Table IV:** World distribution and area of desert bioclimates (adapted from Monod, 1973, and Le Houérou, 1979).

| Region | Semi-arid | Arid (Area in 1000 $km^2$) | Hyper-arid | Total | (%)* |
|---|---|---|---|---|---|
| North Africa | 495 | | 4828 | 5323 | (93%) |
| Asia | 7516 | 7909 | 1051 | 16477 | (38%) |
| Europe | 844 | 170 | 0 | 1015 | (9.5%) |
| Australia | 2157 | 3864 | 0 | 6382 | (80%) |
| North America | 2657 | 1279 | 31 | 3967 | (25%) |
| South America | 1626 | 1217 | 170 | 3014 | (20%) |

* Expressed as a percentage of the total area for the region.

and the Gulf of Guinea during the northern summer. During the northern winter, cold dry air moves from high-pressure centres over central Asia (the winter anti-monsoon) and crosses the China Sea to reach Australia in the southern summer. This initially dry air, accumulates moisture during its long over-ocean trajectory and brings rain to the equatorial margin of the Australian desert in the southern summer. Little of this rain however penetrates into the arid interior.

The northern boundary of the Sahara results from the penetrance of frontal disturbances in winter crossing the Mediterranean when high-pressure air circulation patterns retreat equatorward in the local winter. These winter-rain, summer-drought regions correspond to the classic Mediterranean climate and are associated with characteristic semi-arid and sub-humid vegetation types (the *maquis* and *garigue* of southern Europe, the *macchia* of South Africa, *chaparral* of California and the *mallee* and *mulga* of Australia). Sclerophyllous brushlands also occupy the northern borders of the Sonora and Chihuahua deserts in North America and the southern borders of the Karoo and Kalahari deserts in South Africa. The southern margin of the Australian desert is also a typically Mediterranean region with a winter rainfall régime which in many ways is a mirror image of the North African situation (see Williams, 1979).

All the North American subtropical deserts are terminated poleward by lower temperatures and they pass into colder rain-shadow and continental deserts such as the Mojave, Great Basin and Great Plains deserts (Salt Lake, Snake River, Navajo, Painted, etc.). This is also the case with the deserts of Iran and Afghanistan and the Arabian deserts end northward at the foot of the mountain chains of Turkey and Iran. The cold deserts of Central Asia and North America are so because of the influx of very cold polar continental air masses in winter which are not impeded by intervening mountain ranges and they reduce temperatures to far below freezing. The Sahara is to a large extent

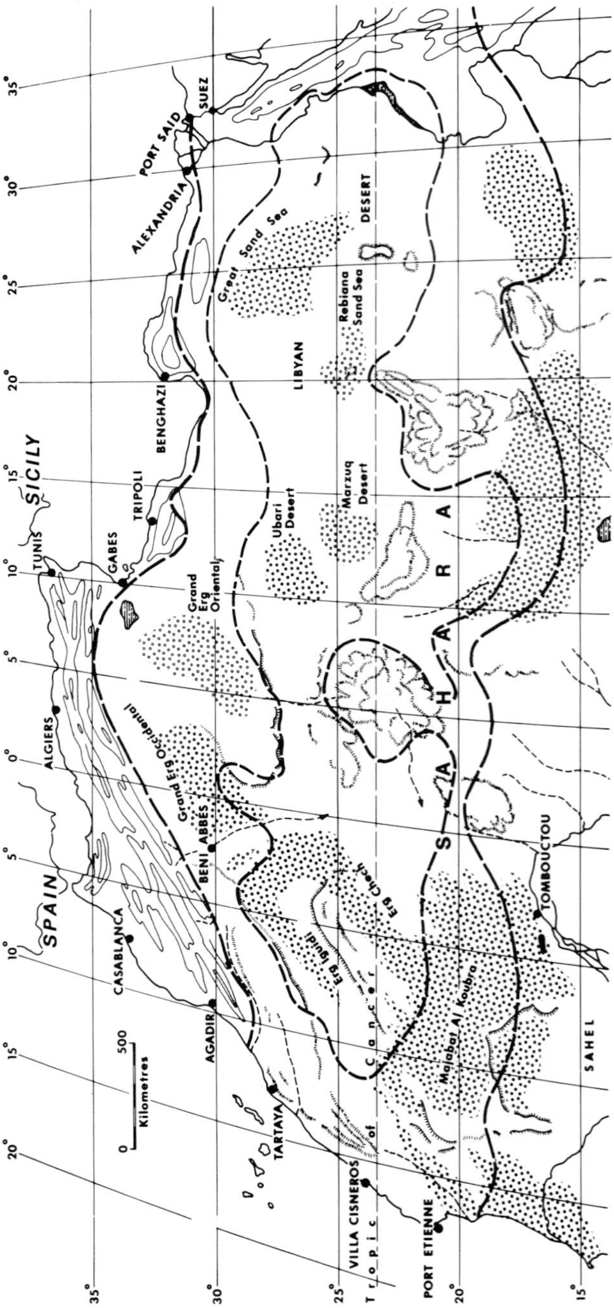


**Fig. 1.4** Map of the Sahara desert (adapted from Monod, 1973 and Grenot, 1974).

protected from this by the buffering influence of the Mediterranean Sea but the same is not true for the deserts of Iran, Baluchistan and Pakistan which are all extremely cold in winter.

Cool coastal deserts represent peculiar situations which are restricted to the west coasts of Africa (Namib), South America (Atacama) and North America (Baja California). These are essentially the extension of subtropical deserts but cold inshore currents (Benguela, Humboldt and the California respectively) induce cold upwellings which serve to chill moist air moving slowly landward from very high pressure cells centred over ocean desert areas (these extend far westward from all the major continents, see Fig. 1.1). Temperature inversions are formed and characteristic 'advectional fogs' carry moisture inland at any time of the day to the coastal desert regions. This fog is usually the sole source of water in these regions (rain does occur but much more rarely) and may also occur at night due to high radiation or when air rises on encountering mountain slopes (orographic fog) as in the Atacama desert of Peru.

The Sahara (Fig. 1.4) is the largest single desert on the earth's surface and also the most uniformly-arid (Grenot, 1974). Table V lists mean monthly and annual rainfall for a number of sites in North Africa which is low and irregular, averaging less than 100 mm per year, so that drought in these regions is virtually permanent. Dubief (1971) has shown that the western 'oceanic' Sahara is nevertheless wetter than the eastern Sahara where rainfall approaches zero. Fig. 1.5 gives Gaussen-Walter ombrothermic diagrams for a variety of sites in the Sahara from which their extreme aridity can be gauged (e.g. Helwan in Egypt and In Salah in Algeria.)

The North American deserts (Fig. 1.6) are, as already noted, much less arid than the Sahara and are comparatively well-vegetated (Johnson, 1968). Only a small region (e.g. Death Valley in California) classifies climatically as hyper-desert and rainfall is generally more regular and abundant. Yuma for example in Arizona receives both summer and winter rain; short-lived thunderstorms produced as a result of the invasion of maritime tropical air from the Gulf of Mexico in summer and occasional migrating frontal disturbances in winter. Despite this, rainfall can be quite variable and varied from 7 mm to 290 mm between 1905 and 1953 with an overall average of 86 mm (Logan, 1968). Fig. 1.7 gives ombrothermic diagrams for a number of localities in the United States and Mexico.

The Australian deserts (Fig. 1.8), the Simpson or Arunta, Great Victoria, Sandy and Gibson, are similar to the North American deserts in terms of aridity but this is compensated for somewhat by the vast area of the arid zone, covering virtually 80% of the continent. A characteristic feature of the Australian arid zone is the broad transition from arid to semi-arid and then sub-humid conditions as one moves out from the central arid core of the continent. As with the North American, the Australian deserts are well-vegetated and the lack of hyper-arid zones can be attributed to a number of

**Table V:** Rainfall in the desert zone (mm) (from Le Houérou, 1979).

| | J | F | M | A | M | J | J | A | S | O | N | D | Year |
|---|---|---|---|---|---|---|---|---|---|---|---|---|---|
| Egypt | | | | | | | | | | | | | |
| Port Said | 18 | 12 | 9 | 5 | 3 | 1 | 0 | 0 | 0 | 3 | 11 | 16 | 78.0 |
| Suez | 5 | 3 | 3 | 2 | 2 | 0 | 0 | 0 | 0 | 1 | 4 | 4 | 24.0 |
| Cairo | 5 | 4 | 4 | 2 | 2 | 0 | 0 | 0 | 0 | 2 | 2 | 4 | 25.0 |
| Dakhla | 0 | 1 | 0 | 0 | 0 | 0 | 0 | 0 | 0 | 0 | 0 | 0 | 1.0 |
| Libya | | | | | | | | | | | | | |
| Beni Walid | 6.7 | 3.7 | 8.3 | 6.0 | 5.6 | 0.3 | 0.0 | 0.0 | 7.0 | 8.5 | 7.0 | 14.4 | 71.0 |
| Rhadames | 6.0 | 5.0 | 4.0 | 13.0 | 1.0 | 0.0 | 0.0 | 0.8 | 0.0 | 2.0 | 7.0 | 4.0 | 43.0 |
| Sebah | 2.0 | 2.0 | 0.8 | 0.6 | 0.2 | 0.0 | 0.0 | 0.0 | 0.0 | 0.7 | 0.2 | 0.0 | 8.0 |
| Kufra | 0.3 | 0.0 | 0.0 | 0.0 | 0.2 | 0.0 | 0.0 | 0.0 | 0.1 | 0.3 | 0.0 | 0.0 | 0.9 |
| Tunisia | | | | | | | | | | | | | |
| Tozeur | 10.0 | 8.0 | 13.0 | 10.0 | 5.0 | 3.0 | 0.2 | 2.0 | 7.0 | 9.0 | 12.0 | 10.0 | 89.0 |
| Fort Saint | 6.7 | 3.0 | 4.1 | 1.3 | 2.7 | 1.0 | 0.0 | 0.0 | 1.5 | 2.4 | 4.1 | 4.0 | 30.8 |
| Algeria | | | | | | | | | | | | | |
| Ghardaia | 9.0 | 4.0 | 7.0 | 4.0 | 5.0 | 4.0 | 1.0 | 2.0 | 5.0 | 7.0 | 13.0 | 7.0 | 68.0 |
| Tamanrasset | 3.9 | 1.1 | 0.8 | 1.3 | 6.6 | 3.6 | 3.1 | 9.5 | 7.4 | 1.9 | 1.1 | 1.8 | 42.6 |
| Bechar | 7.0 | 7.2 | 12.7 | 7.5 | 2.3 | 2.9 | 0.4 | 3.6 | 6.6 | 13.8 | 11.6 | 9.6 | 85.2 |
| Djanet | 4.7 | 1.7 | 1.2 | 1.8 | 3.5 | 3.0 | 0.1 | 0.4 | 3.8 | 1.3 | 1.9 | 1.6 | 25.0 |
| In Salah | 2.9 | 0.8 | 2.1 | 1.1 | 0.3 | 0.1 | 0.0 | 0.7 | 0.1 | 1.3 | 2.3 | 3.9 | 15.6 |
| Tindouf | 4.4 | 2.0 | 5.2 | 2.1 | 0.1 | 0.0 | 0.2 | 5.2 | 8.5 | 9.8 | 8.9 | 6.1 | 52.5 |
| Aoulef | 0.4 | 0.4 | 1.1 | 0.1 | 0.5 | 0.0 | 0.0 | 1.1 | 0.1 | 2.0 | 2.0 | 1.5 | 9.2 |
| Morocco | | | | | | | | | | | | | |
| Bou Denib | 5.4 | 4.2 | 11.1 | 8.2 | 4.2 | 4.7 | 0.9 | 6.9 | 10.6 | 17.2 | 12.7 | 13.8 | 99.9 |
| Ouarzazate | 4.3 | 6.9 | 11.5 | 6.5 | 4.5 | 1.0 | 1.1 | 7.8 | 21.6 | 13.3 | 11.3 | 17.7 | 107.5 |
| Zagora | 0.8 | 2.4 | 5.4 | 7.0 | 0.6 | 1.6 | 0.5 | 4.0 | 11.8 | 8.2 | 5.6 | 11.8 | 59.7 |

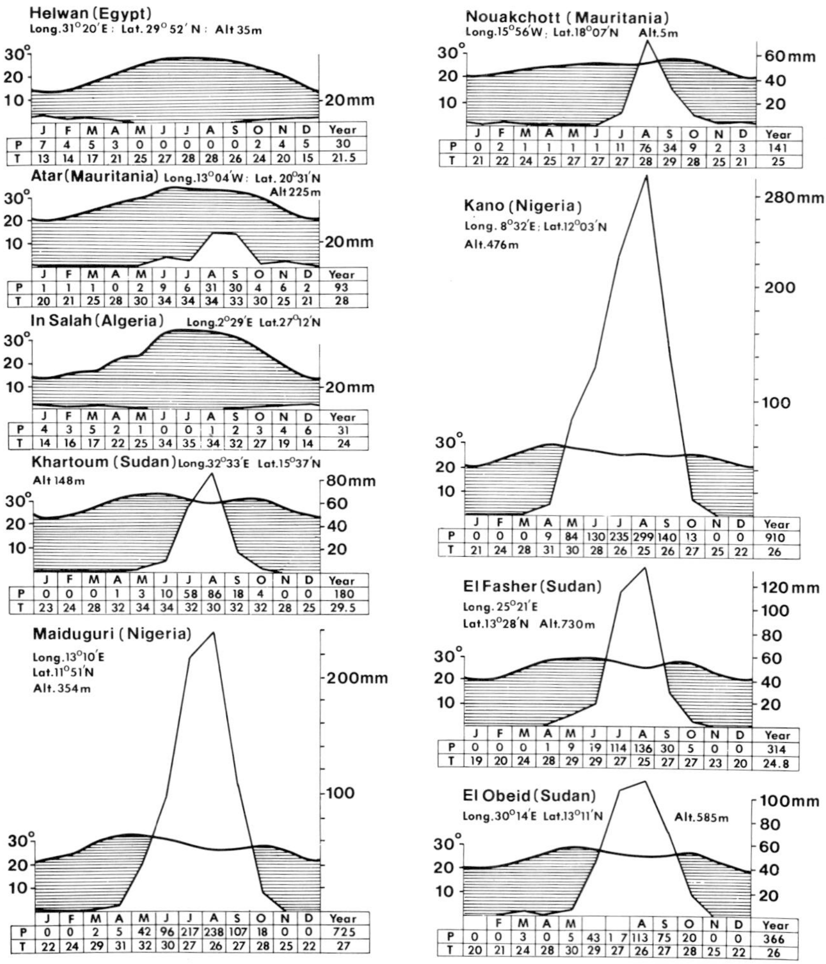


**Fig. 1.5** Ombrothermic Diagrams, developed by Bagnouls & Gaussen (1957), with annual precipitation (P) graphed on a scale equal to 2t (where t = mean monthly maximum temperature) for various stations in the Sahara desert. Dry periods indicated by hatching. (from MAB, 1979).

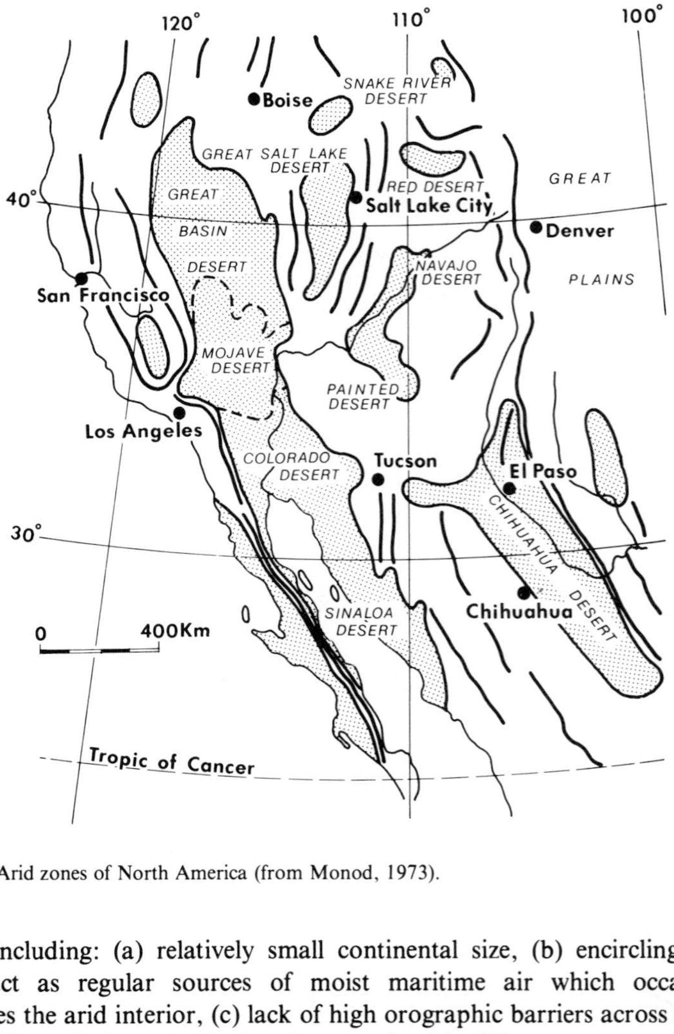


**Fig. 1.6** Arid zones of North America (from Monod, 1973).

factors including: (a) relatively small continental size, (b) encircling oceans which act as regular sources of moist maritime air which occasionally penetrates the arid interior, (c) lack of high orographic barriers across the path of prevailing winds, and (d) the absence of cold offshore currents along the west coast of the continent.

Rainfall patterns have been analysed by Fitzpatrick (1964) who concluded that summer rainfall and a cool-dry winter characterise regions north of the Tropic of Capricorn. As one moves southward, the incidence of winter rain increases but still remains low, except along the south coast, in western Victoria and the south-west of Western Australia. These latter areas experience a typical Mediterranean climate with winter rain and a long summer drought, the two seasons being separated by short climatically-unstable periods. Ombrothermic diagrams for a number of sites within Australia are shown in Fig. 1.9.

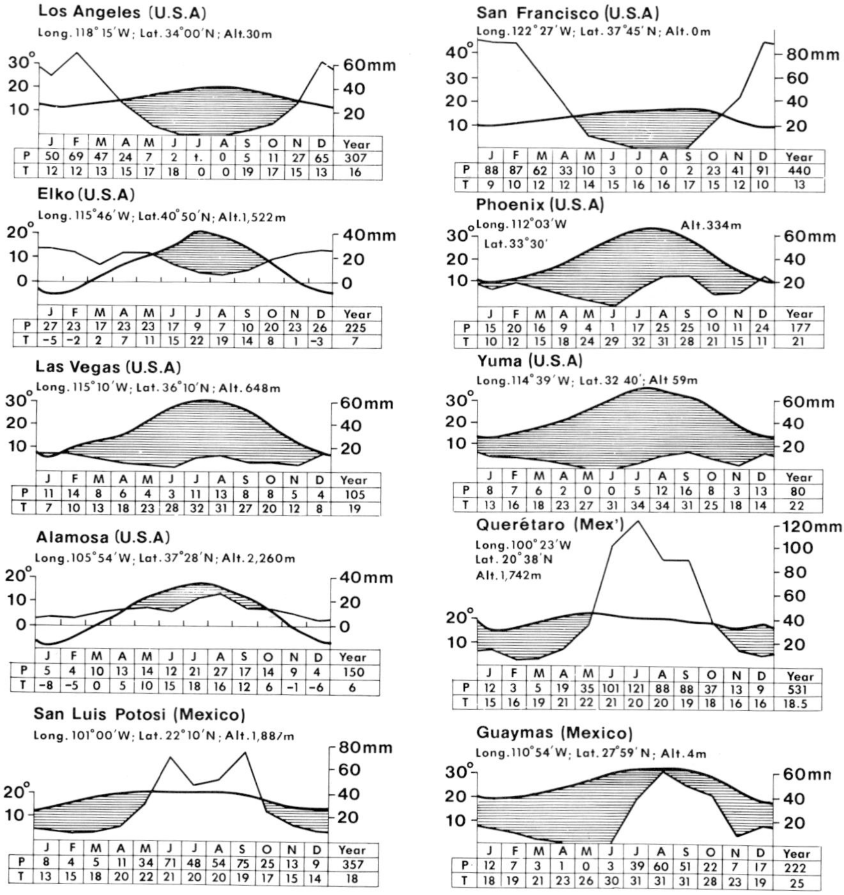


**Fig. 1.7** Ombrothermic Diagrams for a number of stations in the United States of America (from MAB, 1979).

Soil fertility varies markedly between deserts of the world and has a marked effect on any attempts at rain-fed agriculture. Typically the soils of Australia and Africa, being of vastly-greater geological age than those of other countries (Smith, 1968), are poor or deficient in essential elements such as nitrogen and phosphorus (see Fig. 1.10) and this compounds the difficulties associated with agricultural endeavours in these regions.

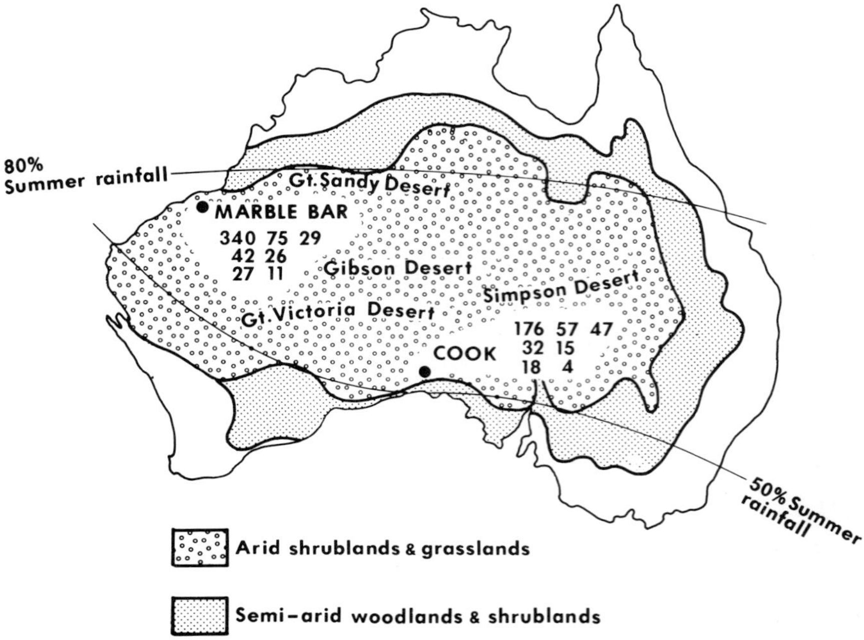


**Fig. 1.8** Arid shrublands and grasslands with surrounding semi-arid crescent of eucalypt low woodlands, shrub woodlands and shrublands. The percentage of rainfall falling in the five summer months is indicated and the figures for Marble Bar and Cook refer to: line 1—mean annual rainfall (mm); percentage rain October-March; mean number of rain days. Line 2—mean monthly maximum and minimum temperatures (°C) for the hottest month; Line 3—mean monthly maximum and minimum temperatures (°C) for the coldest month (from Williams, 1979).

# IV. ORIGINS AND EVOLUTION OF DESERT HERPETOFAUNAE

## A. The Sahara

Most of the Saharan desert is of relatively-recent origin and studies by botanists and geologists have served to establish the various humid and dry phases experienced since the close of the last glaciation (Joleaud, 1938; Alimen, 1963; Balout, 1952; Chavaillon, 1964; Quézel, 1965; Conrad, 1971). Following the Würm glaciation approximately 12 000 years ago, the Sahara was invaded by a typically-Mediterranean flora and fauna, remnants of which still persist in some areas. About 2800 years BP the climate was mild and humid with Atlantic-type weather patterns and plants then found in the Saharan region included olive trees, juniper pines and cypresses with cedar, oak, walnut and heather in the

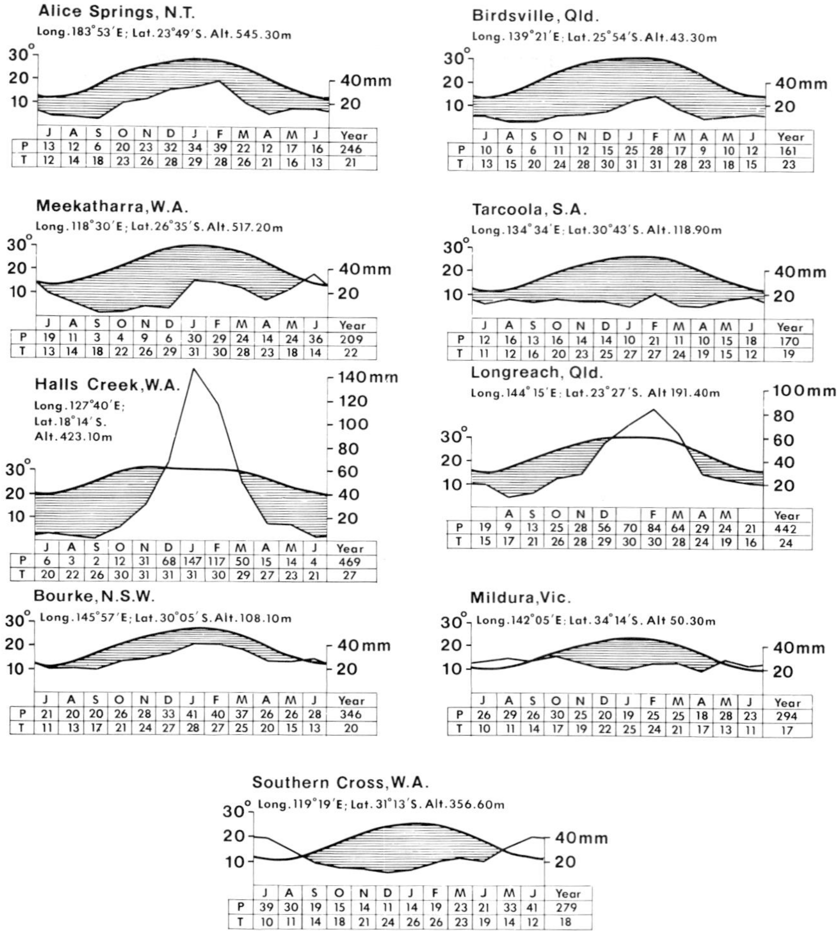


**Fig. 1.9** Ombrothermic Diagrams from arid and semi-arid regions of Australia (from MAB, 1979).

mountains of the High Atlas (see Table VI). Between 2800 and 2700 years BP however there was an abrupt change in climate. It became colder and more contrasted and this coincided with the almost total disappearance of the Mediterranean flora which was replaced by an *Acacia* forest migrating northward from the Sudan. By 2500 BP 'desertification' of the Sahara had already commenced and this has proceeded with alarming rapidity to the present day. Large mammals characteristic of the African savannas (giraffe,

**Table VI:** Postglacial changes in climate and vegetation in the Sahara (from Quézel, 1965).

| Chronology | Postglacial phases in Europe | Vegetation in Northern Europe | Climate in Northern Europe | Prehistoric industries in Central Sahara and $^{14}C$ dates | Vegetation in Central Sahara | Paleosoils in Central Sahara | Climate in Central Sahara |
|---|---|---|---|---|---|---|---|
| – 10,000 BP | Upper Würm | Glacier | Glacial | ? | Steppic. Fragmented Mediterranean vegetation | Alluvial | Arid and steppic in type |
| – 8,000 | Subarctic period—Allerod oscillation | Mountain avens, birch, polar willows | Tundra | Mousterian | Mixed forest of cedars, oak, alder, lime, maple in mountainous regions and Aleppo pines at low altitudes | Diatomaceous paleosoils of forest type plus marshy soils | Temperate and humid climate with dry summers particularly at low altitudes (Mediterranean sub-humid to humid) |
| – 6,000 | Pre-Boreal phase | Birch, aspen, appearance of pines | Cold and humid | | | | |

| | | | | | | | |
|---|---|---|---|---|---|---|---|
| – 5,000 | Boreal phase | Pines, appearance of elm, lime and oak | Gradual warming, climate still continental | Neolithic Méniet – 3,500 Adrar Bous – 3,200 Rock bovidae – 3,000 Sefar – 3,000 | Aleppo pines, olive trees, juniper, cypresses, nettles, with cedars, oak, walnut, mastic and heather | Diatoms | Climate hot and dry (Mediterranean semi-arid) |
| – 2,800 | Atlantic period (optimal climate) | | Rapid re-warming. Mild and humid Atlantic climate | | | Swamp soils | Mountain regions still wet |
| | Sub-Boreal period | Sessile oak, beech, and Scotch pine | Climate colder and more contrasted | – 2,700 Rock-based, Libyan-Berber | Almost total disappearance of the Mediterranean flora with the invasion of the Sahara by *Acacia* forest | No paleosoils | Replacement of Mediterranean climate with a Sahelian climate |
| – 500 | Sub-Boreal period | Progressive disappearance of sessile oak | Climate humid and cold | | Rapid and progressive desertification of the Sahara | No paleosoils | Desert climate |

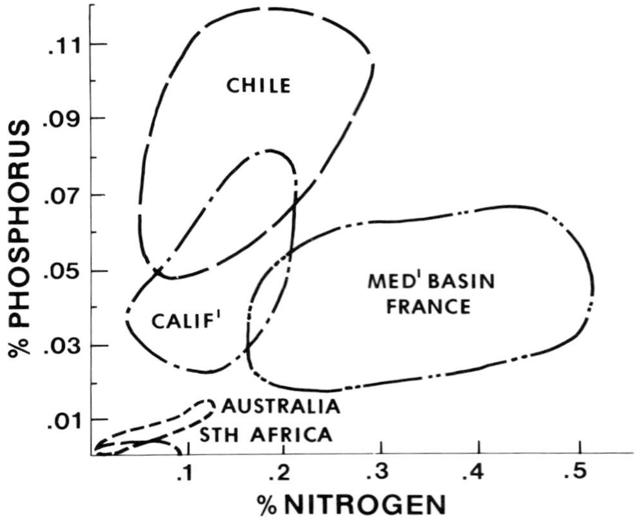


**Fig. 1.10** Levels of phosphorus and nitrogen (total %) in the soils of five regions with Mediterranean climates. Modified from Rundel (1979) with information related to soils of the Mediterranean Basin (from de Castri, 1981).

elephant, rhinoceros etc.) disappeared first during the Christian era and the Arab invasion in the 7th century has been associated with the loss of most of the rest of the fauna. Leopards, lions and cheetahs were still found 100–150 years ago in the Sahara (Le Houérou, 1979) but the hyena, addax, gemsbok, bubal hartebeest, dama gazelle, barbary sheep, nubian ibex and the dorcas and dune gazelle (*Gazelle leptoceros*) are all either now extinct or extremely rare in the Sahara. Some idea of the rapidity of this change can be gauged from the fact that Tunisia, which is now 76.1% desert, was one of the major granaries of the Roman Empire. Massinissa for example, King of Numidia (Eastern Algeria and Western Tunisia), exported more than 2000 t of wheat annually to Rome and Greece between 200 and 170 BC (Camps, 1960).

The characteristic herpetofauna which one encounters today in Morocco, Algeria, Tunisia, Libya and Egypt is in fact a mélange of Mediterranean and tropical relictual elements with a central core of species having Saharo-Sindien affinities. A similar situation exists with the plants and Le Houérou (1959) lists 1200 species with six families accounting for 50%–60% of the flora (Compositae, Graminae, Caryophyllaceae, Cruciferae and Umbelliferae). The rate of endemism is typically high (20 genera and 200 species many of which are Mediterranean relicts found only in mountainous regions) and many genera are monotypic (Jaccard's factor—No. genera/No. species is about 0.5). In the central Sahara the Saharo-Sindien element is dominant and Ozenda (1958) lists 490 species in the central Sahara, with 28% Mediterranean, 47% Saharo-Sindien, 17% tropical and 8% cosmopolitan.

The Saharo-Sindien domain with its characteristic plant and herpetofauna covers a vast area which lies approximately between latitudes 20°N and 30°N and extends from the western Sahara (Morocco) via the central Sahara (Algeria, Libya and Egypt) through Arabia to Iran and Afghanistan and includes the deserts of Baluchistan, West Pakistan and the Thar. Saint Girons (1952), in an early study, identified 14 species of snakes with Saharo-Sindien affinities, 5 of which occur throughout the whole region and 4 of which are restricted to the western sector. Gauthier (1967) and Bons (1967) both give lists of what they consider to be Saharo-Sindien elements in the saharan herpetofauna of Algeria (region of Béni-Abbès) and Morocco respectively and Table VII gives the Saharan herpetofauna in the region of Béni-Abbès in central Algeria. Saharo-Sindien elements clearly predominate (73% of the 33 species) followed by species with Mediterranean affinities (18%) and then species with tropical origins (9% and restricted to the snakes). Many of these tropical species are only found in extremely localised situations around oases and oueds (e.g. *Naja haje*) where supplementary water is available.

Within the Sahara the most specialised ecosystem is undoubtedly that of the 'erg'—the vast mobile sand dunes which harbour a characteristic flora and fauna. About 100 species of plants are found in the erg and a number of these are endemic to this biotope (Grenot, 1974). The Grand Erg Occidental of Algeria covers a vast area of approximately 80 000 $km^2$ and Vernet & Grenot (1972) studied the reptiles associated with this specialised habitat in the region of Béni-Abbès. Contrary to what one might think, *a priori*, the erg is in fact more humid than the other characteristic Saharan biotopes (reg, djebel and hamada) and moisture is usually found at depths between 5 cm and 80 cm which supports the very sparse vegetation. Vernet & Grenot (*op cit*) list 12 species (8 lizards and 4 snakes—see Table VII) which are almost totally restricted to the dunes and which, together with a small number of associated insects, birds and mammals, form a virtually-independent ecosystem (see Fig. 1.11). Six of these same psammophilous species are also found in the northern sand dunes of the Sinai desert (Werner, 1982) and the presence of sand-swimming species such as *Scincus scincus* in both attests to the age of these dunes.

Little can be said at this stage concerning the origins of the North African Saharan herpetofauna, other than that it obviously did not evolve *in situ*. Reptilian fossil material from a Miocene site in Morocco (Béni-Mellalsee H Hoffstetter, 1961) falls clearly into either paleoarctic or tropical categories and attests to wide-ranging and relatively-uniform climatic conditions throughout

**Table VII:** Composition of the Saharan herpetofauna in the region of Bénni-Abbès (Algeria)*.

| Lizards | Geographic Affinity |
|---|---|
| Agamidae | |
| *Agama bibroni* | M |
| *Agama mutabilis* | S |
| *Agama tournevillei*** | S |
| *Uromastix acanthinurus* | S |
| Gekkonidae | |
| *Ptyodactylus hasselquisti* | S |
| *Tropiocolotes tripolitanus* | S |
| *Stenodactylus stendodactylus* | S |
| *Stenodactylus petriei*** | S |
| *Tarentola mautitanica* | M |
| *Tarentola ephippiata* | S |
| *Tarentola neglecta*** | S |
| Lacertidae | |
| *Acanthodactylus scutellatus* | M |
| *Acanthodactylus pardalis* | M |
| *Acanthodactylus inornatus*** | S |
| *Acanthodactylus boskianus* | S |
| *Acanthodactylus longipes* | S |
| *Eremias guttulata* | S |
| *Eremias rubopunctata* | S |
| *Eremias pasteuri*** | S |
| Scincidae | |
| *Sphenops boulengeri*** | S |
| *Chalcides ocellatus* | M |
| *Scincus scincus*** | S |
| Varanidae | |
| *Varanus griseus*** | S |
| Snakes | |
| Leptotyphlopidae | |
| *Leptotyphlops macrorhynchus* | T |
| Colubridae | |
| *Malpolon moilensis*** | S |
| *Psammophis sibilans* | T |
| *Psammophis schokari*** | S |
| *Spalerosophis diadema* | S |
| *Natrix maura* | M |
| *Lytorhynchus diadema*** | S |
| Viperidae | |
| *Cerastes cerastes* | S |
| *Cerastes vipera*** | S |
| Elapidae | |
| *Naja haje* | T |

* Compiled from Gauthier (1967), Bons (1967), Saint Girons (1952), Grenot & Vernet (1972a & b), Vernet & Grenot (1972).
** Species associated with the Erg (mobile sand dunes)
M = Mediterranean S = Saharo-sindien T = Tropical

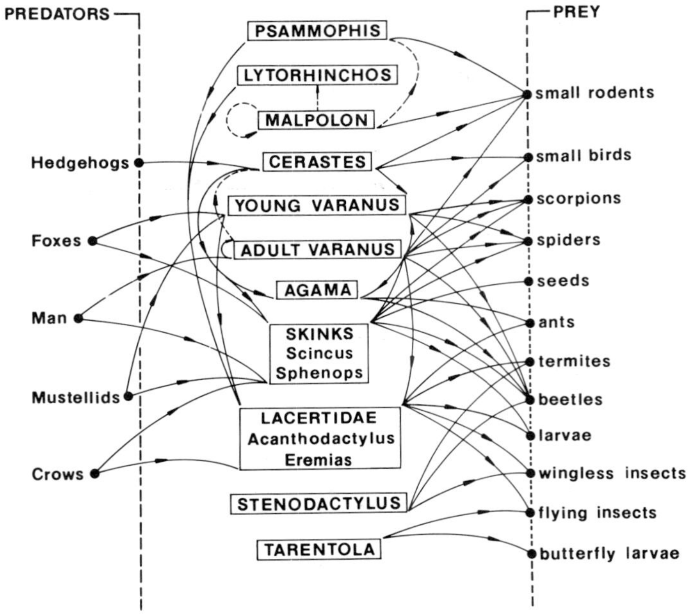


**Fig. 1.11** Trophic structure of the reptilian community found in the Grand Erg Occidental of the algerian Sahara (from Vernet & Grenot, 1972).

the African continent at that epoch (Lavocat, 1961). The Saharo-Sindien herpetofauna must obviously have been located farther eastward at this time—lizards such as *Varanus* are known from the Pliocene (2–12 MYBP) and it is probable that specialised herbivores such as lizards of the genus *Uromastix* had already evolved by this period. Changing climatic conditions in the Pleistocene which heralded the extinction of the unique Ethiopian Tertiary mammalian fauna modified the North African environment in such a way as to make it much more eminently suitable for the Saharo-Sindien herpetofauna (See Saint Girons, 1982a).

## B. Australia

Any consideration of the origins and evolution of the Australian herpetofauna needs to be viewed in the context of the major migration northward experienced by the Australian tectonic plate since late Mesozoic times. The breakup of Gondwanaland commenced at the beginning of the Cretaceous (135 MYBP, Powell, Johnson & Veevers, 1981) and Australia separated from Antarctica in two distinct phases (Crook, 1981). New Zealand was isolated first in the late Palaeocene (80–60 MYBP), followed by the separation of the Australian and Antarctic continents towards the end of the Palaeocene (60–53 MYBP). The Australian continent then migrated northward to collide with the Sunda Arcs about 15 MYBP in the middle Miocene. Between 10 MYBP and the present, Australia has moved farther northward and is now centred over approximately latitude 25° S and consequently experiences, in large part, an arid subtropical climate. Kemp (1981) has attempted to reconstruct Tertiary climates in Australia and widespread warm and humid conditions in the Palaeocene coincided with the first appearance of some of the major elements of the Australian flora (e.g. Myrtaceae, Proteaceae, *Nothofagus*—see Martin, 1981). By the early Eocene Australia had moved north to about latitude 45° S and was substantially free of the Antarctic continent, except in the region of the South Tasman Rise. Warm shallow seas covered much of the southern continent (McGowran, 1978) and tropical rainforest vegetation was widespread. The middle Eocene saw an abrupt change in climate with substantially lower temperatures and the loss of many of the 'tropical' elements of the vegetation. *Nothofagus* trees (Southern Beech) were widespread suggesting a climate somewhat similar to that of present Tasmania.

A trend towards increasing aridity became inevitable with the separation of the South Tasman Rise from Antarctica in the mid-Oligocene (*ca.* 33 MYBP) and the beginnings of circum-polar oceanic circulation. According to Frakes & Kemp (1972) and Kemp (1981) this process would have been completed about 23 MYBP, at the Oligocene/Miocene boundary when Drake Passage between South America and Antarctica opened, allowing the formation of the Circum-Polar Current. This current has the effect of climatically-isolating Antarctica from other land masses and significantly reduces the efficiency of meridional heat transport between the Equator and the South Pole. The first evidence of sea-level ice is found in the late Oligocene (Kenneth & Shakleton, 1976) and all evidence suggests that there was a major fall in environmental temperatures at high latitude in the Oligocene although the extent of the southern ice-cap remains conjectural. What is clear however is that the growth of the East-Antarctic ice-sheet intensified in the early Miocene with something close to the present ice volume being reached by late Miocene times.

The climate of the early Miocene is thought to have been mild, with warm seas again covering much of the southern continent, but this gave way to a much more contrasted climate with well-defined wet and dry seasons. Rainfall during this period was high and there is widespread evidence of weathering profiles associated with duricrust formation in the Miocene which had important consequences for the flora and fauna. Many of the elements essential for biological activity were leached from the parent rock during this period and either bound within the weathering profile or deposited superficially during the process of 'laterisation'. Australian soils are thus notoriously deficient in most essential nutrients, particularly nitrogen and phosphorus, (see Specht, 1981 and Fig. 1.10). Beadle (1966) has argued that the typical 'woody' nature of much of the Australian flora results from the low phosphorus status of the soils which is rendered even more unavailable by the acidic nature of many of the soils.

Beard (1978) has suggested that arid conditions may have first appeared in the Kimberley region of West Australia as early as the late Eocene with a Mediterranean climate in the Pilbara region by the Oligocene. What is clear however is that dry conditions prevailed by the mid-Miocene and lacustrine strata of central Lake Eyre and Lake Frome show evidence of the presence of grasses (Balme, 1962; Galloway & Kemp, 1981). Bowler (1976) suggests that the trend towards increasing aridity begins in the middle Miocene which coincides with the development of the characteristic xerophytic vegetation of Australia and the replacement in the Pliocene of the older *Nothofagus* and gymnosperm rainforests by genera such as *Eucalyptus*, *Casuarina* and *Grevillea* (see also Beadle, 1981). The end of the Miocene was associated with a dramatic cooling of the climate and an extension of the ice shelves in the Ross Sea 400 km north of their present position (Van Couvering et al., 1977). This is thought to have further reduced rainfall, especially in the north-west and the interior of the continent and sedimentary basins show evidence of major regression between 5–6 MYBP (Quilty, 1977), presumably because of much lower water temperatures. The Quaternary climate of Australia has been marked by rapid climatic fluctuations (Kershaw, 1981) with major eustatic changes in sea level during the Pleistocene which have formed land bridges between Australia and New Guinea and Tasmania a number of times. The climate of Australia is now typically arid over much of its area (Williams, 1979) and dominated by its position in the global high-pressure belt. The climate in the south-west of Australia is currently Mediterranean (Beard, 1981) and that of the north, cyclonic and monsoonal (MAB, 1979). Presumably, in time, if the Australian tectonic plate continues its northward progression, the mesic south of Australia will move into the arid zone and the north will experience a more consistent set of tropical conditions.

Given this background one would expect to find, as with the plants (Beadle, 1981), three distinct elements in the Australian herpetofauna: (a) ancient Mesozoic forms with Gondwana affinities, (b) a Tertiary fauna showing moderate adaptation to xeric conditions, and (c) a modern arid-adapted fauna. Bartholomai & Howie (1970) have shown that the Triassic reptilian fauna of Australia differs little from that of other southern continents and only the aquatic turtles of the family Chelidae and the extinct terrestrial tortoises of the family Meiolanidae appear to correspond to Gondwana relicts. Modern groups of reptiles which first appear in the Tertiary include the crocodiles *Palimnarchus* (Pliocene) and *Crocodylus* (Miocene), the chelid tortoises *Emydura* and *Chelodina* (Pliocene), the varanid genera *Megalania* (Miocene) and *Varanus*

**Table VIII:** Diversity and adaptive radiation of Australian desert lizards (from Pianka, 1981).

| Family and genus | Total number of species | No. of species penetrating desert | Percentage Eremaean |
|---|---|---|---|
| Agamidae | | | |
| *Amphibolurus** | 26 | 19 | 73% |
| *Diporophora* | 10 | 3 | 30 |
| *Tympanocryptis* | 6 | 5 | 83 |
| Gekkonidae | | | |
| *Diplodactylus* | 22 | 16 | 73 |
| *Nephrurus* | 6 | 6 | 100 |
| Pygopodidae | | | |
| *Delma* | 13 | 6 | 46 |
| Scincidae | | | |
| *Ctenotus* | 42 | 23 | 55 |
| *Egernia* | 22 | 8 | 36 |
| *Lerista* | 30 | 14 | 47 |
| Varanidae | | | |
| *Varanus* | 21 | 8 | 38 |

* Perhaps polyphyletic.

(Pliocene) as well as agamid lizards close to *Amphibolurus* (Pliocene), boid snakes similar to *Morelia* (Pliocene) and some Pliocene snakes which resemble elapids (Cogger & Heatwole, 1981). There is no evidence however that these Tertiary forms evolved from Gondwana progenitors, and all, with the exception of the chelid tortoises, appear to have invaded Australia when the tectonic plate collided with that of South-east Asia in the mid-Miocene. These invaders appear to have undergone spectacular radiations in the mid and late-Tertiary to give rise to such modern groups as diplodactyline geckos, elapid snakes, the endemic family of pygopodids ('legless' lizards) as well as major segments of the Agamidae, Varanidae and Scincidae (see Table VIII).

Land-bridges formed with New Guinea and south-east Asia in the Pleistocene appear to have been a major factor in structuring the modern Australian herpetofauna which shows strong affinities with that of Asia and New Guinea (see Table IX). Two-way migration appears to have occurred between Australia and New Guinea in the Pleistocene with the pygopodids invading New Guinea and the carettochelids and some colubrid snakes invading Australia. Various hypotheses have been advanced to account for the diversification of various groups of Australian reptiles and a number of these deserve comment.

Kluge (1967a) postulated that the diplodactyline geckos evolved in the tropics of South-east Asia in the late Mesozoic and entered Australia via the Sumatran Tract of the Indo-Australian Archipelago as long ago as the Palaeocene. He suggested that two tribes (the Carphodactylini and Diplodactylini) differentiated as early as the Palaeocene/Eocene and that the then-vast inland lake system (Dieri and Torrens), together with the Flinders Ranges, would have formed an effective barrier separating eastern and western populations and promoting subsequent speciation. The Carphodactylini apparently radiated in association with the tropical flora of Australia at that time (2 genera, *Carphodactylus* and *Pseudothecadactylus* are still tropical) and only the advanced species of *Nephrurus* and *Phyllurus* have invaded the arid regions. The Diplodactylini on the other hand are a primarily-Eyrean group and Kluge (*op cit.*) speculates that differentiation into the various genera (*Crenodactylus, Diplodactylus, Lucasium, Oedura, Rhynchoedura*) occurred during the late Tertiary when the climate was markedly seasonal. The Gekkoninae entered Australia most certainly via land-bridges in the Pleistocene (Kluge, 1967b) and represent the 'modern expanding dominants' (*Cyrtodactylus, Lepidodactylus* and *Hemidactylus*) all of which have representatives in New Guinea, the Pacific Islands and South-east Asia. The gekkonine genera *Heteronotia* and *Phyllodactylus* each contain only two species and only the genus *Gehyra* has radiated extensively within Australia.

The geckos are obviously an extremely old group of animals and both their suggested ancestry to the Pygopodidae—(the only extant reptilian family thought to have evolved in Australia—(Underwood, 1957); Keast, 1959)—and their presence in New Caledonia and New Zealand suggest an early dispersal from South-east Asia. Whether they entered Australia as early as the Palaeocene is doubtful however as recent studies on the movement of the Australian tectonic plate (Powell, Johnson & Veevers, 1981) have shown that southern Asia has remained in its present latitude since the mid-Cretaceous and only undergone a small anticlockwise rotation of about 45°. Powell & Johnson (1978) suggest therefore that Australia and Sundaland did not come into contact until the mid-Miocene (*ca.* 15 MYBP) when the first invasion of Asian reptiles into Australia is thought to have occurred.

The Chelidae however are thought to be Gondwana relicts and the family is now found in South America, Australia and New Guinea. Burbidge, Kirsch & Main (1974) speculated that the ancestors of the chelids were present in Australia in the Cretaceous when it was still part of the Antarctic continent and enjoyed a warm, tropical climate. They speculate that the extensive Cretaceous inland sea would have isolated primitive chelids (represented by the fossil *Chelycarapookus*) into eastern and western populations, the latter now represented by *Chelodina oblonga* and the very rare side-necked tortoise, *Pseudemydura umbrina*. These western species were subsequently restricted to

**Table IX:** Characteristics of the modern Australian reptilian fauna. (from Cogger & Heatwole, 1981).

| Family | World, approx. No. of genera | World, approx. No. of species | Australia, No. of genera | Australia, No. of species | Australia, No. of endemic genera | Australia, No. of endemic species | Shared with New Guinea No. of genera | Shared with New Guinea No. of species | Genera shared with other parts of Asia | Genera shared with areas other than Asia or New Guinea | Worldwide or nearly worldwide species | General distribution |
|---|---|---|---|---|---|---|---|---|---|---|---|---|
| Crocodylidae | 3 | 13 | 1 | 2 | 0 (0) | 1(50) | 1(100) | 1 (50) | 1(100) | 1(100) | 0 | Africa, North America, South America, Asia, Indo-Australasia |
| Cheloniidae & Dermochelyidae | 5 | 7 | 5 | 6 | 0 (0) | 1(17) | 5(100) | 6(100) | 5(100) | 5(100) | 6 | Tropicopolitan |
| Chelidae | 10 | 27 | 4 | 15 | 1(25) | 11(73) | 3 (75) | 4 (27) | 1 (25) | 0 (0) | 0 | Australia, New Guinea, South America |
| Carettochelyidae | 1 | 1 | 1 | 1 | 0 (0) | 0 (0) | 1(100) | 1(100) | 0 (0) | 0 (0) | 0 | Australia, New Guinea |
| Gekkonidae | 82 | 650+ | 16 | 63 | 11(69) | 57(91) | 5 (31) | 6 (10) | 2 (13) | 2 (13) | 3 | Tropicopolitan |
| Pygopodidae | 8 | 30 | 8 | 29 | 7(88) | 28(97) | 1 (13) | 1 (3) | 0 (0) | 0 (0) | 0 | Australia, New Guinea |
| Agamidae | 30 | 280 | 10 | 52 | 5(50) | 49(94) | 5 (50) | 1 (2) | 3 (30) | 0 (0) | 0 | Asia, Africa, Australia, New Guinea |

| | | | | | | | | | | | | |
|---|---|---|---|---|---|---|---|---|---|---|---|---|
| Scincidae | 40 | 600+ | 19 | 188 | 10(53) | 178(95) | 7 (37) | 10 (5) | 4 (21) | 2 (11) | 1 | Cosmopolitan |
| Varanidae | 1 | 24 | 1 | 19 | 0 (0) | 15(79) | 1(100) | 3 (16) | 1(100) | 1(100) | 0 | Africa, Australia, New Guinea, Asia |
| Typhlopidae | 4 | 200 | 1 | 22 | (?)0 (0) | 17(77) | (?)1(100) | 4 (18) | (?)1(100) | 0 (0) | 0 | Cosmopolitan |
| Boidae | 22 | 57 | 4 | 10 | 1(25) | 6(60) | 3 (75) | 4 (40) | 1 (25) | 0 (0) | 0 | Tropicopolitan |
| Acrochordidae | 2 | 2 | 2 | 2 | 0 (0) | 0 (0) | 2(100) | 2(100) | 2(100) | 0 (0) | 0 | Southeast Asia, Indo-Australasia |
| Colubridae | 180+ | 2000+ | 8 | 11 | 0 (0) | 1 (9) | 6 (75) | 8 (73) | 6 (75) | 0 (0) | 0 | Cosmopolitan |
| Elapidae | 30 | 300+ | 25 | 63 | 20(80) | 59(94) | 7 (28) | 4 (6) | 1 (4) | 0 (0) | 0 | Africa, North America, South America, Asia, Australia |
| Hydrophiidae & Laticaudidae | 15 | 52 | 12 | 32 | 2(17) | 11(34) | 8 (67) | 13 (41) | 8 (67) | 2 (17) | 1 | Indo-Pacific region |
| Total | — | — | 117 | 516 | 57(49) | 434(84) | 56 (48) | 68 (13) | 36 (31) | 13 (11) | 11(2) | |

Percentage of Australian genera or species in parentheses.

the south-west corner of the continent by the expansion of aridity in the Tertiary from a north-west Australian centre which also fragmented the distribution of the eastern *Chelodina* and the genera *Emydura* and *Elsyea*. Pleistocene land connections with New Guinea permitted the passage of chelids and there are now species-pairs in several groups on either side of the Torres Strait (Cogger & Heatwole, 1981).

The Varanidae is another ancient family with a wide distribution in Africa, Asia, the East Indies, New Guinea and Australia where it is represented in the Miocene by the giant *Megalania*. Storr (1964) postulated an Asian origin for *Varanus* with the subgenus *Odatria* originating within Australia and giving rise to a group of small secretive monitors. King & King (1975) on the basis of chromosomal morphology however placed the *gouldii* group of species in the *Odatria* group and suggested that *Megalania* may also have been a member of this group. They suggest that *Varanus* entered Australia in the mid-Miocene, along with other Asian reptiles, and that *Varanus timorensis* came from Australia during the Pleistocene at the same time as the *indicus* group entered the east and the north of Australia via land-bridges.

The Scincidae form an extraordinarily diverse and numerous group of lizards within Australia and their taxonomy is still somewhat confused (see Cogger & Heatwole, 1981). Horton (1972) has nonetheless suggested a model for speciation within the wide-ranging genus *Egernia* which is closely related to the Asian genus *Mabuya*. He suggests that the progenitor(s) entered Australia via the Torres Strait during a period of low sea level in the Pliocene and formed a 'species pool' in the Queensland rainforest. Pleistocene pluvials are then invoked to spread these forms initially southward on either side of the Great Dividing Range and then interpluvial arid phases isolate the original Torresian populations as well as those in Queensland from those that have penetrated farther southward (see Heatwole, 1976 for details). Subsequent pluvials and interpluvials are then employed to account for invasion and speciation within the arid central areas of Australia along the lines suggested originally by Main, Lee & Littlejohn (1958) for the Australian amphibian fauna. One criticism levelled at this scenario is that the rates of evolution within the genus *Egernia* would need to be exceptionally high to account for the current species' diversity. Also 'classic' Australian concepts of a succession of 'glacial pluvials' together with a mid-recent hypisthermal major period of aridity have recently been called into question by Galloway & Kemp (1981).

The Agamidae (dragon lizards) are an Old-World group of lizards which have radiated extensively within Australia, and primarily within the arid regions. Cogger (1961) advanced two explanations to account for the explosive radiation of this group: isolation of taxa in xeric refugia by anastomosing stream systems in the Pleistocene, and selection for heterochronic growth processes with subsequent neoteny (as in *Amphibolurus fordi*, for example).

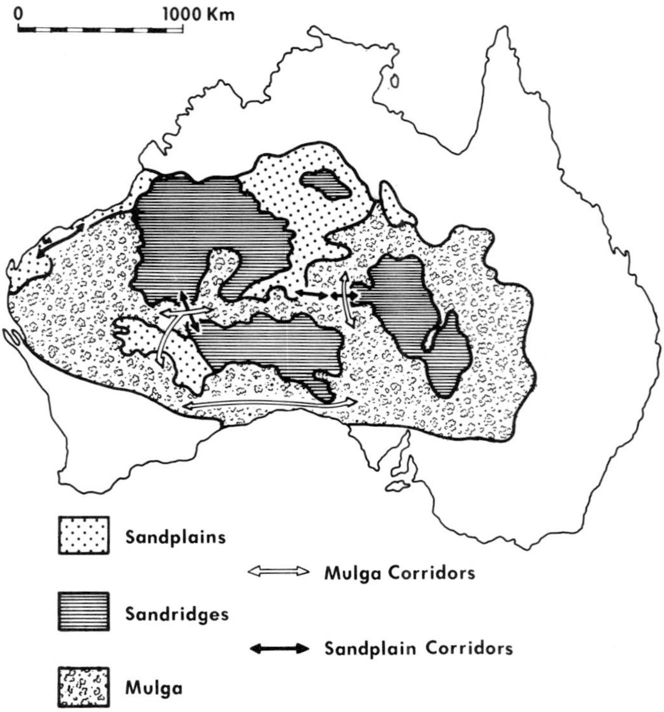


**Fig. 1.12** Map showing the approximate extent of sandplains, sandridges and shrub habitats. Shrub and sand corridors indicated by arrows. Note especially the two long east-west shrub corridors: the northern one is the Giles Corridor while the southern one runs through the Nullarbor Plain and its bordering zone of myall. Only the major habit occurring in any given area is shown; small-scale maps such as this cannot show adequately the habitat heterogeneity within subregions which in some cases is quite pronounced (from Pianka, 1981).

Four invasions of Asian stock commencing in the mid-Tertiary were envisaged. (a) the first giving rise to the bizarre and highly specialized 'mountain devil' (*Moloch horridus*); (b) a Pliocene invasion giving rise to the amphibolurid radiation (e.g. *Amphibolurus, Diporophora, Chlamydosaurus, Tympanocryptis* and *Lophognathus*); (c) and (d) two late Pleistocene invasions bringing *Physignathus lesueurii* and *Gonocephalus* both of which are still restricted to rainforest areas in northern Queensland. Pianka (1969, 1972 and 1981) divided lizard species living in the arid interior of Australia into three groups on the basis of habitat: (a) sand-ridge or dune species (*Amphibolurus fordi, Diporophora winneckei, Lophognathus longirostris, Ctenotus brooksi, dux* and *leae, Diplodactylus stenodactylus* and *Nephrurus laevissimus*); (b) mulga

(*Acacia*) shrub species (*Amphibolurus recticulatus* and *scutulatus, Varanus caudolineatus, Diporophora amphiboluroides, Ctenotus leonhardii, Egernia depressa, Sphenomorphus richardsoni, Diplodactylus pulcher* and *Nephurus vertebralis*); and (c) sand-plain-spinifex (*Triodia*) species (*Varanus eremius, Moloch horridus, Ctenotus calurus, grandis, helenae, pantherinus, pianki, quattuordecimlineatus, Egernia striata, Pygopus nigriceps* and *Nephrurus levis*). The mulga-shrub habitat is widespread in the Australian desert as shown in Figure 1.12 and effectively isolates large sand-ridge areas and spinifex sand-plains which form parts of the Great Sandy Desert (sand-dunes), Tanami desert (spinifex), Simpson (sand-dunes) and Great Victoria deserts (part sand-dune and part spinifex). Pianka (1972) identified a corridor of mulga-shrub country which maintains continuity between the eastern and western sections of this habitat and named it the 'Giles Corridor' and suggested this as a narrow conduit through which gene flow occurred. In this hypothesis, habitat-specialisation is followed by reduced mobility and, ultimately, speciation in response to fluctuations in the size and shape of these habitats with variation in climate, particularly during the Pleistocene. What is not obvious at this stage however is whether these changes in habitat constitute real barriers to gene flow or represent instead the ecological preferences of the adult species. *Amphibolurus ornatus* for example is a species which is restricted to exfoliating granite tors in Western Australia but morphological differentiation between widely-ranging populations is slight and the juvenile form is highly dispersive and readily traverses other habitats (Bradshaw, 1971). Other species such as *Amphibolurus nuchalis* occur throughout the area occupied by the three major habitats described above, and seem in no way limited by them. The grouping together of these particular species, according to very broad habitat definitions, may or may not prove useful in the long run in identifying modes of speciation in widely-divergent groups of reptiles. What is not in doubt however is the extraordinary richness, in terms of species, of these central arid regions of Australia.

In an effort to clarify the situation, Cogger & Heatwole (1981) summarised the conventional wisdom concerning the distribution of reptiles within Australia and then undertook a quantitative analysis of species' distributions at the level of family. Similarity coefficients were computed and from these broad areas within which certain species occur were defined and then plotted on the map of the continent. From this study Cogger & Heatwole (*op. cit.*) were able to draw a number of major conclusions, some of which agree with and others of which modify substantially previous opinion. One can summarise these findings as follows:

(a) In the case of pygopodids and geckos the first major phenetic division isolates the fauna of the south-west corner of the continent from the remainder.

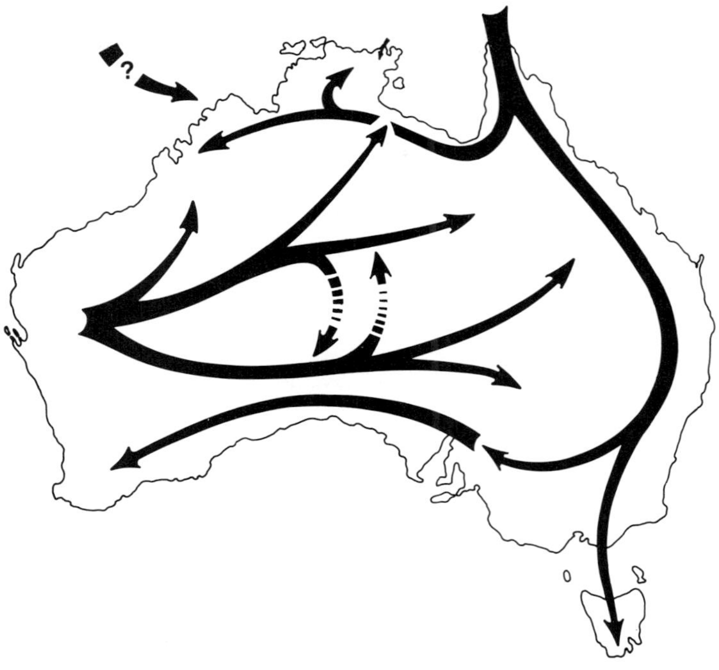


**Fig. 1.13** Major dispersal routes postulated for the Australian reptilian fauna (from Cogger & Heatwole, 1981).

Furthermore, the divisions within the Pygopodidae are virtually confined to the southern half of the continent suggesting that these lizards represent a relictual fauna which was little influenced by the changes seen during the Pleistocene.

(b) In the case of both the elapid snakes and the skinks, the first major phenetic division isolates the fauna of the Cape York Peninsula from the remainder, indicating the importance of recent immigration in these groups.

(c) A broad western Torresian or Timorean subregion is indicated for the skinks, varanids, gekkonids and elapid snakes but is entirely lacking for the pygopodids.

(d) Only in the Agamidae does the first phenetic division isolate the central Eyrean component indicating a major radiation in this region. In all other families except the Pygopodidae this region also exists and is further subdivided into centres corresponding to the north-eastern (Pilbara), southern (Nullarbor) and central-eastern (Simpson) desert, suggesting foci of speciation.

**Table X:** Numbers of species of lizards in different families found in sympatry on desert study areas on three continents (from Pianka, 1981).

| Lizard family | North America | Kalahari | Australia |
|---|---|---|---|
| Agamidae | | 1 (1) | 2– 8 (11) |
| Chameleontidae | | 1 (1) | |
| Gekkonidae | 1 (1) | 4–7 (7) | 5– 9 (13) |
| Helodermatidae | 1 (1) | | |
| Iguanidae | 3–8 (9) | | |
| Lacertidae | | 3–5 (7) | |
| Pygopodidae | | | 1– 2 (3) |
| Scincidae | | 3–5 (6) | 6–18 (28) |
| Teiidae | 1 (1) | | |
| Varanidae | | | 1– 5 (5) |
| Xantusidae | 1 (1) | | |
| Totals | 4–11 (13) | 12–18 (22) | 18–40 (60) |

Total number of different species in each family is listed in parentheses.

There seems little doubt that the various classifications offered by Cogger & Heatwole (1981) reflect real differences in continental radiation patterns of the Australian herpetofauna and they also enable the identification of the major dispersal routes within the continent which are shown in Figure 1.13. What is immediately apparent from this figure is the importance, as a major centre of speciation and dispersal, of the central and Pilbara regions of Western Australia. This is the area which is postulated to have been arid since mid-Miocene times and its importance as a xeric refuge during the Pleistocene cannot be over emphasised. Clearly, much work remains to be done on the evolution of the Australian herpetofauna but its primarily Asian origins in the Tertiary with subsequent extinctions and accessions in the Pleistocene seem to be well established.

## C. North America

These deserts are all located in the central and south-western regions of the United States and Mexico (see Figure 1.6) and the associated herpetofauna includes a number of families not found in the deserts of the Old World—the Helodermatidae, Iguanidae, Teiidae and Xantusidae (see Table X). Savage (1960) in an early analysis of the zoogeography of the Baja California region classified the modern North-American herpetofauna into a number of assemblages (see Figure 1.14) and recognised a single unit for the desert and

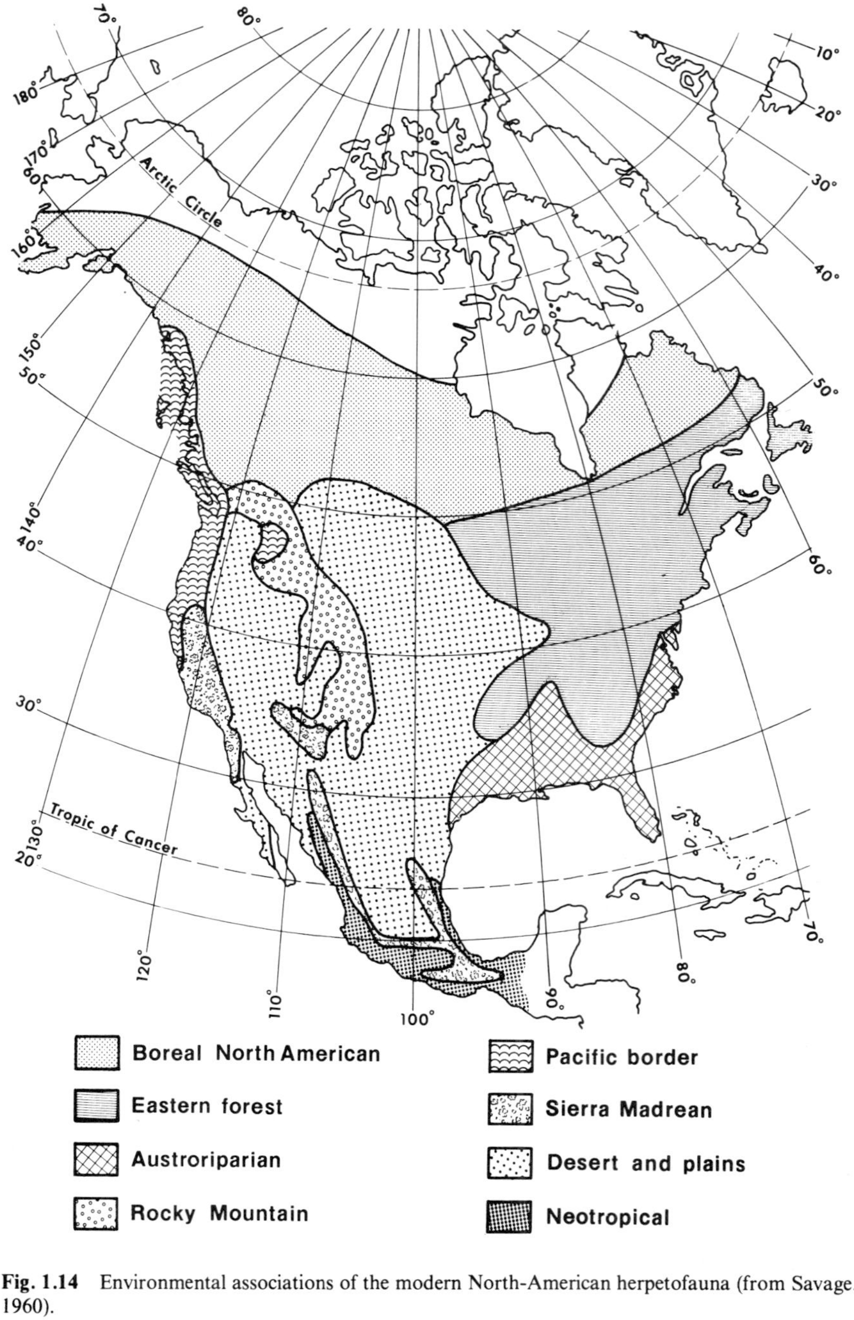


**Fig. 1.14** Environmental associations of the modern North-American herpetofauna (from Savage, 1960).

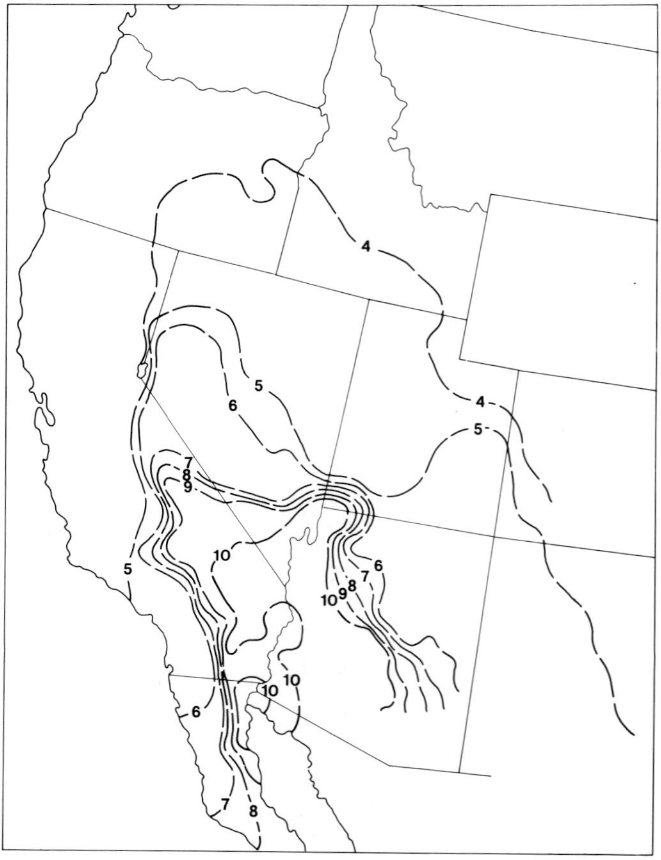


**Fig. 1.15** Contour map of the number of species of North-American flatland desert lizards, based upon eleven of the twelve species. The eleven-species isopleth has been omitted (from Pianka, 1967).

Great Plains biome. The species composition of this group of arid-adapted species is now well known and fine-grain species density maps have been constructed by Pianka (1967). As seen from Figure 1.15, high species diversity is associated with the Sonora and northern Chihuahua deserts. To a large extent the fossil herpetofauna of these regions is not well known and attempts to retrace the recent evolution of desert reptiles are forced to rely on palaeobotanical evidence (e.g. Savage 1966; Raven & Axelrod, 1975; Macmahon, 1978). The studies of Axelrod (1950, 1958) resulted in the concept of three major geofloras undergoing evolution and dispersion in the Americas during the Tertiary with one of these, the Madro-Tertiary, containing plants which were to become increasingly adapted to arid conditions. A neotropical flora and fauna appears to have occupied present desert areas during much of the early Tertiary and there is evidence of long-term environmental trends towards increasing aridity with so called 'subdesert' conditions prevailing towards the end of the Miocene (Johnson, 1968). Deserts expanded markedly during the Pliocene to form the Mohavia which included the present Mojave, Colorado and Sonora deserts, and the northern Great Basin Desert which corresponds roughly to the present Great Basin and Plains desert. During the Pleistocene, extensive glaciation in the north of the American continent restricted deserts to small regions in northern Mexico and the extreme south-western United States. Deserts such as the Chihuahua of Mexico are thought to be of recent post-glacial origin (i.e. 10–12 000 yr BP, see Wells, 1977) and the most common plant covering much of the present Mojave, Colorado and Chihuahua deserts, creosote bush (*Larrea divaricata*) apparently only entered North America from South America at this time (Johnson, 1968) and now occurs as a single species with a number of chromosomal races (Wells, 1977).

Although it is generally agreed that the herpetofauna of the New World deserts is 'younger' than that of the Old World (Mayhew, 1968) it is also assumed that substantial elements of the current desert herpetofauna originated in the Pliocene or late Miocene, even if it is somewhat difficult to identify these same elements. Pianka (1967) notes that lizards of the genera *Cnemidophorus, Uta, Crotaphytus* and *Sceloporus* occur widely today in savanna grasslands on the edge of deserts and speculates that they may have existed in the Great Basin by the middle Pliocene. This seems unlikely in the case of *Cnemidophorus* however which is the only Teiid genus to have invaded North America from South America and Presch (1980) suggests a more recent invasion following the formation of the Panamanian land-bridge. Fossil evidence of the Gila Monster (*Heloderma*) from Colorado (Gilmore, 1928) proves that at least this family was present during the Oligocene. Savage (1960) speculates that ancestors of desert lizards such as *Dipsosaurus dorsalis* (desert iguana), *Sauromalus obesus* (chuckwalla) and *Sceloporus orcutii* survived glacial maxima during the Pleistocene in thorn-scrub refugia in the Baja California and southern Mexico (see Figure 1.16).

More recent biogeographic and phenetic studies of the American desert herpetofauna (e.g. Tihen, 1964; Blair, Hulse & Mares, 1976; Morafka, 1977; Grenot & Serrano, 1982; Barbault & Halffter, 1981; Lambert & Reid, 1981) unfortunately give no clue as to the age of the assemblages, other than to suggest that they have evolved together in the face of common environmental exigencies over appreciable periods of time. Recent studies (see later) have shown that the lizard fauna of the North American deserts is noticeably less diverse than that of Old-World deserts and this could, in part, reflect their more

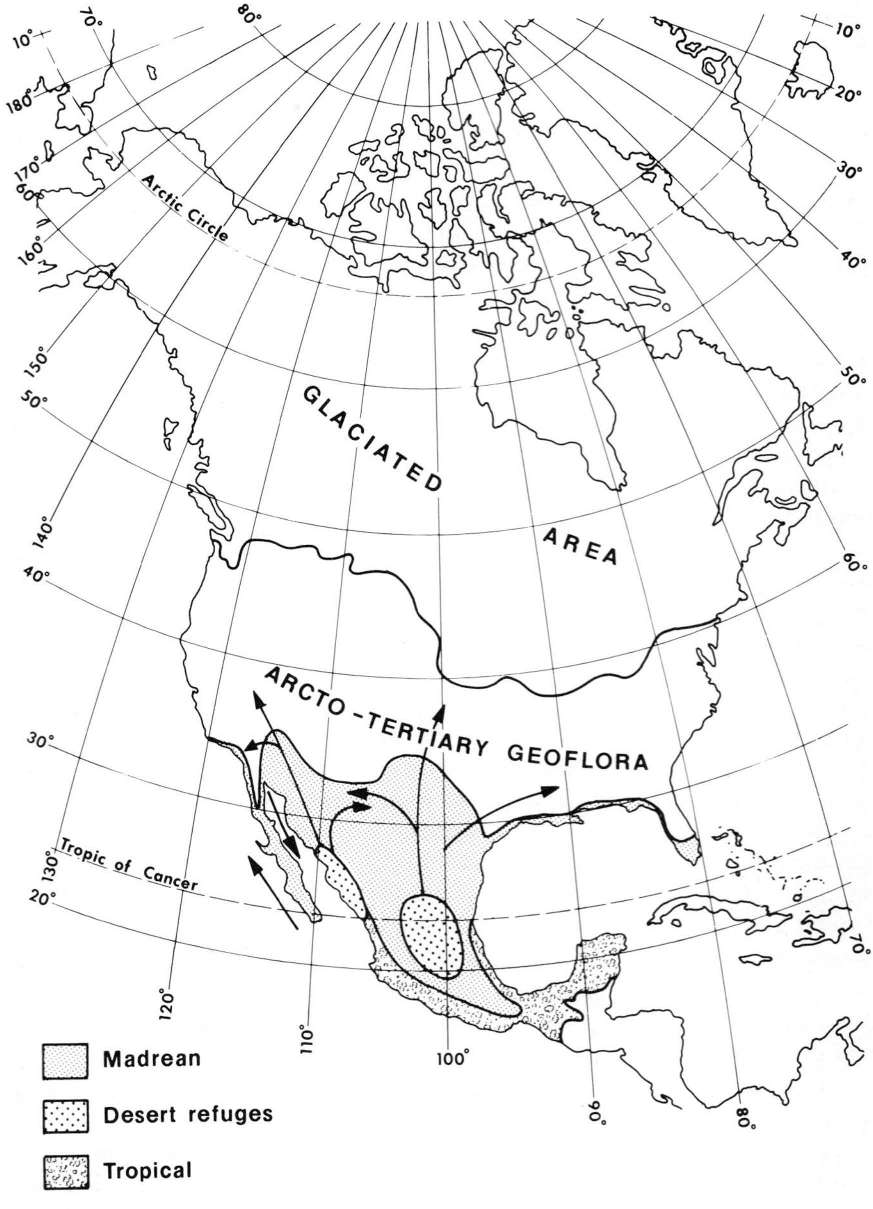


**Fig. 1.16** Distributional pattern of herpetofaunal elements during North American glacial maxima to indicate probable location of refugia for desert forms. Solid arrows indicate the general direction of dispersion for desert forms as glaciers retreated and arid conditions made more northern areas available for occupancy (from Savage, 1960).

recent origin. Future studies will undoubtedly add much to our knowledge of the evolution of this herpetofauna.

## V. COMMONALITIES

The deserts of North Africa, Australia and North America differ substantially in the species-richness of their reptilian herpetofaunae, Australian deserts having densities of up to 40 species/$km^2$, the Kalahari 18/$km^2$ and the North American deserts 11/$km^2$ (Pianka, 1966, 1969, 1971, 1975, 1981b; Keister, 1971; Kitchener *et al.*, 1980; Heatwole & Butler, 1981—see Table X). At the generic level, Australian deserts are also richer, with 23 genera, than the Kalahari with 13 genera and the North American deserts (12 genera) but the reasons for these differences are obscure. Pianka (1973, 1974, 1975), Huey & Pianka (1977a) and Schall & Pianka (1978) analysed niche-breadth relationships of desert lizards on the three continents but found no clear trends. The diversity of food eaten by an 'average' species is just as great in Australia as in North America and average niche overlap was not greater in more diverse saurofaunas, but rather varied inversely with lizard diversity. The higher diversity of the Australian saurofauna is thus not due to either narrower niches or greater niche overlap but to a greater 'recognition' of habitats by these species (Pianka, 1978; Pianka & Schall, 1981). Other workers have speculated on the reasons for the greater ecological diversity of lizards in Australia and Morton (1979) has shown that the high diversity of all vertebrate insectivores is a general phenomenon in Australian deserts and may be a reflection of higher insect diversity on this continent (see Table XI). The Australian desert also covers an area approximately 3 to 4 times as large as the North American deserts, is geologically much older, and might therefore be expected, *a priori*, to house a richer and more diverse fauna (see Fig. 1.17).

Main (1978) has given an account of the various environmental exigencies which characterise the Australian desert and render survival there difficult for animals. To some extent these features are characteristic of most deserts: absence of water, absence of vegetational cover, extremes of temperature and seasonal deficiences of nitrogen and energy in the plants. The soils of the Australian deserts are particularly poor in plant nutrients, due to their long history of weathering in the Tertiary, and the peculiarly unique flora of Australia, which defies conventional international classification (Williams, 1979) has evolved partly in response to this. Plants in the Australian desert are characteristically high in fibre and 'woody' in response to the low phosphorus status of the soils and thus provide poor nutrition for herbivorous animals.

Milewski (1981, 1983) has recently compared the structure of Mediterranean ecosystems in Australia and southern Africa and he advances the very

**Table XI:** Approximate total numbers of species of various vertebrate taxa in Australia and the continental United States (from Pianka & Schall, 1981).

| Taxonomic group | Australia | United States |
|---|---|---|
| Mammals | 224 | 296 |
| Birds | 650 | 620 |
| Frogs | 133 | 52 |
| Toads | 1 (introduced) | 25 |
| Salamanders | 0 | 99 |
| All Amphibians | 134 | 176 |
| Turtles | 16 | 50 |
| Snakes | 101 | 104 |
| Lizards | 351 | 99 |
| Totals, all terrestrial vertebrates | 1476 | 1345 |
| Area in $10^6$ km$^2$ | 7.69 | 7.83 |

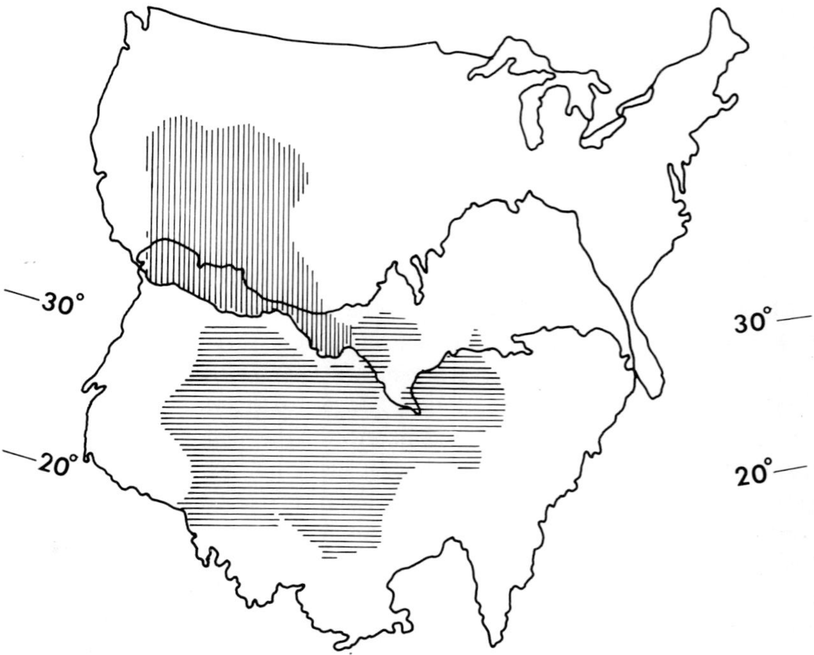


**Fig. 1.17** Equal-area projections of Australia and the United States superimposed. Latitudes and west coasts are aligned and approximate distributions of deserts are indicated by stippling (from Schall & Pianka, 1978).

interesting hypothesis that the low and unreliable productivity of the Australian soils is more suitable for ectotherms such as reptiles and their invertebrate prey than endothermic birds and mammals, with their higher daily energy requirements, and it is this unique infertility of the soils which accounts for this peculiar species richness. Other desert soils are not necessarily so deficient in phosphorus and nitrogen as those of Australia and South Africa however (Rundel, 1979) and this should not be seen as an essential feature of the desert environment. Deserts are climatic phenomena which can occur in regions with markedly-different geology, physiography and geographical location. To a large extent it is these latter factors which will determine the 'type' of desert that is produced (i.e. hot or cold, edaphic, coastal etc.) but the one constant factor is the presence of aridity. In looking for commonality therefore we should concentrate on aridity rather than other, local, factors which at first sight may seem more important and it is to aridity that we might anticipate finding parallel or convergent adaptations in the desert reptiles of the world.

# 2

# The Maintenance of Physiological Homeostasis in the Desert Environment

## I. AVANT PROPOS

The fundamental question of the maintenance of homeostasis in desert environments was first posed by Bartholomew (1964) in a provocative article in which he attempted to gauge the extent to which 'success' in desert-living organisms was consequent upon the evolution of superior regulatory capacities. Slobodkin & Rappaport (1974) have argued that the 'appropriateness' of an organism's response to a given environmental perturbation can be measured through its subsequent change in state. Adjustments which are 'adequate' to the occasion will entrain no drastic modification of the *milieu intérieur*, but will be associated with small changes in physiological parameters (e.g. increased haematocrit reading in mountainous regions, variations in blood levels of regulatory hormones etc.). Implicit in this argument is the concept of an optimal internal state of the organism which is 'protected' by the appropriate behavioural and physiological responses when confronted with environmental perturbations whether of a regular or irregular nature. Perturbations which surpass the capacities of the organism to respond appropriately engender modifications of the *milieu intérieur* which are of a pathological type (elevation or depression of body temperature, disturbance of water and electrolyte balance etc.) and are normally taken as indicators of stress.

The term 'stress' in the physiological literature is usually employed in two somewhat different senses. Either it indicates some environmentally imposed demand on the animal (e.g. heat or cold stress) or it describes the resultant change in physiological condition of the animal, consequent upon an environmental change which exceeds its regulatory capacities. The distinction between the two usages is not particularly clear in many cases (e.g. the 'stress' of reproduction) and it would seem useful to limit the use of the term to a clearly-definable set of circumstances. Of the two usages the latter (which owes more to Selyé than Cannon) seems the more useful in that it enables one to detect stress in animals by a significant departure of the *milieu intérieur* from whatever is considered to be its normal state. Animals exposed to temperature extremes are thus not necessarily stressed by this; stress only intervenes if the intensity of the exposure exceeds the thermoregulatory capacities and the body temperature departs significantly from some preferential level.

Elevated corticosteroid levels in the plasma (often used as an index of 'stress') are thus part of the normal adaptive response of the animal to the initial phases of many environmental changes (low temperatures, starvation etc. in mammals) and can only be used as an index of stress if accompanied by other physiological modifications of a profound nature. One must thus distinguish between variations in regulatory factors which are a normal part of the organism's homeostatic response, and changes in state or condition which are inherently inimical for the long-term survival of the organism. Elevated plasma levels of, say, antidiuretic hormone thus could not be taken, in isolation, as a measure of stress consequent upon water deprivation; stress would only be indicated if the total body water content of the animal was also depressed, indicating significant dehydration. In this simple example five distinct operational levels can be distinguished: (a) environmental perturbation (lack of water), (b) physiological stimulus (increase in plasma osmotic pressure due to imbalance between water influx and efflux), (c) hormonal response (increased rate of secretion of antidiuretic hormone from the pars nervosa of the pituitary gland), (d) effector response (decreased rate of urine production by the kidney), and (e) physiological state (total water content of the body which may or may not decline depending upon the severity of water deficiency in relation to the animal's capacities for water conservation through a modification of renal function). The significant parameter which the animal is attempting to control in this example is not osmotic pressure or antidiuretic hormone concentration, nor even the rate of filtration and reabsorption of water in the kidney, but the total body water content which must remain within certain definable limits if the animal is to survive.

If we attempt to apply the schema outlined in Table XII to the particular case of reptiles, a number of problems become immediately apparent. In many cases the identity of the relevant hormones is not known, let alone the

**Table XII.** Hierarchical nature of regulatory interactions between an organism and its immediate environment.

| Level | Nature of change | Example |
|---|---|---|
| Primary | Perturbation | Deficiency of water |
| Secondary | Stimulus | Elevated plasma osmolality |
| Tertiary | Hormonal response | Release of ADH |
| Quaternary | Effector response | Decreased urine production |
| Pentary | Resultant state | ±total body water content |

regulatory functions they subtend, and the existence of many so-called 'basic' physiological processes (in fact only well-established in laboratory mammals) is still a matter for debate (e.g. gluconeogenic and 'diabetic' effect of corticosteroids; action of aldosterone on sodium reabsorption; rôle of thyroid in regulation of oxidative metabolism; regulation of blood glucose concentration etc.). The peptide arginine vasotocin, for example, was first isolated from the reptilian pituitary gland over 20 years ago, but it has only recently been shown to function as a physiological antidiuretic hormone, and the rôle of mesotocin, which also occurs in the reptilian pars nervosa, is completely unknown.

To proceed, one needs to posit the existence of a hierarchy of interactions such as that shown in Table XII in reptiles, whilst being quite clear that the evidence for this is at best fragmentary. This enables one to postulate a series of physiological parameters which would be expected to be regulated, especially in desert environments characterised by aridity and extremes of temperature. Primary amongst these must be body temperature,* and water and electrolyte balance (the volume and distribution of water within the body and the concentration of electrolytes such as sodium, potassium and chloride) which, *a priori*, appear as the factors most likely to be affected in small diurnal vertebrates by the exigencies of the desert milieu. It is quite conceivable that, in some reptiles, these parameters are not regulated (e.g. there is now considerable evidence to show that many tropical lizard species do not regulate their body temperature by behavioural means—Huey, 1982) and that survival may depend instead on extreme tolerance, blood sodium levels being a good case in point. What is possible at this stage, however, and what will be attempted here, is an examination of the nature and extent of the variation in these basic parameters in desert reptiles, compared with the variation commonly

* De Witt & Friedman (1979) argue cogently that it is not temperature *per se* that is regulated in reptiles but rather a physiological rate process whose rate is an exponential function of temperature.

encountered in laboratory situations where the animals are not deprived of adequate sources of food, water and heat.

An absence of regulatory processes should manifest itself in the form of a very high variance in laboratory controlled situations where animals are exposed to variations in, for example, electrolyte status, coupled with an absence of change in tertiary hormonal factors normally thought to regulate these processes. Plasma aldosterone levels, for example, show almost no change with wide alterations in plasma sodium concentration in the scincid lizard *Tiliqua rugosa* (Bradshaw & Grenot, 1976) which raises the question of whether this electrolyte is regulated at all in this particular species. 'Stress' should be discernible in field situations where significant changes in state or body composition are evident, despite the triggering of homeostatic regulatory mechanisms. Such a situation could be the finding of body temperatures close to the Critical Thermal Maximum (CTMax) despite the deployment of high-temperature avoidance behavioural patterns; or cases of chronic dehydration accompanied by elevated plasma antidiuretic hormone levels and anuria.

Such information does exist in the reptilian literature, but it is scanty and in most cases very fragmentary. Most of the data have also not been collected with the above questions in mind (i.e. are the animals exhibiting homeostatic or alternative strategies) and it will be necessary to view them in a slightly different context if they are to be of assistance in reaching conclusions. The first section of this chapter will therefore deal with mechanisms and efficiency of thermoregulation in desert reptiles. This will be followed by a similar analysis of water and electrolyte metabolism. The third section will review current knowledge of the hormonal control of water and electrolyte balance in reptiles, with particular attention given to the rôle played by pituitary and adrenal hormones in modulating renal and extrarenal sites of exchange.

## II. THERMOREGULATION IN DESERT REPTILES

A massive amount of information has now been accumulated on both the mechanisms and the efficiency of thermoregulation in, particularly, desert species of lizards and the subject has been reviewed frequently (see Schmidt-Nielsen & Dawson, 1964; Templeton, 1970; Cloudsley-Thompson, 1972; Dawson, 1975; Heatwole, 1976; Saint Girons, 1980). Other reptilian groups are much less well known and it is only relatively recently that environments other than the desert (e.g. the tropics) have attracted the attention of researchers. Huey & Slatkin (1976) and Huey (1982) have recently reviewed reptilian temperature regulation in relation to ecological concepts and Avery (1982) has documented field studies. From these reviews it is possible to glean a number of general conclusions relative to desert species.

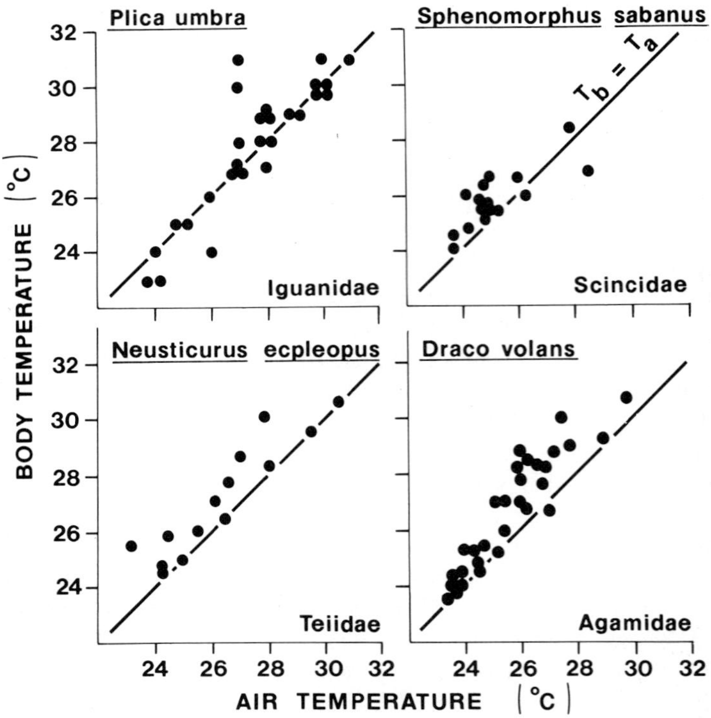


**Fig. 2.1** Body temperatures during activity, correlated with associated air temperatures, of four species of diurnal lizards from several families which appear to be thermoconformers (from Huey, 1982).

## A. General Conclusions

### *1. Temperature Control*

Early reviews of reptilian thermoregulation (e.g. Schmidt-Nielsen & Dawson, 1964; Templeton, 1970) emphasised the high precision of temperature control by non-physiological means displayed by lizards inhabiting desert regions, but subsequent studies with tropical species have served to temper this view somewhat. Within the family Iguanidae for example, which is by far the best studied, five species are known to maintain very high activity temperatures in excess of 38°C for extended periods (*Callisaurus draconoides, Crotaphytus wislizenii, Dipsosaurus dorsalis, Sauromalus obesus* and *Uma notata*). Many species of the genera *Anolis, Ctenosaura, Liolaemus, Petrosaurus, Phrynosoma,*

*Sceloporus* and *Uta* are also known to maintain somewhat lower, but, nonetheless, quite elevated body temperatures when active, but extensive studies of the temperature relations of, particularly, members of the genus *Anolis* in Central America, have shown that activity temperatures may be quite variable and often less than 30°C. A number of diurnal lizard species are now known to be 'thermoconformers' with body temperatures indistinguishable from that of the surrounding environment when active (see Fig. 2.1) and many fossorial snakes, amphisbaenians, sea snakes, as well as freshwater and marine turtles appear not to thermoregulate (Avery, 1982). By way of contrast, only one species within the Old World family Agamidae has been shown to lack thermoregulatory capacities—the flying lizard *Draco volans* (Alcala, 1967)—but this apparent difference with the iguanids may only reflect the fact that tropical genera such as *Calotes, Cophotis, Gonocephalis* and *Lophognathus* have yet to be studied in detail.

## 2. Mean Body Temperature

The mean body temperature (MBT) of animals active in the field was taken to be an unvarying and conservative feature of the thermal biology of different reptilian taxa, but detailed field studies of lizards such as *Dipsosaurus, Amphibolurus* and *Phrynosoma* have shown clearly that the MBT represents a compromise between physiological preferences and ecological opportunities. These physiological preferences are better reflected by the 'preferred body temperature' (PBT) determined in a thermal gradient, but this is also known to be influenced by a multitude of factors including habitat, elevation, time of day, season and physiological condition (see Huey 1982). Huey & Slatkin (1976) concluded from a cost–benefit analysis of lizard thermoregulation, however, that 'evolution of optimal temperatures is conservative within genera and that flexibility of thermoregulatory behaviour minimizes any resulting ecological disadvantages'. Only within the large Central American genus *Anolis* is there any evidence of marked lability in thermal optima and this clearly parallels the remarkable speciation of this genus in the tropics.

## 3. Preferred Body Temperature

Although some minor exceptions exist (see Dawson, 1975) there is now overwhelming evidence that most physiological processes in reptiles proceed at optimal rates at temperatures close to the preferred body temperature (see Fig. 2.2) and thus provide a rationale for behavioural thermoregulation. There is still debate, however, on the question of whether some species possess more than a single preferendum and many reptiles select higher body temperatures after feeding, both in the laboratory and in the field (Lang, 1979; Regal, 1966;

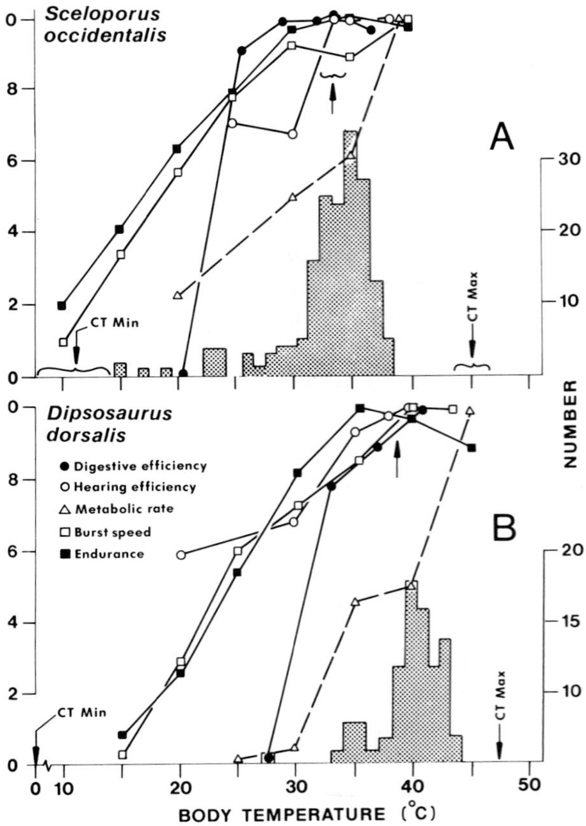


**Fig. 2.2** Relative performance of several physiological systems, and the relative metababolic rate for the lizards *Sceloporus occidentalis* (A) and *Dipsosaurus dorsalis* (B). Arrow indicates selected body temperatures (from Huey, 1982).

Bradshaw, Gans & Saint Girons, 1980) and the apparent preferred body temperature may also differ during the day and the night (Regal, 1967; Spellerberg, 1974; Saint Girons, 1980). There may also be great disparity between mean body temperatures of geckos at night when active and their mean body temperature during the day when in refuge (Licht *et al.*, 1966). Some geckos are also known to bask during the day (Werner & Goldblatt, 1978; Bustard, 1967) and select high preferred body temperatures in a photothermal gradient (Bradshaw *et al.*, 1970; Dial, 1978; Arnold & Gallagher, 1977; Werner & Whitaker, 1978). This paradoxical behaviour has been studied by Mutton (1970), in the Australian genera *Diplodactylus* and *Gehyra*, who found that nocturnal species of these geckos were unable to withstand prolonged exposure to ultra-violet light which produced blindness, and, ultimately, death from sunburn. Partially 'diurnal' species such as *Diplodactylus spinigerus* have no exposed blood vessels on the iris and possess a heavily-melanised peritoneum which protects them from these effects of short-wave radiation.

### 4. *Temperature Regulation in other Reptiles*

The question of temperature regulation in reptiles other than lizards which also occur in desert regions has not been studied in great depth. Avery (1982) points out that no comprehensive study of the thermal biology of any boid snake has been published, and few studies of colubrid snakes in which "knowledge of thermal relationships extends beyond a non-systematic collection of body temperature measurements over short periods" have appeared. There is no doubt however that many snakes thermoregulate behaviourally although most of those studied are not desert species (see Heatwole & Johnson, 1979; Stewart, 1965; Webb & Heatwole, 1971; Lillywhite, 1980; Saint Girons & Bradshaw, 1981). Desert dwelling species such as *Spalerosphis cliffordi* (Dmi'el & Borut, 1972) in Israel and the North American rattle-snake *Crotalus mitchelli* and desert sidewinder, *Crotalus cerastes* (Moore, 1978) seem to differ little in their thermal preferences, however, and typically maintain low preferred body temperatures close to 30°C. Without exception, desert snakes revert to nocturnality during the summer months when diurnal temperatures are typically high, thus keeping body temperatures within the modest limits set by their physiology (Saint Girons, 1980). Desert tortoises seem to be similar to snakes in this respect with low-regulated body temperatures (Mackay, 1964; McGimmin & Voigt, 1971; Regan, 1974) and summer–nocturnal activity (Avery, 1982).

## B. Behavioural Mechanisms

The precise sequence of behavioural attitudes which are employed by desert lizards in elevating their body temperature above that of their surroundings, and then maintaining it within tolerable limits, and close to that preferred for activity, has been described in many species and an example will suffice to illustrate the principles involved. The Ornate Dragon lizard (*Amphibolurus ornatus*) is a small (about 20 g, see Plate II) agamid lizard which is restricted to a habitat of large granite tors (Plate III) in the semi-arid region of Western Australia. These Precambrian granites exfoliate constantly under the influence of marked diurnal fluctuations in surface rock temperature of the order of 40°C and the resultant thin slabs of rock which gradually become detached from the granite matrix provide the sole form of cover for this lizard (see Plate IV). Dragon lizards retreat beneath these rocks at night, during the heat of the day and when menaced by potential predators. Morphologically, this species is

**Plate II** The Ornate Dragon Lizard, *Amphibolurus ornatus*; an adult male photographed at Baker's Hill in Western Australia.

**Plate III** An extensive exfoliating granite tor at Baker's Hill in Western Australia which forms the sole habitat for the agamid lizard, *Amphibolurus ornatus*.

**Plate IV** Detail of exfoliating granite which provides *Amphibolurus ornatus* with refugia that are used both at night and during the hottest periods of the day (see also Fig. 2.3).

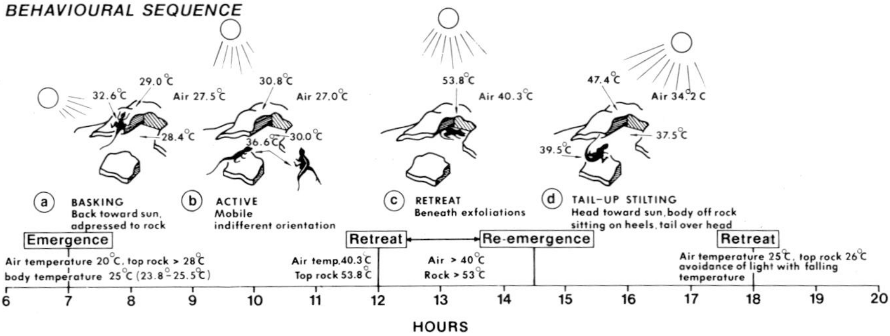


**Fig. 2.3** Diagrammatic representation of behavioural thermoregulation in the agamid lizard *Amphibolurus ornatus*. Timing and duration of behaviour patterns will depend upon characteristics of the day (from Bradshaw & Main, 1968).

uniquely adapted to this particular habitat with a body strongly flattened in the dorso-ventral plane so that the thickness never exceeds 1 cm and permits them to occupy the small space beneath the flattened slabs of rock. The typical thermoregulatory sequence of *Amphibolurus ornatus* during a hot summer's day is shown diagrammatically in Fig. 2.3. A number of discrete behaviour patterns which serve either to elevate the body temperature above that of the surroundings (emergence, basking, orientation of body perpendicular to sun's rays) or decrease the body temperature or retard its rate of elevation ('stilting', orientation of body parallel to sun's rays; 'shuttling', retreat beneath rock) can be discerned. These latter 'high-temperature avoidance' behaviour patterns are typically displayed in the later, hotter part of the day when increasing environmental temperatures threaten to raise the body temperature above the preferendum. The preferred body temperature of this species as determined in a photo-thermal gradient in the laboratory is 36.6°C (Bradshaw & Main, 1968) and this is also the mean body temperature of animals 'active' (i.e. by definition, not involved in thermoregulatory behaviour) in the field. The mean body temperature of 'stilting' animals (i.e. the body lifted up away from the ground and with the tail swung over the head) was 39.5°C, however, which is almost 3°C above the preferendum and they will tolerate body temperatures up to 40°C before they retreat beneath the rocks. Fig. 2.4 shows that these sheets of rock can be particularly effective in providing the animal with an adequate thermal refuge during the hottest part of the day, particularly if they are still attached to the granite matrix, in which case the temperature rarely exceeds 35°C. Plates V–X show photographically the daily thermoregulatory sequence of another Australian agamid, *Amphibolurus nuchalis*, at Shark Bay.

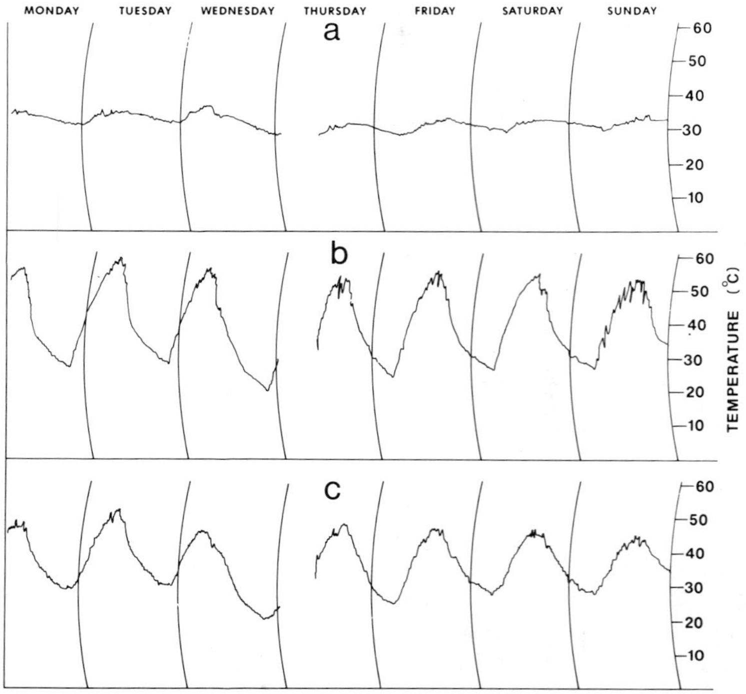


**Fig. 2.4** Continuous temperature records from three microenvironments at Bakers Hill in Western Australia during one week of January 1965 (summer). (a) beneath a large 'fixed' rock used as a refuge by the lizards; (b) on top of a 'fixed' rock; (c) beneath a small 'movable' rock not frequented by lizards during periods of high temperature (from Bradshaw & Main, 1968).

The association of discrete patterns of behaviour with narrow ranges of body temperature and a characteristic set of environmental temperatures suggests that their function is indeed thermoregulatory and examples of this can be seen in the case of *Amphibolurus ornatus*. 'Shuttling' behaviour, for example, results in a fall of the mean body temperature from 39.5° to 37.9°C despite a significant increase of the substrate temperature from 47.4°C to 49.8°C (Bradshaw & Main, 1968) and 'stilting' behaviour holds the body temperature at 39.5°C despite a 6.5°C increase in rock surface temperature. This maintenance of mean body temperature within very narrow limits (38°–40°C) despite progessively increasing environmental temperatures is seen particularly clearly in Fig. 2.5. The use of the mean body temperature in these early studies, although instructive, is nonetheless somewhat misleading as particular behaviours are displayed over a range of body temperatures and are

**Plate V** *Amphibolurus nuchalis* at Shark Bay, 'basking' with its back perpendicular to the sun's rays in the early morning (note shadow), soon after emerging from its burrow. (Air temperature = 28.9°C.)

**Plate VI** *Amphibolurus nuchalis* at Shark Bay at the end of a 10–15 minute basking period, just prior to commencing activity with a body temperature close to 37°C. (Air temperature = 28.8°C.)

**Plate VII** An extraordinary photograph of a female *Amphibolurus nuchalis* at Shark Bay, balanced on its hind legs in a completely vertical position at mid-day, thereby reducing to an absolute minimum the area of its body exposed to the sun which is directly overhead (note shadow). The air temperature at the time was 41°C.

**Plate VIII** Male *Amphibolurus nuchalis* at Shark Bay exhibiting 'shuttling behaviour' and here seen cooling itself temporarily in the shade of an *Acacia sclerosperma* bush. (Air temperature = 42.4°C.)

**Plate IX** An 'elevated' male *Amphibolurus nuchalis* on a small *Arthrocnemum* shrub at Shark Bay, showing marked pallor due to aggregation of melanin in the dermal melanophores. This reduces the albedo of the animal and enhances thermoregulation by reflecting much of the incident radiation. (Air temperature = 40.3°C.)

**Plate X** An 'elevated' male *Amphibolurus nuchalis* in late afternoon perched on a fencepost. This animal shows extreme pallor and the air temperature at the time was 41.8°C.

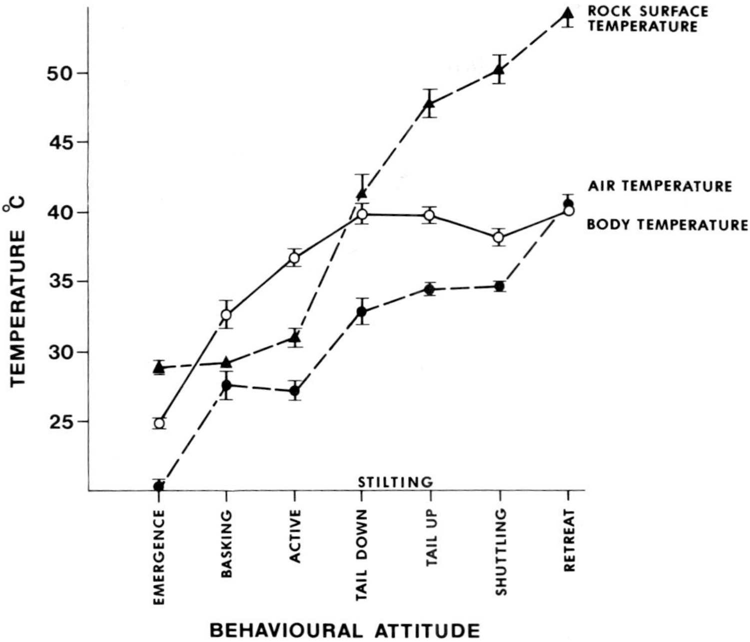


**Fig. 2.5** Relationship between mean body temperature associated with various thermoregulatory behaviour patterns of the lizard *Amphibolurus ornatus* and associated environmental temperatures (plotted from data in Bradshaw & Main, 1968).

better characterised by this measure rather than by the mean, which is a measure of central tendency. There has also been considerable discussion in the literature of the question of whether, at the neurological level, a single preferred body temperature is indeed regulated. Some models (McGinnis 1966, McGinnis & Dickson 1967; DeWitt, 1967a,b) postulate a single thermal set-point whereas Heath (1970) proposed a coupled on–off model which has received some experimental support (see Berk & Heath, 1975a; Garrick *et al.*, 1977; 1979; Kluger *et al.*, 1973; Cabanac, 1979). The frequency of distribution of body temperatures at which the lizard *Dipsosaurus dorsalis* will actively seek heat in a shuttle box situation (LBTS) and also move to a cooler situation (HBTS) is shown in Fig. 2.6 and it is clear that they do not overlap.

Myhre & Hammel (1969) and Cabanac *et al.* (1967) studied control of body temperature in the lizard *Tiliqua scincoides* and identified two sets of temperature-sensitive neurones in the preoptic region of the brain. With decreasing brain temperature, cold-sensitive neurones slowly increased their rate

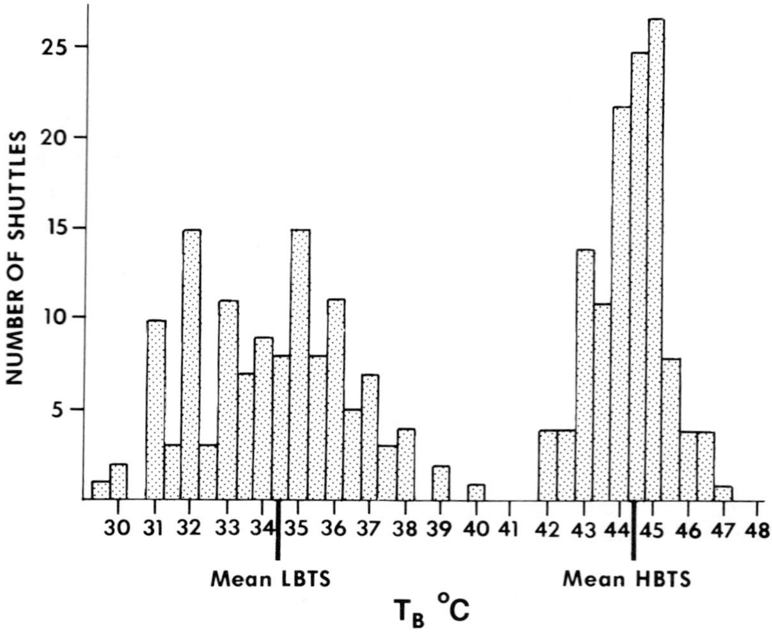


**Fig. 2.6** The frequency of individual body temperaures recorded from a single *Dipsosaurus dorsalis* in a 'shuttle box' situation showing absence of overlap between high and low temperatures. $T_B$ = body temperature; LBTS = low body temperature at shuttling; HBTS = high body temperature at shuttling (from Berk & Heath, 1975b).

of firing (low $Q_{10}$ neurones) whilst with increasing brain temperature the warm-sensitive neurones (high $Q_{10}$ neurones) increased their rate of firing. Activity curves for these two sets of cold and warm-sensitive neurones were found not to overlap in the range of body temperatures of active *Tiliqua scincoides*, suggesting the presence of an upper and a lower set-point rather than a single preferendum. Kluger *et al.* (1973) and Berk & Heath (1975a) were also able to show that posterior hypothalamic lesions which also touch the mammillary body reduce the frequency of cycling between hot and cold environments by prolonging the hypothermic phase. Massive destruction of the mammillary bodies had the effect of significantly lowering the LBTS from 37.3°C to 32.6°C (see Table XIII). It must be remembered, however, that the mammillary bodies are regions in the reptilian brain, through which several major nerve tracts pass, and the effects of these lesions thus cannot be ascribed to any identifiable nuclei. To date, the most useful model of reptilian behavioural thermo-regulation is the 'three neurone' model proposed by Berk & Heath (1975b)

**Table XIII:** Alterations in thermoregulatory behaviour of the lizard *Dipsosaurus dorsalis*** following damage to the mammillary bodies (modified from Kluger, Tarr & Heath, 1973).

| Condition | Ratio cycles/ hour | Abnormal push-up | Average high-set point (°C) | Average low-set point (°C) | Mammillary destruction (%) | N |
|---|---|---|---|---|---|---|
| Controls | 1.49 ± 0.32 | 0 | 41.2 ± 0.39 | 37.3 ± 0.49 | 0 | 5 |
| | (0.83 – 2.43) | | (40.4 – 42.4) | (35.5 – 39.4) | | |
| Lesioned Animals | | | | | | |
| a.......... | 0.49 ± 0.03* | + | 40.9 ± 1.06 | 35.5 ± 0.64 | 25 | 3 |
| | (0.45 – 0.54) | | (38.8 – 42.2) | (34.0 – 37.2) | (10–20) | |
| b.......... | 0.08 ± 0.04* | + | 39.4 ± 0.72 | 32.6 ± 1.44* | 80 | 5 |
| | (0.02 – 0.22) | | (36.6 – 40.8) | (30.1 – 37.2) | (50–100) | |

** Values are mean ± S.E.; range in parentheses.
* Significant at $P < 0.05$, Student 't'.

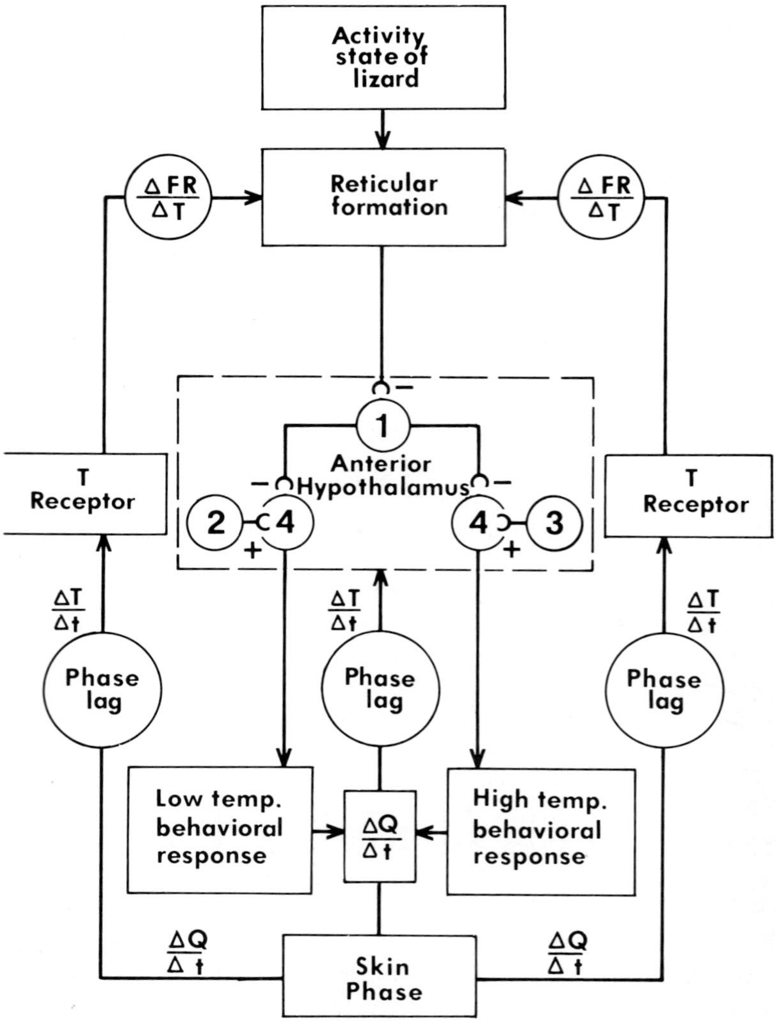


**Fig. 2.7** Hypothetical neural network of thermoregulation. (1) = temperature insensitive neurone; (2) = low $Q_{10}$ cold neurone; (3) = high $Q_{10}$ warm neurone; (4) = motorneurone.

$$\frac{\Delta Q}{\Delta t} = \frac{\Delta \text{ Heat Content}}{\Delta \text{ Time}} \qquad \frac{\Delta FR}{\Delta T} = \frac{\Delta \text{ Firing Rate}}{\Delta \text{ Temperature}} \qquad \frac{\Delta T}{\Delta t} = \frac{\Delta \text{ Temperature}}{\Delta \text{ Time}}$$

(+) = excitatory synapse; (−) = inhibitory synapse; T receptor = peripheral thermal receptor. (from Berk & Heath, 1975b).

shown in Fig. 2.7. DeWitt & Friedman (1979) have recently suggested that the characteristic negatively-skewed temperature distribution of reptiles indicates the regulation, not of body temperature *per se*, but of some physiological rate process whose rate itself is an exponential function of temperature. This rate process has yet to be identified in any species (Firth & Turner, 1982) and it is interesting to note that the numerical model of Barber & Crawford (1977) can account for negative skewness in body temperatures if one assumes a lower sensitivity for the lower set-point (LBTS) than the upper set-point.

The question of whether lizards regulate a single 'preferred' body temperature or instead are behaviourally motivated by separate upper and lower set-points is germane to this analysis because, on the former theory, animals would be 'stressed' as soon as their body temperature exceeded the preferendum (e.g. DeWitt, 1967b, concluded that lizards 'permit' their body temperature to rise above the preferendum when engaged in territorial behaviour and predator defence, but maintain the preferred body temperature during feeding and courtship). With the second model one could only invoke stress if the body temperature exceeded the upper set-point (HBTS) which, for example, in *Dipsosaurus dorsalis* has a mean of 41.7°C (Beck & Heath, 1975b), some 7°C below the Critical Thermal Maximum (CTMax) for this species (Whitfield & Livezy, 1973).

Relatively sophisticated analyses of lizard thermoregulatory behaviour and the development of equilibrium heat-transfer models predicting the body temperature of an animal under a variety of different conditions developed originally from the work of Norris (1967) and Bartlett & Gates (1967). These early models were generalised by Porter & Gates (1969), but they were somewhat limited by the need for local measurements of the microclimate and a number of restrictive assumptions about microclimatic variables. Porter *et al.*, (1973) presented a coupled transient microclimate/transient ectotherm model for the desert iguana, *Dipsosaurus dorsalis*, which performed very robustly as did the related model of Spotila *et al.*, (1973), developed for large reptiles (see also Bakken & Gates, 1975). James & Porter (1979) further investigated the robustness of this model by first analysing the relationship between a number of discrete behaviour patterns and microclimate variables in the tropical African rainbow lizard *Agama agama* in the field in Ghana. Principal component analysis and canonical correlation analysis were used to identify microclimatic variables useful in predicting behaviour and to permit an ordination of behavioural events in terms of the thermal environment. In the case of *Agama agama*, two particularly important variables were soil surface temperature and net radiation which, when plotted together, describe a climate space within which all the behavioural interactions occur in a regular pattern as shown in Fig. 2.8. Porter & James (1979) then tested the applicability of the transient model, originally developed for a temperate-desert lizard, to this tropical

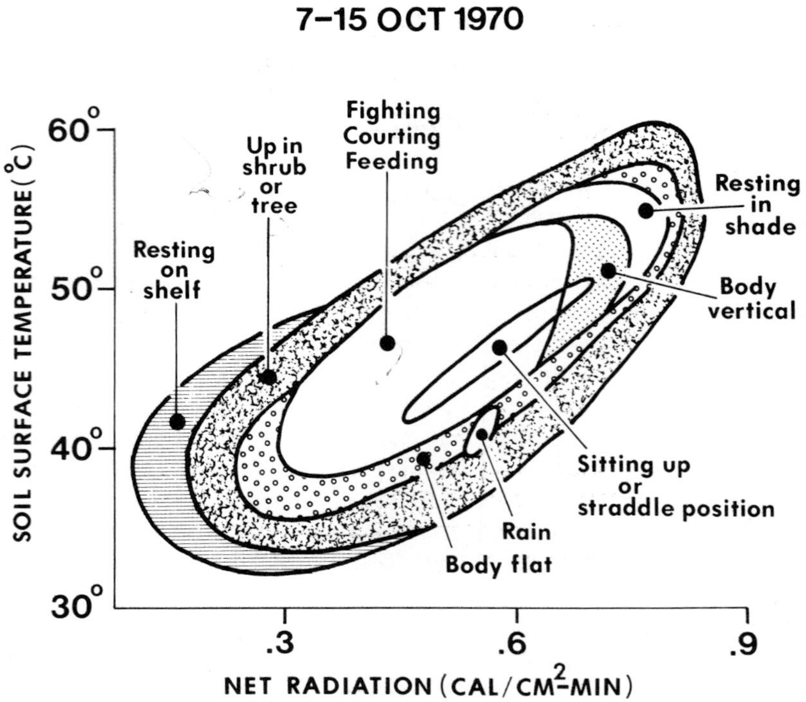


**Fig. 2.8** Summary of behavioural data for the lizard *Agama agama* during the week of 7–15 October 1970 plotted in the time independent space of values for soil surface temperature and net radiation. Behavioural events for the nine days were superimposed and lines drawn to define their spatial pattern. To read the radiation in $10^{-2}$ watts, insert $2.1 \times 10^{-2}$, $4.2 \times 10^{-2}$ and $6.3 \times 10^{-2}$ along the X-axis (from James and Porter, 1979).

species. The fit is surprisingly good although to some extent, as Porter & James (1979) admit, there is little detailed information on the physiology of this species (see Harris, 1964) with which to compare the predictions of the model. The model predicts the effect on total energy demands of changes in location (e.g. sitting in shade rather than being exposed fully to the sun reduces daily energy requirements from 11%–19%, regardless of the size of the animal and its height above the ground—see Fig. 2.9), posture, change in metabolic rate and variations in preferred body temperature. To date this is the most sophisticated model that has been proposed to account for lizard thermoregulatory behaviour and its broad applicability to both desert and tropical species should facilitate the testing of current ideas derived from theoretical cost–benefit analyses of thermoregulation (see Huey & Slatkin, 1976; Tracy, 1982).

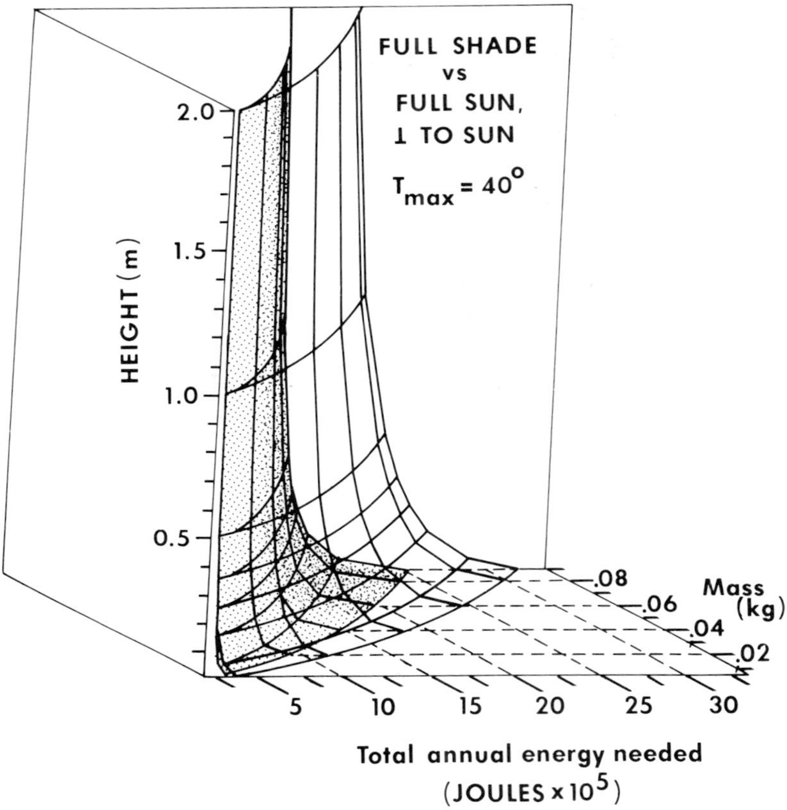


**Fig. 2.9** The overlap of metabolic cost to the lizard *Agama agama* choosing to remain in full sun, normal to the sun ('hottest'), and full shade ('coolest'). The shaded surface is full shade. Maximum tolerated temperature is assumed to be 40°C for both (from Porter & James, 1979).

## C. Physiological Mechanisms

Physiological processes could influence the primarily behavioural thermoregulation of reptiles in a number of ways and these include: (i) increased heat production through metabolism; (ii) variations in rate of heat exchange with the environment through circulatory adjustments and changes in dermal vascularity; (iii) systematic changes in colour and hence reflectivity of the skin; and (iv) evaporative cooling. All of the above are known to occur to some extent in reptiles (see Table XIV) but the contributions due to metabolism and

**Table XIV:** Major differences in components for temperature control among birds, mammals, and reptiles (from Bartholomew, 1982).

| | Birds | Mammals | Crocodilians | Turtles | *Sphenodon* | Lizards | Snakes |
|---|---|---|---|---|---|---|---|
| Endogenous heat production | | | | | | | |
| High resting metabolic rate (BMR) | + | + | − | − | − | − | − |
| Shivering thermogenesis | + | + | − | − | − | − | − |
| Locomotor thermogenesis | + | + | + | + | + | + | + |
| Chemical thermogenesis | −(?) | + | − | − | − | − | − |
| Integumentry control of heat flux | | | | | | | |
| Pilomotor or filomotor activity | + | + | − | − | − | − | − |
| Integumentary colour changes | − | − | − | − | ? | + | ? |
| Evaporative cooling | | | | | | | |
| Panting, gaping | + | + | + | + | ? | + | + |
| Saliva spreading | − | + | − | + | − | − | − |
| Urinating on body | + | − | − | + | − | + | − |
| Sweating | − | + | − | − | − | − | − |
| Regional heterothermy | | | | | | | |
| Vascular shunting | + | + | + | + | + | + | + |
| Cardiac shunting | − | − | + | + | + | + | + |

evaporative cooling appear to be, on the whole, fairly negligible. Bartholomew (1982) points out that the 'factorial metabolic scope' (maximum divided by resting metabolic rate) of lizards with preferred body temperatures between 35°C and 40°C is such that some can, however, briefly attain rates of heat production which approximate the basal rates of similar-sized birds and mammals. Thermal conductance of, particularly, small reptiles is so high however that such additional heat would be lost almost as rapidly as it is produced and it is only in large reptiles, with a high thermal inertia, that such heat production could be of significance. The Indian phthon *Python molurus* is known to be capable of elevating its body temperature by as much as 4.7°C above that of the external environment when incubating its eggs by spasmodic contractions of its body musculature (Hutchinson *et al.*, (1966) but to date this appears to be an isolated instance (Bennett & Dawson, 1975). Calculations suggest that endogenous heat production may contribute to differences observed in rates of heating and of cooling in the lizard *Tiliqua scincoides* (Bartholomew *et al.*, 1965) and a number of varanids (Bartholomew & Tucker, 1964) and may retard cooling in the marine iguana *Amblyrhynchus cristatus* (Bartholomew & Vleck, 1979). Smith (1975, 1979), working with the American alligator, arrived at a similar conclusion, but this has yet to be demonstrated in animals exposed to ecologically realistic thermal surroundings. Purely on the basis of large body size and thermal inertia, one might expect to find some evidence of endothermy in large crocodiles but Terpin *et al.*, (1979) concluded after a careful study with *Alligator mississippiensis* that there was no evidence of elevation of body temperature through endogenous heat production.

Gaping or 'panting' at high body temperatures is commonly observed in lizards constrained in laboratory situations and a number of studies have attempted to assess the extent of the evaporative cooling emanating from this behaviour. Panting in the chuckwalla *Sauromalus obesus* can reduce body temperature by as much as 4°C below ambient for several hours during exposure to dry air at 47°C (Case, 1972) and can lower the temperature of the brain even further (Crawford, 1972). Similar temporary reductions in body temperature have been recorded in desert lizards such as *Dipsosaurus dorsalis* (Templeton, 1960), *Crotaphytus collaris* (Dawson & Templeton, 1963), *Uromastix acanthinurus* (Grenot & Loirat, 1973), and also *Amphibolurus barbatus* (Warburg, 1965). The amount of heat dissipated in this way is quite small, amounting to no more than 50% of the metabolic rate (see Fig. 2.10) and it seems intuitively obvious in the case of desert animals that the short-term advantage to be gained from evaporative cooling would be far outweighed in the long term by the associated loss of vital body water.

Various authors (e.g. Bradshaw & Main, 1968; Cogger, 1974; Porter & James, 1979) have also remarked on the fact that animals are rarely observed

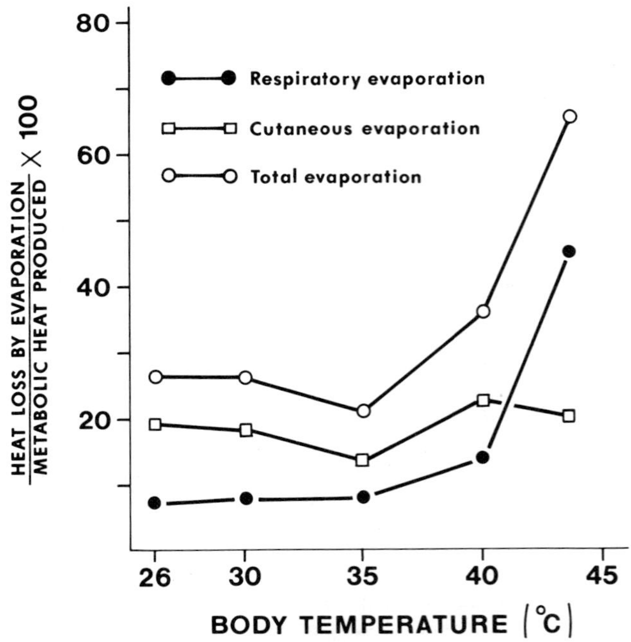


**Fig. 2.10** Percentage of metabolic heat production of the lizard *Sauromalus obesus* which is dissipated by respiratory, cutaneous and total evaporation at various environmental temperatures (from Crawford & Kampe, 1971).

panting in the field and normally retreat to some thermal refuge before body temperature reaches the panting threshold. Panting under field conditions may thus be useful as an index of thermal stress (elevated body temperature associated with high-temperature avoidance behaviour patterns).

Circulatory adjustments which facilitate thermoregulation have now been well-documented in a number of reptilian species following the early demonstration of Bartholomew & Tucker (1963) that the Australian lizard *Amphibolurus barbatus* heats more rapidly than it cools. Heart rate is higher during heating in the majority of lizards and alligators studied (see Fig. 2.11), but interestingly does not change in varanids (Bartholomew & Tucker, 1964) nor in *Iguana iguana* (Spray & Belkin, 1972). Changes in vasomotor activity and intracardiac shunts, which produce an increase in cardiac output by a right-to-left cardiac shunt which allows venous blood to bypass the pulmonary circuit and enter the left systemic arch (White, 1976), would appear to be more important factors in the regional control of body temperature (Bartholomew,

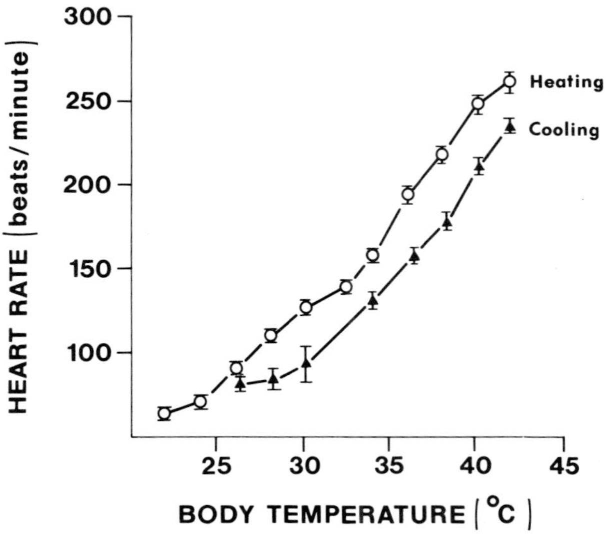


**Fig. 2.11** The response of heart rate (beats/min) to change in body temperature, recorded during heating and cooling in the agamid lizard *Amphibolurus nuchalis*. Each point represents mean ± standard error with N > 5 (from Rice & Bradshaw, 1980).

1982). Cowles (1958) was one of the first to appreciate that changes in subcutaneous blood flow may also play an adaptive rôle in reptilian thermoregulation, although the techniques at his disposal did not enable him to marshal definitive evidence to support his contention. Morgareidge & White (1969) used the $^{133}$Xe clearance technique to estimate dermal blood flow and found evidence for changes emanating from local heating of the skin. They showed that these responses were unrelated to changes in either heart rate, core temperature or any portion of the skin other than that undergoing heat flux.

Weathers (1971) extended these findings in his study of the desert iguana *Dipsosaurus dorsalis* and suggested that local changes in subcutaneous blood flow are a direct result of heat on the vascular smooth muscle. Baker *et al.* (1972) presented evidence to suggest that the peripheral circulatory response to local heating is dependent upon the thermal load and concluded that control over heat transfer in lizards depends upon a selective redistribution of blood between the shell and the core. Changes in blood viscosity produced by regional temperature variation may also contribute to the modification of rates of heating and cooling by the cardiovascular system (Firth & Turner, 1982). It is generally assumed that the functional significance of these physiological modifications of rates of heating and cooling is manifested in a prolongation of the time that animals can maintain favourable body temperatures in the field. To date this has not been tested in the field and Bartholomew (1982) points out that, in the lizards which have been studied, there does not appear to be any obvious correlation between the capacity to control rate of change in body temperature and either habitat or the preferred body temperature.

The contribution to physiological regulation of the body temperature made by systematic changes in colour, and hence reflectivity of the skin, has been more difficult to assess in reptiles, even though from simple physical principles it seems intuitively obvious that dark-coloured animals should heat more rapidly when exposed to radiant energy. Cole (1943) appears to have been the first to attempt the simple experiment of tethering a light and dark-coloured lizard together in the sun and measuring their subsequent rates of temperature increase and, in fact, the darker-coloured animal did heat more rapidly. Unfortunately the 'experiment', although apparently unequivocal, is deceptively simple and in no way excludes the operation of factors other than skin reflectivity, such as changes in dermal vascularity, which could equally well account for the difference. The majority of the studies which have been concerned with measuring differences in rates of heating and cooling in reptiles have also relied upon either constant-temperature chambers for heating, where the primary modes of heat transfer occur by convection and long-wave length radiation, or water baths (Bartholomew & Lasiewski, 1965) where heat exchange occurs solely by conduction. Under these circumstances changes in the reflectivity of the skin should have little effect on the rate of heat exchange of the animal and, unfortunately, in those studies where incandescent heat lamps have been employed to heat lizards (e.g. Weathers, 1971; Baker *et al.*, 1972), changes in dermal reflectivity were not monitored. The interrelationship between changes in dermal vascularity and reflectivity during heating and cooling have thus far only been investigated in the desert lizard *Amphibolurus nuchalis* by Rice & Bradshaw (1980). These authors found that dermal reflectance increased in a linear fashion from approximately 10% at a body temperature of 20°C to almost 32% at 40°C in a similar manner to that described by Pearson (1977) for *Liolaemus multiformis*. Dermal vascularity was also found to increase as a function of body temperature as shown by the decreasing half-life ($t_{1/2}$) of injected $^{85}Kr$ which fell from 20 minutes to less than 10 minutes over the same temperature range (see Fig. 2.12). Local heating of the skin was also found to increase the clearance rate of $^{85}Kr$, even though heart rate and deep body temperature were unaffected, indicating the capacity for autoregulation of dermal blood vessels, also described in the green iguana *Iguana iguana* by Baker *et al* (1972).

Injections of synthetic melanocyte-stimulating hormone ($\alpha$-MSH) were used to darken animals and Rice & Bradshaw (1980) established that this did

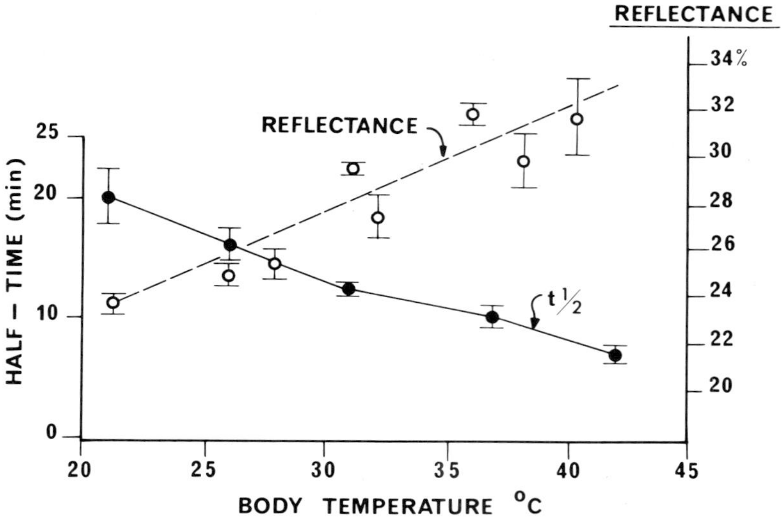


**Fig. 2.12** Temperature-dependent changes in reflectance and dermal vascularity (measured as the half-time for removal of subcutaneously-injected $^{85}Kr$) of the lizard *Amphibolurus nuchalis* (from Rice & Bradshaw, 1980).

not alter in any way changes in dermal vascularity. Without exception, however, α-MSH-injected *Amphibolurus nuchalis* subjects heated at faster rates than control subjects and reached higher equilibrium body temperatures with the same thermal input from a heat lamp (see Fig. 2.13). Surgical hypophysectomy similarly did not affect $^{85}Kr$ clearance rates, but these individuals, which exhibit a permanent pallor after the operation, heated at a slower rate than the control subjects and reached even lower equilibrium body temperatures (α-MSH-injected = 40.2°C; control subjects = 37.2°C; hypophysectomised subjects = 33.7°C; (see also Figure 2.13). These differences in rate of heating appear to result uniquely from the differences in reflectivity of the skin and thus provide, for the first time, an unequivocal demonstration of the importance of changes in reflectance in reptilian thermoregulation. Significantly, rates of cooling measured in α-MSH-injected and hypophysectomised individuals did not differ from rates of cooling in control subjects (see Table XV), indicating that the effect of changes in reflectivity is on heating rates alone. This is almost certainly due to the fact that, during cooling, the primary modes of heat transfer are by long-wavelength radiation and convection, with the animals acting as black body radiators, and colour would thus have little effect on cooling rate. Instantaneous rates of temperature change (IRT) were

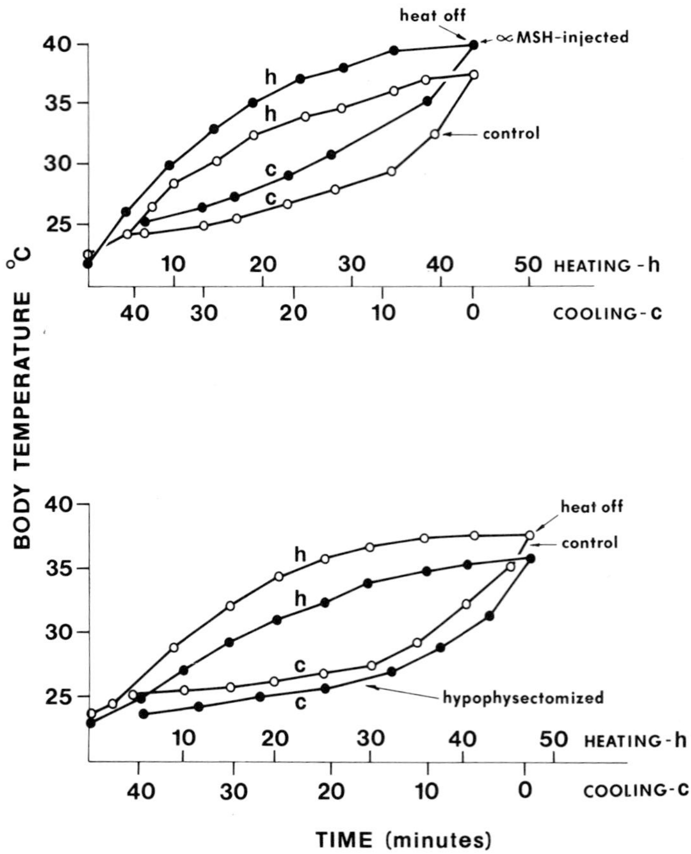


**Fig. 2.13** Heating and cooling curves of intact, $\alpha$-MSH injected and hypophysectomised *Amphibolurus nuchalis*, showing both thermal hysteresis and the effect of permanent changes in dermal reflectance (from Rice & Bradshaw, 1980).

also measured in dead *Amphibolurus nuchalis* subjects, rendered either dark or pale, by Rice & Bradshaw (1980) (see also Table XV), thus permitting an assessment of the extent to which physiological (vascular) and physical (reflectivity) changes operate in concert. The IRT varied by 67.9% in live animals, which is over twice the range seen in dead animals (29.5%). This reflects the degree to which physiological processes, almost certainly vascular, augment the extend to which purely reflectance changes can alter the rate of heating. In the living animal, the two effects are additive and synergistic, as local heating of peripheral vessels leads to their dilatation which enhances

**Table XV:** Instantaneous rate of temperature change (in °C/min) at 30°C during heating and cooling of live and dead *Amphibolurus nuchalis* (from Rice & Bradshaw, 1980).

| | α-MSH-injected | | Control | | Hypophysectomised |
|---|---|---|---|---|---|
| (a) Heating: | | | | | |
| Body temperature | | | | | |
| Live | 1.09 ± 0.09 (8) | P < 0.001 | 0.66 ± 0.08 (10) | P < 0.001 | 0.35 ± 0.03 (11) |
| | P < 0.001 | | | | NS |
| Dead | 0.61 ± 0.04 (6) | | P < 0.02 | | 0.43 ± 0.06 (3) |
| (b) Cooling: | | | | | |
| Body temperature | | | | | |
| Live | 0.69 ± 0.16 (4) | NS | 0.69 ± 0.07 (5) | NS | 0.87 ± 0.14 (4) |
| | P < 0.001 | | | | P < 0.02 |
| Dead | 0.31 ± 0.01 (6) | | NS | | 0.23 ± 0.08 (3) |

Mean ± S.E. (N)

peripheral flow and subsequent transfer of heat to the core. This effect is most clearly seen where $^{85}$Kr washout rates were measured, not at stable body temperatures, but during heating and cooling. Krypton-85 clearance was faster during heating than cooling in control animals ($t_{1/2}$ = 8.03 min *vs* 12.98 min), but is reduced in hypophysectomised individuals and does not differ significantly between heating and cooling ($t_{1/2}$ = 13.65 and 11.47 min respectively). Injection of $\alpha$-MSH into hypophysectomised animals during heating dramatically accelerated the clearance of $^{85}$Kr and reduced the $t_{1/2}$ to levels seen in intact animals. Thus, despite the fact that vascular mechanisms were intact in the hypophysectomised individuals (as shown by their normal rates of clearance of $^{85}$Kr at stable body temperatures) these are not activated in the dynamic heating situation in the absence of $\alpha$-MSH. Reduced reflectivity at low body temperatures thus appears to be critical for the initiation of vascular responses in *Amphibolurus nuchalis* which result from local heating of the skin and subsequent autoregulation of dermal vessels.

Differences between heating and cooling rates in lizards thus appear to facilitate thermoregulation and probably result from a combination of factors which are almost certainly not constant from one species to another. Endogenous metabolism would appear to account for the differences observed in rates of heating of *Tiliqua scincoides*, but can only account for about 25% of the observed difference in *Amphibolurus barbatus* and *Varanus gouldii*, both of which are species showing marked heterochromatism at high environmental temperatures. Temperature-dependent changes in reflectance have been shown to be very important in the thermoregulation of *Amphibolurus nuchalis* and almost certainly this will also be found to be the case in other desert species which demonstrate marked changes in albedo at high temperatures. The question of the ecological significance of colour differences and changes in reflectivity have been discussed in detail by Norris & Lowe (1964) and by Norris (1967) and their conclusion that the colour of lizards represents a compromise between the exigencies of temperature regulation and concealment is one with which few biologists would take issue. Studies carried out on species which lack heterochromatic ability, such as the small lizard *Lacerta dugesii* which occurs in a number of different colour morphs (Crisp *et al.*, 1979) show that these differences have little effect on heat balance and almost certainly function to enhance crypsis and predator avoidance. Pronounced blanching of the skin at high temperatures on the other hand appears to be a characteristic of desert-living species of lizards (Atsatt, 1939) which, by reducing input of radiant energy from the visible and near infrared spectrum, facilitates thermoregulation. The effectiveness of this physiological adaptation would appear to be tempered, however, by the fact that peripheral circulation continues to increase as the body temperature rises (at least up to 40°C in *Amphibolurus nuchalis*) and there is no evidence of peripheral vasoconstruction at high body temperatures which might retard heat transfer to the core of the animal.

## D. Hormonal Aspects of Thermoregulation

Various hormones have been suggested to play a rôle in reptilian thermoregulation, but there is little evidence to support this (see Firth & Turner, 1982). Bradshaw & Main (1968) suggested the blanching of the skin of desert lizards at high temperatures might be due to peripheral vasoconstriction limiting the supply of $\alpha$-MSH to dermal melanophores, but the study of Rice & Bradshaw (1981) with *Amphibolurus nuchalis* has shown that this is not the case. Control of heterochromatic responses in reptiles may be either nervous and/or hormonal depending upon the species (see Waring, 1963) but there is good evidence that when hormonal, only a single hormone, $\alpha$-MSH, is involved (Bradshaw & Waring, 1969); Bagnara & Hadley, 1973) and there is only evidence for a second 'melanophore-concentrating hormone' (MCH) in fish (Baker & Rance, 1983). The work with *Amphibolurus nuchalis* suggests that $\alpha$-MSH-induced darkening of the skin at low body temperatures is very important in triggering peripheral vascular responses which are an essential component in the rapid heating of basking animals. This needs to be confirmed by measurements of circulating levels of $\alpha$-MSH over a range of body temperatures which have not been reported in any reptile.

Thyroxine ($T_4$) is known to increase the energy metabolism of lizards kept at temperatures close to their preferendum (Wilhoft, 1966a; Licht, 1974) and also increases heart rate in *Varanus albigularis albigularis* (Louw *et al.*, 1976). Thyroidectomised *Sceloporus cyanogenys* regulated their body temperatures at a lower level than control subjects (Wilhoft, 1966b) and this effect was rectified by injections of $T_4$. Levels of $T_4$ and $T_3$ have again not been measured in reptiles, however, nor have the effects of the potent tri-peptide pyro-glu-his-pro-amide (thyrotrophic-hormone releasing factor–TRH) been assessed.

The parietal eye and associated pineal complex has attracted a great deal of attention and research in reptiles, but its significance in thermoregulation is still obscure (see Ralph *et al.*, 1979; Ralph, 1984). As early as 1902, Langlois in a series of experiments with the North African lizards *Uromastix acanthinurus* and *Varanus arenarius* (*griseus*) found that the panting response to high temperatures depended upon light falling directly upon the parietal eye and could be inhibited by interposing an opaque screen. Glaser (1958) and Stebbins & Eakin (1958) first suggested that the parapineal organ and parietal eye function as a radiation dosimeter in lizards after observing increased exposure to the sun and bright lights of parietalectomised animals (Stebbins & Wilhoft, 1966), but this has not been supported by electrophysiological studies (see

Dodt, 1973). Nonetheless, reproductive cycles are accelerated and thyroid activity increased in the lizard *Sceloporus occidentalis* by shielding or removing the parietal eye (Stebbins & Cohen, 1973) and parietalectomised or nerve-sectioned *Anolis carolinensis* and *Sceloporus magister* select higher body temperatures than control subjects in thermal gradients (Hutchison & Kosh, 1974; Roth & Ralph 1976; Engbreston & Hutchison, 1976). The effects of heat and light were separated by Roth & Ralph (1977) in a shuttle-box study with *Anolis carolinensis* and parietalectomised animals were found to select higher body temperatures, even in the absence of a light stimulus.

Melatonin (N-acetyl-5-methoxytryptamine) is secreted by the mammalian pineal and the enzyme which catalyses the terminal step in its biosynthesis, hydroxyindole-O-methyl transferase (HIOMT), has been detected in the pineal and lateral eyes of a number of lizards (see Quay, 1965, 1969). Plasma melatonin levels have also been measured in the Australian lizard *Tiliqua rugosa* by radioimmunoassay (Kennaway *et al* 1977) and were found to be higher during the scotophase than the photophase. Injections of melatonin in the lizard *Amphibolurus muricatus* lowered the panting threshold of animals heated with an incandescent lamp, as did parietalectomy, (Firth & Heatwole, 1976), but the reverse effect (i.e. an elevated panting threshold) was observed when parietalectomised animals were heated non-radiantly. Some of the effects of intraperitoneal injections of melatonin on behavioural temperature regulation in the lizard *Crotaphytus collaris* are shown in Fig. 2.14 from which the extreme variability of the response on different days can be gauged. It would be premature to conclude at this stage, on the basis of the experimental evidence available, that the parietal eye and melatonin are vitally involved in reptilian thermoregulation. All of the studies in lizards, however, point to the parietal eye as an organ which functions to prevent excessive and debilitating elevations of the body temperature and further studies of changes in plasma melatonin levels are needed. One of the few studies to show a clear effect of hormones on thermoregulatory behaviour is that of Garrick (1974) on the viviparous lizard *Sceloporus cyanogenys* in which pregnancy was found to depress the preferred body temperature and injections of progesterone were able to duplicate this effect.

## E. Maintenance of Thermal Homeostasis in the Field

Due to the paucity of information on other groups, an analysis of thermal homeostasis can only be attempted with lizards, and then only with a very restricted number of species that have been adequately studied both in the field and the laboratory. Bradshaw & Main (1968) scored field body temperatures of a number of *Amphibolurus* species by the extent to which they exceeded the

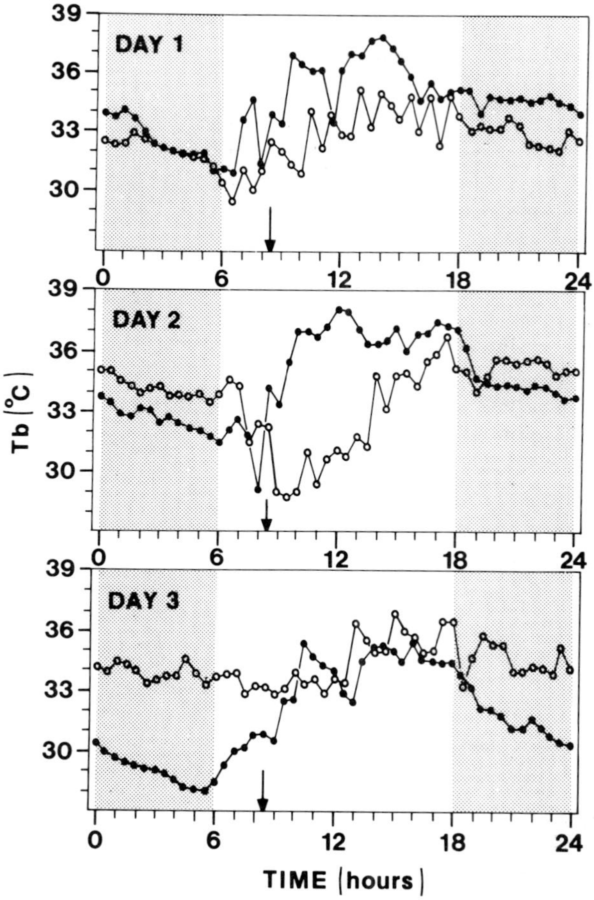


**Fig. 2.14** Mean body temperature selected by *Crotophytus collaris* during three days in a thermal gradient. Closed circles indicate $T_b$'s of lizards injected with melatonin: open circles are mean $T_b$'s of lizards receiving a control injection. Each point equals the mean of six individuals. The arrow indicates the time of injection. The scotophase is indicated by stippling (from Cothran & Hutchinson, 1979).

preferred body temperature (1°C = 0, 1°–2°C = 1, 2°–3°C = 2 etc.) and found that desert species such as *Amphibolurus nuchalis* and *A. caudicinctus* accumulated significantly higher mean scores than non-desert species (see Table XVI). This was interpreted as reflecting the greater dependence of desert species on high-temperature avoidance behaviour patterns to maintain thermal homeostasis in the hotter environments which they occupy. It was also shown

**Table XVI:** Comparison between the score/animal for different species of *Amphibolurus* and the highest body temperatures recorded in the field (from Bradshaw & Main, 1968).

| Species | N | Score/Animal ± S.E. | Highest field body temperature record |
|---|---|---|---|
| *A. nuchalis* | 47 | 1.150 ± 0.24 | 43.8°C |
| *A. caudicinctus* | 19 | 0.843 ± 0.37 | 41.4 |
| *A. ornatus* | 69 | 0.695 ± 0.14 | 41.8 |
| *A. minor* | 15 | 0.267 ± 0.18 | 38.6 |
| *A. minimus* | 33 | 0.000 | 37.4 |

Treating the score/animal as a mean and using Student $t$ test, the significance of the differences are as follows:

*A. nuchalis*/*A. caudicinctus* NS
*A. nuchalis*/*A. ornatus* sig. $t = 2.36$, 114 DF, $P = 0.01$
*A. caudicinctus*/*A. ornatus* NS
*A. caudicinctus*/*A. minor* sig. $t = 2.10$, 32 DF, $P = 0.05$
*A. ornatus*/*A. minor* NS
*A. minor*/*A. minimus* sig. $t = 2.32$, 46 DF, $0.02 < P < 0.05$
*A. ornatus*/*A. minimus* sig. $t = 3.58$, 100 DF, $P << 0.01$

NS, Not significant; DF, degrees of freedom.
(See text for derivation of score/animal.)

that this daily exposure to body temperatures above the preferred body temperature entrained an increased thermal resistance which was lost within a short period when field acclimated animals were maintained under constant conditions in the laboratory. Bradshaw & Main (1968) did not consider that *A. nuchalis* and *A. caudicinctus* were ever exposed to 'thermal stress' in the field as high-temperature avoidance behaviour patterns elicited at high body temperatures (e.g. 'elevated' which occurs at a body temperature of 39°C, around 2°C above the preferred body temperature) were considered to be part of the normal thermoregulatory répertoire of these lizards, and desert species are only forced, by the exigencies of their environment, to resort to them more frequently than species living in cooler regions.

If one examines the normal sequence of events which occurs when lizards are heated above the preferred body temperature, a number of stages can be discerned as shown in Table XVII. Heatwole & Firth (1982) determined the Voluntary Maximum Temperature (VMax) for the non-desert agamid lizard *Amphibolurus muricatus* which varied from 34.5°C to 38.4°C, depending upon the season. The VMax and the Panting Threshold (PT) were significantly correlated in this species and data plotted from the study of Whitfield &

**Table XVII:** Behavioural and physiological correlates of heating above the PBT in lizards.

| Temperature | Definition | Examples *D. dorsalis* | Examples *A. nuchalis* |
|---|---|---|---|
| Preferred body temperature (PBT) | Mean BT maintained preferentially in a photo-thermal gradient | 38.5°C[1] | 36.4°C[2] |
| Maximum voluntary temperature (VMax—same as HBST) | BT at which animal seeks shade or underground retreat | 41.7[3] | ? |
| Panting threshold (PT) | BT at which mouth opens and is associated with increased ventilation rate | 44.5[4] | 44.1[5] |
| Critical thermal maximum (CTMax) | BT at which animal loses righting reflexes but from which recovery is possible | 48.6[4] | 48.5[6] |
| Lethal temperature (LT) | Lowest 'high' BT from which recovery is impossible | 50.2[8] | 49.0[7] |

**Authority**

1. De Witt (1967)
2. Licht *et al.* (1966)
3. Berk & Heath (1975b)
4. Whitfield & Livezey (1973)
5. Heatwole (1970)
6. Heatwole (1976)
7. Bradshaw & Main (1968)
8. McGinnis & Dickson (1967)

Livezey (1973) in Figure 2.15 for a number of North American lizards show that PT and the Critical Thermal maximum (CTMax) are also positively correlated. Thus, although various studies have shown that CTMax and PT may be influenced by a wide range of environmental factors (including photoperiod, thermal acclimation, season, rate and mode of heating, time of day, dehydration: see Huey, 1982; Heatwole *et al.*, 1973; Heatwole *et al.*, 1975; Parmenter & Heatwole, 1975) they appear to change in concert, which suggests that all of the critical temperatures listed in Table XVII are linked within a given species and move upwards or downwards together with environmental acclimation.

Representative activity temperatures of reptiles in relation to critical temperatures were collated by Huey (1982) and these are shown in Table XVIII from which it is clear that most reptiles enjoy wide temperature safety-margins when active in the field. The difference betwen the CTMax and mean body temperatures for diurnal lizards for example is 10.4 ± 0.56°C and the difference between the CTMax and the highest body temperature ever recorded in the field for 38 species was 6.4 ± 1.03°C. If the PT is taken, as suggested previously, as an obvious index of thermal stress (i.e., significantly elevated

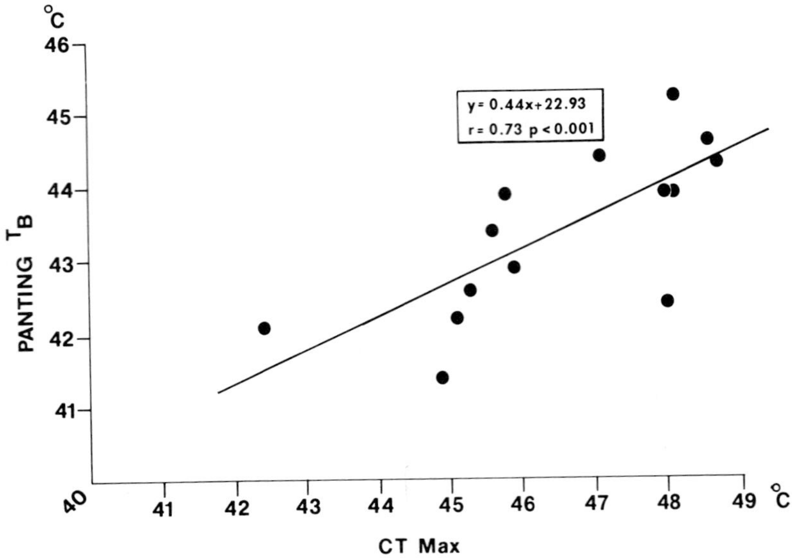


**Fig. 2.15** Correlation between the body temperature at which panting commences and the CTMax recorded in 14 species of North-American lizards (modified from Whitfield & Livezey, 1973).

body temperature coupled with the deployment of emergency thermoregulatory responses) then it seems clear that *Amphibolurus nuchalis* is probably never exposed to thermal stress in its natural environment. The highest body temperatures ever recorded for this species in the field are 43°C (Licht *et al.*, 1966), 43.8°C (Bradshaw & Main, 1968), and 43.5°C (central Australia—Heatwole, 1970), which are all below the PT for this lizard of 44.1°C. Heatwole (1970) records one body temperature of 46°C from an *Amphibolurus nuchalis* in the Simpson Desert which was panting prior to capture and the animal presumably had been forced from its burrow which at that time had a temperature of 44.9°C. Bradshaw & Main (1968) described a similar occasion at Port Hedland in the north-west of Western Australia when extremely high environmental temperatures forced two species of lizards above ground, an agamid, *Diporophora bilinea* and a gecko, *Diplodactylus michaelsoni* (see Figure 2.16). The mean body temperatures of these two species at this time were 44.6°C and 42.6°C respectively and must have been very close to their PTs although neither species was panting when captured. Significantly, no *Amphibolurus nuchalis* were observed above ground at this time when the air temperature was 45.5°C and one individual dug from a burrow in fact had a body temperature of 39.2°C. This highlights the

**Table XVIII:** Representative activity temperatures of reptiles in relation to critical temperatures (modified from Huey, 1982).

| Group | MTB–CTMin $\overline{X} \pm SE$ (N) | Min $T_b$–CTMin $\overline{X} \pm SE$ (N) | CTMax–MTB $\overline{X} \pm SE$ (N) | CTMax–Max $T_b$ $\overline{X} \pm SE$ (N) |
|---|---|---|---|---|
| Crocodilians | | 23 (1) | | 1–2 (1) |
| Turtles | | | 13.7 ± 1.10 (15) | 7.4 ± 1.24 (12) |
| *Sphenodon* | 7.4 (1) | 2.7 (1) | | |
| Lizards: | | | | |
| diurnal species | 26.7 ± 1.53 (9) | 20.3 ± 3.17 (9) | 10.4 ± 0.56 (38) | 6.4 ± 1.03 (38) |
| nocturnal/fossorial species | | | 13.0 ± 1.36 (3) | 6.5 ± 1.17 (3) |
| Snakes | | | 13.1 ± 0.66 (19) | 6.8 ± 0.66 (18) |

Symbols as follows: MTB = mean $T_b$ of reptiles in field; Max $T_b$, maximum $T_b$ recorded in field; Min $T_b$, minimum $T_b$ recorded in field; CTMax, Critical Thermal Maximum; CTMin, Critical Thermal Minimum; N, number of species. CTMax–MTB, "Ecological Thermal Safety Margin" and CTMax–Max $T_b$, "Minimum Thermal Safety Margin" of Heatwole (1976).

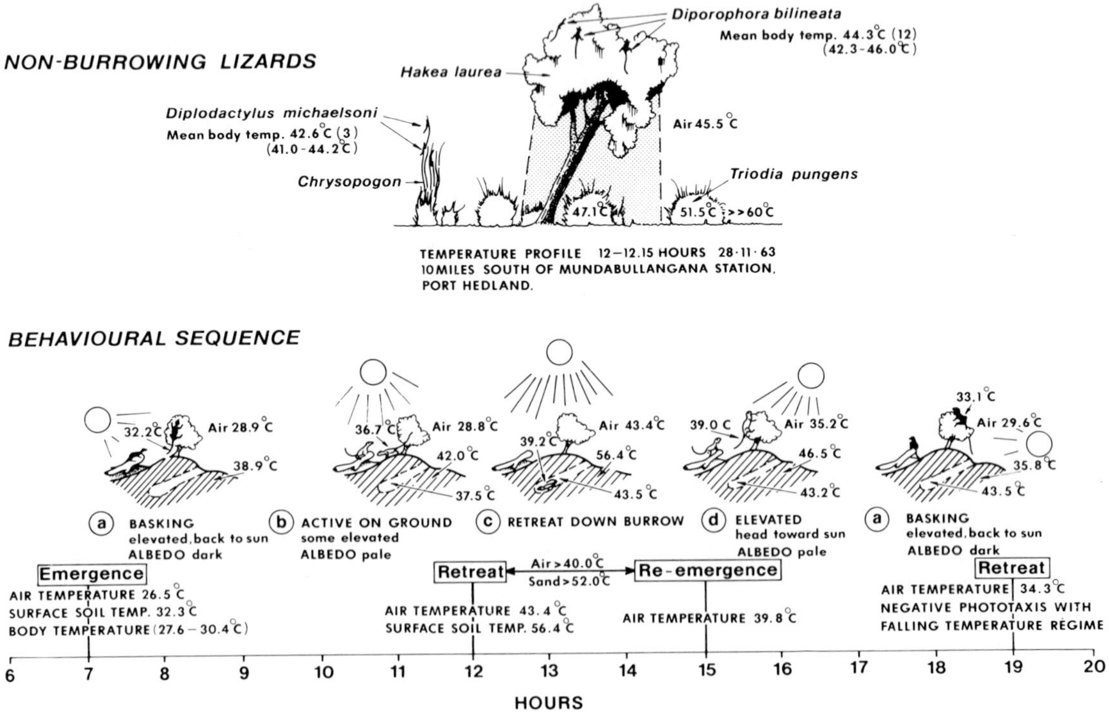


**Fig. 2.16** Diagrammatic representation of behavioural thermoregulation in the Australian agamid lizard *Amphibolurus nuchalis* and temperature relations for non-burrowing lizards, *Diplodactylus michaelsoni* and *Diporophora bilineata*, during the period when *A. nuchalis* retreats within its burrow (from Bradshaw & Main, 1968).

importance of the burrow in this species as a thermal refuge, the utilisation of which enables it to avoid elevated body temperatures which would entail excessive loss of body water.

Cogger (1974) arrived at a similar conclusion when studying the small desert-living *Amphibolurus fordi*. He records the VMax for this species as 41°C and reduced activity with shade-seeking behaviour commenced at a mean body temperature of 39.6°C. Cogger (1974) mentions that *Amphibolurus fordi* invariably pants at body temperatures above 45°C–46°C but that no individual was ever observed to do so in the field during two years of intensive study and the highest body temperature record in the field was 41°C, the same as the VMax.

Close inspection of field body temperature records from other species of lizards leads to a similar conclusion that they rarely, if ever, experience temperatures as high as their PT and thus, by the definition advanced here, could not be considered to suffer 'thermal stress'. McGinnis (1966) records the maximum body temperature of *Sceloporus occidentalis* in the field as 38.9°C, well below the PT for this species of 42.9°C (Whitfield & Livezey, 1973). The body temperature measured by telemetry of *Varanus griseus* in semi-liberty in the Algerian Sahara did not exceed 35°C and averaged 32°C during the day (Francaz *et al.*, 1976) well below the PT of 42°C for this species (Vernet, 1977) and the Lethal Temperature of 44.8°C (Grenot, 1968a). Sokolov *et al.* (1975) similarly record a maximum body temperature of 38.4°C for *Varanus griseus* also studied in the field by telemetry and Stebbins & Barwick (1968) found that the mean body temperature of a single free-ranging *Varanus varius* during summer in a semi-arid region of eastern Australia was 35.5°C with a maximum register of 38°C.

The body temperatures of free-ranging *Varanus gouldii* were recorded by King (1980) in animals living in both mesic as well as arid situations and found to be significantly higher in the drier localities (38.5 ± 0.6°C and 37.0 ± 0.4°C in arid and semi-arid localities versus 35.1 ± 0.4°C in cooler regions). When body temperatures were measured under similar environmental conditions however (air and black-bulb temperatures) in arid and mesic localities, there was no significant difference between the means (35.5 ± 0.6°C versus 35.6 ± 0.3°C) suggesting that, as with *Amphibolurus nuchalis*, arid-living *Varanus gouldii* are forced to spend longer periods of the day at elevated body temperatures which are nevertheless well below temperatures which could be considered stressful. The maximum body temperature recorded by Mayhew (1963) for *Sceloporus orcutti* in the field was 38°C and, although the PT does not seem to have been measured for this species, it probably lies between 41° and 43°C as for other species of this genus (Whitfield & Livezey, 1973).

Huey & Pianka (1977b) recorded mean body temperatures throughout the year for a large number of lizard species in the Kalahari Desert in southern

Africa and, although PTs and CTMax data are not available for these species, it is notable that the mean body temperature for the summer months was in no instance significantly greater than the overall mean representing all seasons. Spellerberg (1972) studied the critical temperatures and temperature tolerances of 33 species of south-eastern Australian reptiles and concluded that the CTMax showed little correlation with species' distribution and aridity of habitat and that the CTMin was of greater ecological significance in limiting the survival of these species. Reptiles from the more arid parts of south-east Australia were found generally, however, to have a higher CTMax and this has been confirmed by Greer (1980) in his study of the CTMax of a large number of scincid lizards. Of the skinks, however, only two genera are known to pant (*Tiliqua* and *Menetia*), and then in a peculiar manner, which does not suggest that the response is primarily of significance in thermoregulation (see also Heatwole *et al.*, 1973). Greer (1980) found that the CTMax of diurnal surface-active species was higher by several degrees than that of nocturnal and crepuscular forms (see Figure 2.17) and generally somewhat lower than that of desert agamids and iguanids which fall in the range from 46°–48°C.

At this stage the nature of the data preclude any final conclusions on the question of thermal stress in desert-living species and more work is needed in this area. The studies carried out on desert and non-desert species of the genus *Amphibolurus* in Australia suggest that there is little apparent variation in thermoregulatory behaviour between species when precise sequences are analysed, and that selection has not favoured the development of superior regulatory capacities in the desert-living members of this genus (Bradshaw, 1981). A similar conclusion appears to apply to scincid genera but many desert species of the large genus *Ctenotus* have yet to be studied in any detail. Webb *et al.* (1972) reported that *Oedura cogger*, a desert-adapted gecko, had a higher PT than *O. tryoni* which lives in cooler regions, but this could merely be a reflection of differences in acclimation régime. Heath (1965), for example, found that the PT of the desert horned toad *Phrynosoma platyrhinos* was 44.1°C, significantly higher than that of *P. cornutum* and *P. coronatum*, which are both non-desert species, but Ballinger & Schrank (1970) subsequently showed that acclimation to a temperature of 40°C for 60 hours raised the CTMax of *P. cornutum* to 45.5°C. Comparisons between the maximum body temperatures recorded in the field and the PTs for different species (see Table XIX) suggest that few desert lizards are ever subjected to thermal stress and this is certainly even more evident in the case of snakes living in arid habitats (Lillywhite, 1980). More research is clearly needed and it will be of interest to see whether the PT proves to be a useful index of the occurrence of thermal stress in non-scincid lizards.

Currently the only instances of a breakdown in thermal homeostasis which have been documented are a direct result of other detrimental changes in the

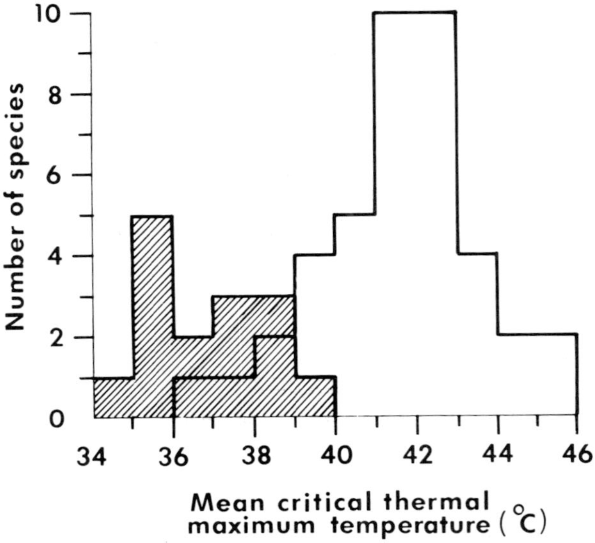


**Fig. 2.17** Histograms showing the number of species with different mean critical thermal maximum temperatures in groups of skinks defined on behaviour and distribution. Open-histogram: diurnal, surface-active species regardless of distribution; hatched-histogram: crepuscular to nocturnal or cryptozooic to fossorial species in genera that have radiated widely in the interior of Australia (from Greer, 1980).

*milieu intérieur* which probably have a wide-ranging effect on the central nervous system. Bradshaw (1970) noted that a combination of hypernatraemia and chronic dehydration in *Amphibolurus ornatus* led to an abandonment of thermoregulatory behaviour in the field. Henzell (1972), working with several species of the scincid genus *Egernia*, also noted that dehydration had a detrimental effect on the precision of thermoregulation, and Rice & Bradshaw (1980) found that loss of condition in *Amphibolurus nuchalis* was associated with a temporary loss of the ability to alter skin reflectance with change in body temperature. Nagy (1973) also records that the chuckwalla, *Sauromalus obesus*, abandons diurnal activity, and, hence, thermoregulation, for long periods in the field when starved and Cogger (1974) found that even mild starvation was sufficient to lower the activity temperatures of *Amphibolurus fordi*. Thermal homeostasis thus seems to be a necessary aspect of overall physiological homeostasis which is normally well-maintained in desert lizards, due to the inherent efficacy of their behavioural and physiological thermoregulatory mechanisms, but which may in turn itself be affected, if other

**Table XIX:** Relationship between maximum body temperatures recorded in the field and panting threshold for a number of species of lizards.

| Species | Max BT recorded in field | Panting Threshold (PT) | Authority |
|---|---|---|---|
| *Amphibolurus nuchalis* | 43.0°C | | 1 |
| | 43.8 | | 2 |
| | 43.5 | | 3 |
| | 46.0 | 44.1°C | 3 |
| *A. barbatus* | 38.6 | 39.5 | 4 |
| *A. fordi* | 41.0 | 45.5 | 5 |
| *Dipsosaurus dorsalis* | 42.0 | 44.5 | 6, 7 |
| *Sceloporus occidentalis* | 38.9 | 42.9 | 8, 7 |
| *S. orcutti* | 38.0 | 41–43 | 9, 7 |
| *Varanus griseus* | 35.0 | 42.0 | 10, 11 |
| | 38.4 | | 12 |
| *V. varius* | 38.0 | ? | 13 |
| *V. gouldii* | 39.8 | ? | 14 |

1. Licht *et al.* (1966)
2. Bradshaw & Main (1968)
3. Heatwole (1970)
4. Warburg (1965)
5. Cogger (1974)
6. De Witt (1967a)
7. Whitfield & Livezey (1973)
8. McGinnis (1966)
9. Mayhew (1963)
10. Francaz *et al.* (1976)
11. Vernet (1977)
12. Sokolov *et al.* (1975)
13. Stebbins & Barwick (1968)
14. King (1980)

systems fail. Considerably more data are needed however to test adequately the preliminary conclusion reached here that desert species of reptiles rarely, if ever, experience what may be defined as 'thermal stress' under natural conditions.

## III. WATER AND ELECTROLYTE BALANCE

'Osmoregulation' and 'water and electrolyte balance' are usually treated synonymously in the literature and taken to signify the "processes by which the amounts of water and specific solutes within the body of an organism are maintained constant, or within tolerable limits" (Shoemaker & Nagy, 1977). Bentley (1959) seems to have been one of the first to appreciate the need for study of the water and electrolyte balance of reptiles, both in the laboratory and in their natural environment, and his paper on the large Australian skink *Tiliqua* (*Trachydosaurus*) *rugosa*, was an important benchmark in the field. He found that the response of *Tiliqua* to the administration of salt solutions was

**Table XX:** Plasma $Na^+$ and $K^+$ concentrations in *Tiliqua rugosa* estimated in August and January–February (from Bentley, 1959).

| | Plasma concentration ( Mmol/ L) | |
|---|---|---|
| | $Na^+$ | $K^+$ |
| August<br>10 animals) | 151.87 ± 1.57<br>(147.82 – 156.52) | 3.46 ± 0.139<br>(2.85 – 4.10) |
| January–February<br>(12 animals) | 195.88 ± 7.10<br>(162.16 – 242.20) | 4.59 ± 0.126<br>(4.01 – 5.64) |

Values are given as means ± S.E.; in parentheses, range.

very reminisent of that of newborn mammals. He ascribed this to their very limited glomerular population, and hence filtering surface, which was about seven times less than that of a similar-sized mammal. In addition, Bentley (1959) measured the concentrations of sodium ($Na^+$) and potassium ($K^+$) in the plasma of animals collected during different seasons in the field, and found marked evidence of hypernatraemia in summer (see Table XX). In interpreting these data Bentley (*op cit*) advanced the hypothesis that arid-living reptiles may "conserve body water at the expense of abandoning maintenance of the constancy of body fluid composition", during periods of water stress, and he noted the extraordinary tolerance of these animals to hypernatraemia, with levels of over 230 mmol/L being produced experimentally without apparent ill effect (see also Borst *et al.*, 1960).

This early work serves to bring into sharp focus the central question concerning water and electrolyte homeostasis in reptiles—whether episodes such as periodic hypernatraemia result from a limited ability to regulate the composition of the body fluids or, instead, should be viewed as 'adaptive' responses to an exigent environment which enhance the individual's chances of survival. With reference to the definition of 'specific adaptations' advanced in the Introduction, these responses would need to be shared only by desert-living species if they are to be considered as such, and one would need to demonstrate that mesic-living species, for example, do not 'abandon' osmoregulation when similarly deprived of water. Such an approach is also important to resolve the question of 'stress': one needs to know, first, whether reptiles regulate plasma sodium levels between narrow limits, as do mammals; then whether such regulation involves the use of hormones, and, last, whether hypernatraemia in the field is associated with elevated levels of these same regulatory hormones. Only then would one be justified in concluding that the animals showed water and electrolyte 'stress'. Unfortunately, as with the question of temperature regulation, the data in the literature have for the most part been collected with

**Table XXI:** Fluid distribution and plasma electrolyte concentrations in lizards (*Amphibolurus ornatus*) in the field during a mild summer (from Bradshaw & Shoemaker, 1967).

| Season | N | Mean fluid volume (mL/100g ± S.E.) | | | | | Mean plasma electrolytes ± S.E. | |
|---|---|---|---|---|---|---|---|---|
| | | TBW | PV | BV | ECFV | ICFV | Sodium (Mmol/L) | Potassium (Mmol/L) |
| Early summer (Nov) | 7 | 73.6 ± 1.04 | 5.4 ± 0.08 | 7.6 ± 0.16 | 25.7 ± 0.82 | 47.9 ± 0.78 | 155 ± 2.4 | 5.5 ± 0.62 |
| Mid-summer (Dec) | 8 | 73.8 ± 1.62 | 7.1 ± 0.35 | 9.3 ± 0.39 | 33.0 ± 1.27 | 41.2 ± 1.63 | 172 ± 1.5 | 5.7 ± 0.15 |
| Significance of difference between means for early and mid-summer | | N.S. | P < 0.01 | P < 0.01 | P < 0.01 | P < 0.01 | P < 0.01 | N.S. |
| Late summer (Feb)— 10 hours after rain | 5 | 73.2 ± 0.66 | 4.8 ± 0.25 | 6.7 ± 0.20 | 25.4 ± 1.05 | 47.9 ± 1.47 | 149 ± 3.9 | 5.0 ± 0.35 |
| Mean values from adequately-hydrated animals | 22 | 73.5 ± 0.30 | 5.1 ± 0.10 | 7.3 ± 0.12 | 25.5 ± 0.43 | 48.0 ± 0.49 | 150 ± 1.3 | 5.0 ± 0.17 |

No significant differences exist between any of the means for November, February, and those from adequately-hydrated animals. TBW = total body water content; PV = plasma volume; BV = blood volume; ECFV = extra-cellular fluid volume; ICFV = intracellular fluid volume; N.S. = not significant.

other motives in mind, and it is only possible to resolve this problem in a very limited number of lizard species whose water and electrolyte metabolism has been studied in sufficient detail, both in the laboratory and in the field.

Seasonal changes in water and electrolyte balance of the small Australian agamid lizard, *Amphibolurus ornatus*, were first reported by Bradshaw & Shoemaker (1967) who measured changes in fluid volumes (blood and extracellular fluid volume and total body water content) as well as the concentrations of $Na^+$ and $K^+$. These authors found evidence of significant hypernatraemia in the field during the summer months, associated with an isosmotic expansion of the extracellular compartment, caused by a movement of water from the cells (see Table XXI). These animals were not dehydrated, however, as shown by their normal total body water content, and it was thus clear that the hypernatraemia was the result of the addition of sodium ions to the body fluids, rather than the loss of ion-free water.

These lizards exist almost exclusively on a sodium-rich diet of ants of the genus *Iridomyrmex*, and, during the summer months, when free water is unavailable, sodium ions are retained in the extracellular fluid due to a progressive inhibition of renal excretion. During brief four to six week periods of drought, this sodium is retained entirely within the extracellular fluid which is in consequence expanded in volume by an isosmotic shift of water from the intracellular compartment. This is only a temporary state of affairs, however, and, as seen in Table XXI, as little as 10 hours after a sudden summer thunderstorm, during which the lizards were observed to drink copiously, plasma sodium concentrations fell to normal levels and the cellular dehydration was rectified. This same pattern of transient hypernatraemia during the summer months is common to all the species of the genus *Amphibolurus* that have been studied to date (see Bradshaw, 1970) and would appear to be inevitable in any reptile lacking a nasal salt gland or other significant avenue of extra-renal excretion, which is forced to subsist on a sodium-rich diet.

Longer periods of drought can provoke significant dehydration in *Amphibolurus ornatus*, with some animals losing over 40% of their original body weight before succumbing. The mean plasma sodium concentration of a sample of animals at Bakers Hill in Western Australia after a 73-day drought was 243 mmol/L and the corresponding potassium concentration was 12.2 mmol/L (see Table XXII). Tissue sodium concentrations of these animals were also elevated, indicating that dietary sodium was no longer exclusively sequestered in the extracellular compartment and, as would be expected from this, the cellular compartment was consequently fully hydrated. Considerable individual variability was evident, however, in the extent to which animals were, in addition, suffering from a loss of body weight and this was found to correlate with the growth rate of the individual. Figure 2.18 shows a weight/length regression for marked individuals of known growth rate at Bakers Hill at the end of a hot, dry summer period and it is evident that, when compared with the regression for fully-hydrated individuals, weight loss is

**Table XXII:** Comparison between the distribution of electrolytes in *Amphibolurus ornatus* before and after rain at Baker's Hill (from Bradshaw, 1970).

| Variable | Pre-rain sample | Significance of difference | Post-rain sample | Significance of difference | Fully hydrated |
|---|---|---|---|---|---|
| Plasma sodium (mmol/L) | 243.0 ± 9.96 | P < 0.001 | 166.0 ± 2.74 | P < 0.01 | 150.0 ± 1.32 |
| Plasma potassium (mmol/L) | 12.2 ± 0.60 | P < 0.001 | 5.6 ± 0.88 | P > 0.1 | 5.0 ± 0.17 |
| Tissue sodium | | | | | |
| (mmol/L tissue water) | 75.5 ± 8.33 | P > 0.5 | 71.2 ± 9.99 | P < 0.01 | 28.5 ± 3.24 |
| (mmol/kg dry wt.) | 216.8 ± 20.52 | P > 0.1 | 259.8 ± 41.33 | P < 0.01 | 98.2 ± 10.31 |
| Tissue potassium | | | | | |
| (mmol/L tissue water) | 147.0 ± 3.69 | P < 0.01 | 128.2 ± 4.67 | P > 0.5 | 123.8 ± 4.86 |
| (mmol/kg dry wt.) | 426.2 ± 16.49 | P > 0.1 | 461.2 ± 12.41 | P = 0.05 | 406.0 ± 16.20 |
| Tissue water (mL/100g) | 74.3 ± 0.63 | P < 0.01 | 78.3 ± 0.36 | P < 0.01 | 76.8 ± 0.24 |
| Urine sodium (mmol/L) | 103.0 ± 9.53 | P < 0.02 | 65.3 ± 6.72 | | |
| Urine potassium (mmol/L) | 38.2 ± 10.35 | P > 0.1 | 27.1 ± 1.96 | | |

All values mean ± S.E.

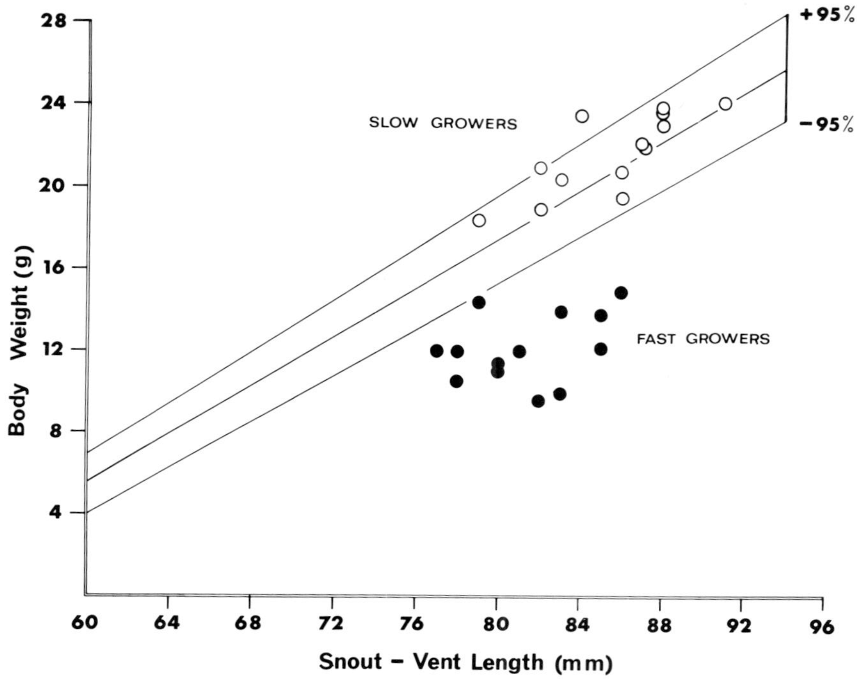


**Fig. 2.18** Body weight/snout-vent length relationship for the agamid lizard *Amphibolurus ornatus* from Bakers Hill in Western Australia. Animals of known growth rate are plotted against the regression for fully-hydrated individuals (± 95% confidence limits) prior to rain in the summer of 1964 and the differential weight loss of the fast-growers is evident (from Bradshaw, 1970).

restricted to 'fast growers'. This unusual situation has been analysed in some detail by Baverstock (1975) and Baverstock & Bradshaw (1975) who found that, under laboratory situations, *A. ornatus* with apparently genetically determined high rates of growth void significantly greater volumes of urine per day and have higher rates of evaporative water loss than 'slow growers' when both are subjected to salt loading (see Figure 2.19).

Despite the loss of weight associated with these processes, 'fast-growing' *A. ornatus* possess considerable powers of regulation and, even in individuals that have lost over 40% of their initial body weight, the haematocrit is not elevated and, as seen in Figure 2.20, plasma and blood volumes are indistinguishable from those of fully-hydrated individuals. Progressive weight loss in the field is associated with a linear decline in total body water content (TBW) which is parallelled by a fall in intracellular water content (ICFV) until there has been an approximate 25% weight loss. This has the effect of

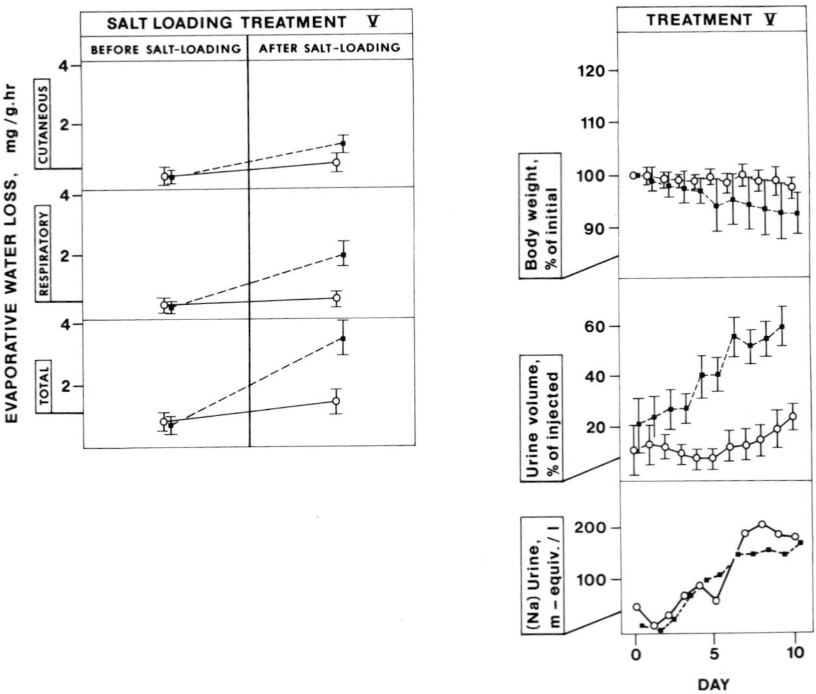


**Fig. 2.19** Changes in rate of evaporative water loss (partitioned into cutaneous and respiratory components), body weight and urine volume and sodium concentration of fast- and slow-growing *Amphibolurus ornatus* with chronic salt loading (150mmol/L NaCl). Open circles—'slow growers'; closed squares—'fast growers' (from Baverstock, 1975).

'protecting' the extracellular compartment, including the plasma and blood volumes, from reduction in size. Weight losses in excess of 20% appear to involve a reduction in ECFV (extracellular fluid volume), but at the expense of the interstitial fluid volume, so that there is no reduction in the circulating fluid volume right up until the point of death. This is an impressive response to dehydration whereby unavoidable fluid losses are limited to the intracellular rather than the extracellular compartment, thereby preserving the volume and viscosity of the circulating fluid in the body. Conservation of plasma volume with dehydration has been described in a number of other desert-inhabiting animals including the camel (Macfarlane *et al.*, 1963), the Mongolian gerbil *Meriones unquiculatus* (Kutcher, 1967), and the desert spiny mouse *Acomys cahirinus* (Horowitz & Borut, 1970, 1975). This ability would appear to result

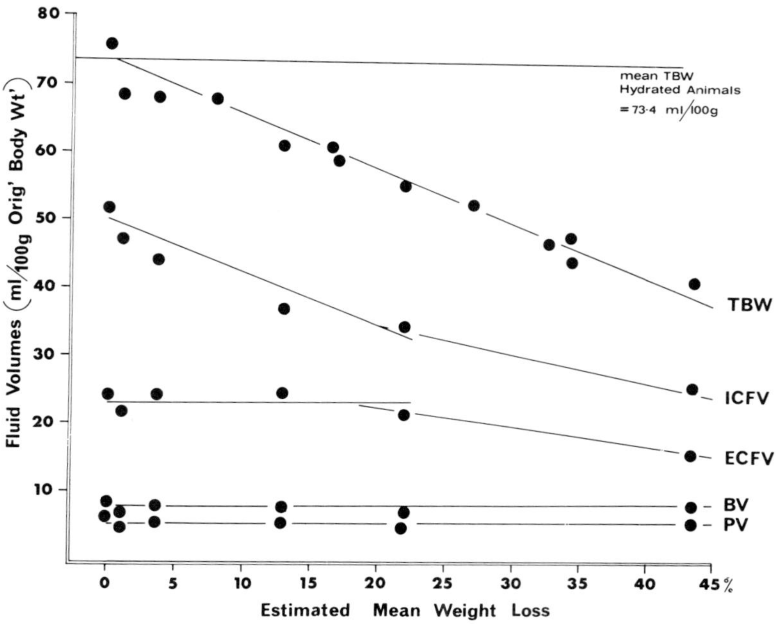


**Fig. 2.20** Changes in fluid volumes with progressive dehydration in a sample of fast-growing *Amphibolurus ornatus* at Bakers Hill in Western Australia. TBW = total body weight content in mL/100g; ICFV = intra-cellular fluid volume; ECFV = extra-cellular fluid volume; PV = plasma volume; BV = blood volume (from Bradshaw, 1970).

from an increase in the colloid osmotic pressure of the plasma and a decrease in the permeability of blood capillaries to serum albumin (Borut *et al.*, 1972; Horowitz & Samueloff, 1979).

The handling of sodium and potassium ions during weight loss in 'fast-growing' *A. ornatus* also shows interesting differences of relevance to the problem of ionic regulation. Plasma sodium concentrations may increase to levels as high as 300 mmol/L in some individuals and tissue sodium concentrations and content also increase significantly (see Table XXII) and water is restored to the intracellular compartment (Bradshaw, 1970). As dehydration progresses, sodium ions are retained within the tissues at ever-increasing concentrations (which may reach levels as high as 100 mmol/L tissue water, over three times the normal concentration), but the potassium concentration in the tissues remains unchanged at about 150 mmol/L throughout the period of weight loss, apparently by persistent tubular secretion via the renal-portal system. 'Slow-growing' *A. ornatus*, which characteristically

**Table XXIII:** Water and electrolyte composition of the diet of *Dipsosaurus dorsalis* (herbivore) and *Amphibolurus ornatus* (insectivore) (from Bradshaw & Shoemaker, 1967; Minnich & Shoemaker, 1970).

| Species | Locality | Water content | Electrolyte concentrations (mmol/L) | | |
|---|---|---|---|---|---|
| | | | Sodium | Potassium | Na/K |
| *Dipsosaurus dorsalis* | | | | | |
| Spring | Palm Desert | 55.6% | 39.1 | 360.0 | 0.11 |
| Summer | Palm Desert | 51.0 | 34.3 | 364.0 | 0.09 |
| Summer | Thousand Palms | 40.1 | 84.7 | 481.0 | 0.18 |
| *Amphibolurus ornatus* | | | | | |
| Winter | Bakers Hill | 77.0% | 200.8 | 102.0 | 1.97 |
| Summer | Bakers Hill | 74.8 | 290.0 | 82.4 | 3.54 |

do not lose weight in summer, thus have only to contend with the problem of sodium loading during even the most severe of droughts, whereas 'fast-growers' suffer the multiple effects of sodium loading, dehydration and potassium depletion. Schmidt-Nielsen (1975) points out that the maintenance of a rather constant potassium concentration of about 150 mmol/L appears to be critical for the integrity of all cells and speculates that protein synthesis is particularly sensitive to changes in intracellular potassium concentration. Recovery from the effects of a prolonged drought is thus much more difficult for 'fast' than 'slow growers', because of the problem of the loss of cell mass in the former. Bradshaw (1971) has documented instances of differential morality in this species.

The water and electrolyte balance of the North-American desert iguana, *Dipsosaurus dorsalis*, was studied in its natural habitat in southern California by Minnich (1970a) and Minnich & Shoemaker (1970). It is a herbivorous species, feeding on the flowers and leaves of a number of plants including *Oenothera clavaeformis, Palafoxia linearis, Cryptantha angustifolia,* the indigo bush, *Dalea emoryi*, and the flowers of the creosote bush, *Larrea divaricata*. The average water and electrolyte composition of the diet of desert iguanas in spring and summer is shown in Table XXIII and compared with that of the Australian lizard *Amphibolurus ornatus*. The difference in the sodium/potassium ratio of the two diets is very noticeable, with potassium rather than sodium being the predominant cation in the diet of the herbivorous desert iguana. Seasonal changes in plasma and electrolyte concentrations are shown in Table XXIV from which it is evident that this species experiences neither hypernatraemia nor hyperkalaemia in the field. Plasma sodium concentrations averaged 131 mmol/L in winter, increasing to a maximum of 168 mmol/L in spring but, because of large individual variations, the difference was not statistically significant. Using data on rates of water turnover of free-ranging animals it is possible to estimate the daily water and electrolyte budget of *D. dorsalis* in the field at this time and budgets have been presented by Bentley (1976) and by Minnich (1979). Table XXV shows such a budget, based on the data reported in Minnich (1970) and Minnich & Shoemaker (1970), and it is important to note that this differs slightly from those already published. Bentley (1976) (see his Table XVI) unfortunately miscalculated potassium loss and thereby grossly over-estimated (by a factor of ten) the importance of the nasal salt gland of this species in potassium excretion. Minnich (1979) (see his Table V) has similarly magnified the importance of the salt gland by attributing to its activity the remainder of the balance which is unaccounted for in the raw data. In fact only 51.5% of the daily intake of sodium ions and 58.1% of the potassium intake could be accounted for after summing all excretory routes and this could either result from an over-estimation of the dietary intake or from some sequestering of salts in the body over the period of measurement.

**Table XXIV:** Electrolyte concentrations (in mmol/L) in the body fluids and urine of desert iguanas captured during different seasons (from Minnich, 1970).

| | | Spring | | Summer (Palm Desert) | | Summer (Thousand Palms) | | Winter | |
|---|---|---|---|---|---|---|---|---|---|
| Fluid | Ion | N | Mean ± S.E. | N | Mean ± S.E. | N | Mean ± S.E. | N | Mean ± S.E. |
| Plasma | $Na^+$ | 21 | 168.0 ± 1.92 | 18 | 163.0 ± 2.60 | 11 | 165.0 ± 2.66 | 4 | 131.0 ± 18.2 |
| | $K^+$ | 21 | 3.40 ± 0.40 | 18 | 3.36 ± 0.22 | 11 | 2.60 ± 0.35 | 4 | 3.44 ± 0.11 |
| | $Cl^-$ | 21 | 120.0 ± 1.64 | 18 | 120.0 ± 2.75 | 11 | 112.0 ± 2.23 | 4 | 91.5 ± 15.8 |
| Muscle | $Na^+$ | 9 | 44.9 ± 2.73 | 16 | 40.5 ± 1.88 | 6 | 40.3 ± 0.94 | 5 | 42.4 ± 2.88 |
| | $K^+$ | 9 | 122.0 ± 3.90 | 16 | 126.0 ± 2.54 | 6 | 122.0 ± 3.00 | 5 | 117.0 ± 4.48 |
| | $Cl^-$ | 9 | 30.0 ± 1.33 | 16 | 25.9 ± 1.18 | 6 | 26.4 ± 1.32 | 5 | 29.9 ± 1.42 |

**Table XXV:** Water and electrolyte budget of the desert iguana, *Dipsosaurus dorsalis*, under field conditions in southern California (water in mL/kg.day, electrolytes in μmol/kg.day) (from Minnich 1970, and Minnich & Shoemaker, 1970).

| Gain | | | | Loss (% Total) | | | |
|---|---|---|---|---|---|---|---|
| Avenue | Water | $Na^+$ | $K^+$ | Avenue | Water | $Na^+$ | $K^+$ |
| Dietary | 26.4 | 916 | 9744 | Urinary | 0.78 ( 2.8) | 188 (39.8) | 3835 (67.7) |
| Metabolic | 3.6 | — | — | Faecal | 18.60 (66.5) | 281 (59.5) | 1691 (29.9) |
| | | | | Salt gland | ? | 3 ( 0.7) | 139 ( 2.4) |
| | | | | Evaporation* | 8.60 (30.7) | — | — |
| Total: | 30.0 | 916 | 9744 | Total: | 27.98 | 472 | 5665 |

Note that the balance is unaccounted for and the lizards were in positive water and electrolyte balance during the period of assessment.

* Estimate from Minnich (1970).

Most of the animals in fact gained weight during the period that water turnover was being measured and were thus in positive water and electrolyte balance (see Minnich & Shoemaker, 1970). What is clear from Table XXV is that faecal water loss is important in this species, accounting for 66.5% of the total, and that a large proportion of dietary potassium is excreted in the form of insoluble urate salts (see Minnich 1972; Minnich & Pieal, 1972). Sodium excretion by this route is less important, however, and 59.5% of this ion is excreted via the faeces. These data also underline the relative unimportance of the nasal salt gland of this species under these field conditions, contributing only 0.7% to sodium excretion and 2.4% to potassium elimination. Templeton *et al.* (1972a) also found in laboratory experiments with *D. dorsalis* that the nasal salt gland was active in potassium excretion during rehydration of dehydration animals, but that it played a negligible rôle in fully-hydrated individuals (see Figure 2.21).

Perhaps not surprisingly, there has been a tendency amongst American workers to interpret the relative constancy of field electrolyte data from species such as *Dipsosaurus dorsalis* in terms of an enhanced regulatory capacity when compared with lizards such as *Amphibolurus ornatus* (see Minnich, 1970; Minnich & Shoemaker, 1972). Homeostatic capacities, however, can only be assessed and compared under rigorous laboratory conditions where electrolyte intakes can be controlled, preferably by injection, and renal and extra-renal excretion rates measured accurately. Such a comparison is possible in this particular case because both species have been studied in the laboratory, under near-identical conditions by Bradshaw (1972) and Bradshaw *et al.* (1972). In these studies renal responses to chronic administration of distilled water, 50 mmol NaCl and 150 mmol NaCl were assessed and changes in adrenocortical activity were monitored through variations in circulating levels of corticosteroids. As shown in Figure 2.22, the renal responses to gross electrolyte depletion and loading in these two species are broadly similar, and homeostatic in tendency, but it is clear that the renal response of *A. ornatus* is somewhat more rapid and ultimately more effective than that of *D. dorsalis*. For example, by Day 6 of a series of daily injections of 150 mmol NaCl, the urine voided by *A. ornatus* equalled this in sodium concentration whereas the sodium concentration of the urine of *D. dorsalis* had only increased to about 125 mmol/L by Day 8, and daily sodium excretion rates averaged less than 50% of those of *A. ornatus*. Table XXVI compares relative changes in plasma sodium concentrations of the two species and it is evident that *D. dorsalis* regulates the *milieu intérieur* less well under these circumstances. Water loading led to a −7.4% change in plasma sodium levels in *A. ornatus* but was accompanied by a −20.6% change in *D. dorsalis*. Sodium loading similarly elevated plasma sodium levels of *D. dorsalis* more than those of *A. ornatus* (+18.4% versus +14.2%). These differences are the result of alterations in the fractional

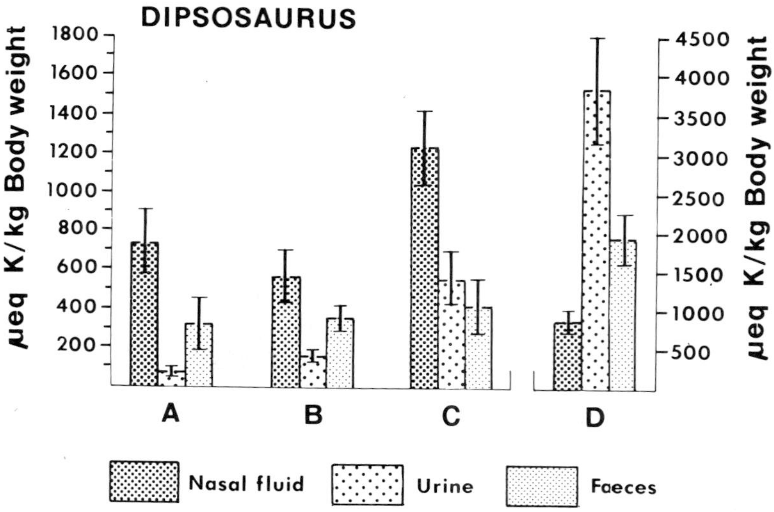


**Fig. 2.21** The relationship of potassium removed in 72 hours per kg body weight by the nasal fluid, urine and faeces of the iguanid lizard *Dipsosaurus dorsalis* during rehydration (A, B and C) and when fully-hydrated. Bars represent the means ± standard error (from Templeton, Murrish, Randall & Mugaas, 1972a).

reabsorption of sodium in the kidneys which determines overall rates of excretion. Fractional reabsorption of sodium ($FR_{Na}$) fell from 99.85 ± 0.06% with water loading, to 88.38 ± 2.01% with salt loading in *A. ornatus*, a relative decrease of −11.49%, whereas the $FR_{Na}$ of *D. dorsalis* only fell by −3.43% under identical circumstances.

These data thus show that the *renal* response to variations in electrolyte status of *D. dorsalis* is less effective, when judged homeostatically, than that of *A. ornatus*, despite the fact that the latter species suffers greater variations in plasma sodium levels in its natural environment. Table XXIII makes it abundantly clear that *A. ornatus'* 'problems' stem from a sodium-rich diet with an $Na^+/K^+$ ratio of from 2–3, while the diet of *D. dorsalis* is virtually sodium-free. Clearly, if *D. dorsalis* were forced to subsist on a diet as rich in sodium as that of *A. ornatus* it would experience even greater variations in sodium concentration than does *A. ornatus*, and the relative constancy of its *milieu intérieur* in its natural habitat is in no way due to inherently superior capacities for regulation. These data also reinforce the conclusion, from field studies, that the nasal salt gland of *D. dorsalis* contributes little to electrolyte homeostasis in this species.

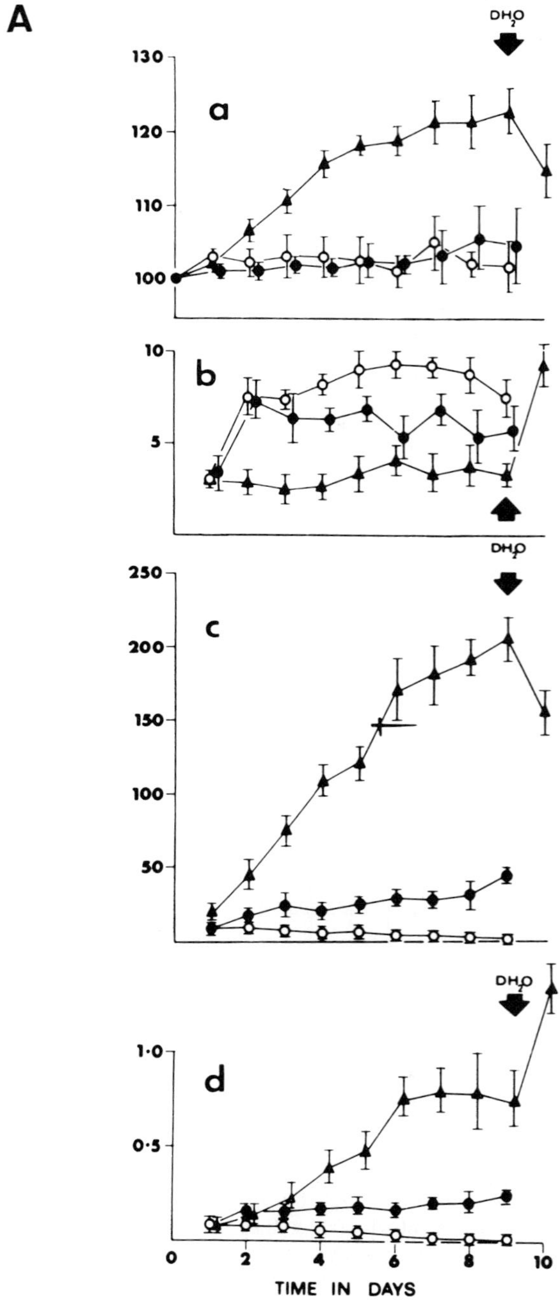
A
DH2O
130
120
110
100
a
10
5
b
DH2O
250
200
150
100
50
c
DH2O
1·0
0·5
d
0
2
4
6
8
10
TIME IN DAYS

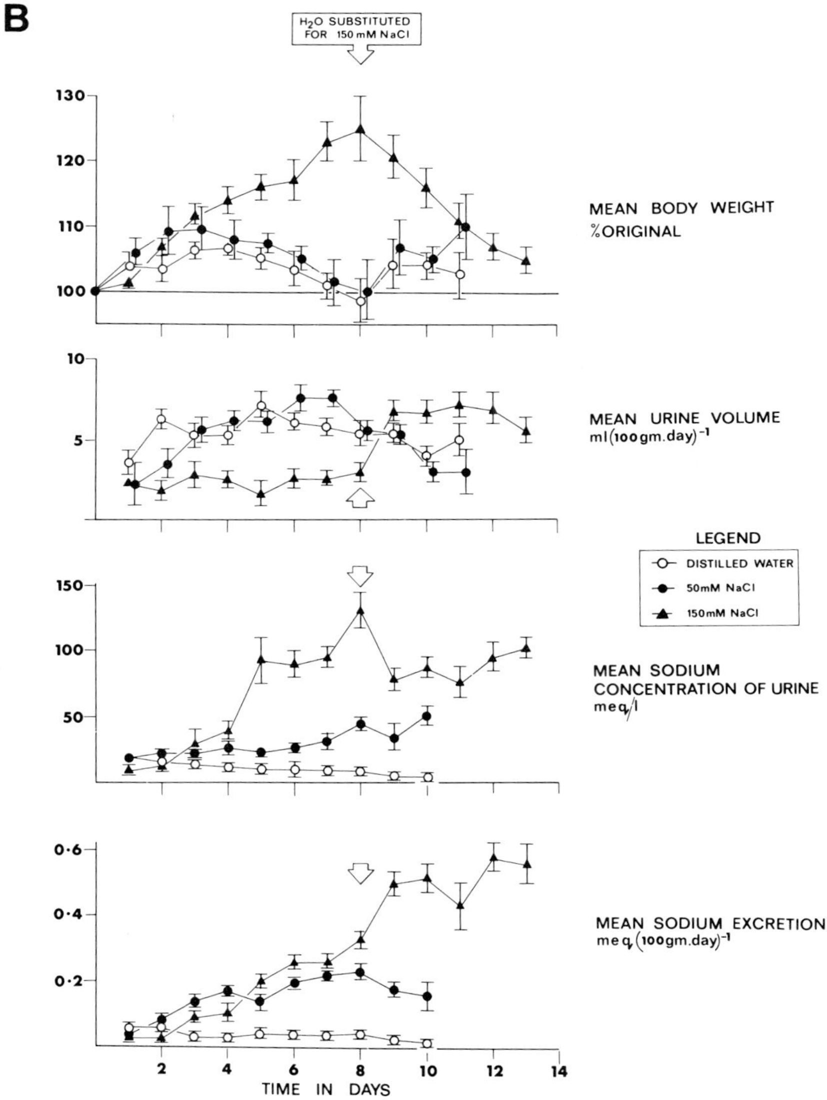


**Fig. 2.22** Changes in (a) mean body weight; (b) urinary volume; (c) urinary sodium excretion and (d) total daily sodium excretion of (A) the Australian agamid lizard *Amphibolurus ornatus* and (B) the American iguanid lizard *Dipsosaurus dorsalis* with progressive water or saline diuresis. All fluid loads = 10mL (100g. day)$^{-1}$ and all points = Mean ± S.E. On day 9, animals previously injected with 150mmol NaCl were injected instead with distilled water ($H_2O$) (from Bradshaw 1972, and Bradshaw, Shoemaker & Nagy, 1972).

**Table XXVI:** A comparison between relative changes in plasma sodium concentrations and fractional reabsorption of sodium in the kidney with sodium loading and depletion in the lizards *Dipsosaurus dorsalis* and *Amphibolurus ornatus* (from Bradshaw, Shoemaker & Nagy, 1972; Bradshaw, 1972).

| Species | Day | Water loading* | Δ% | Salt loading* | Δ% |
|---|---|---|---|---|---|
| *Amphibolurus ornatus* | 0 | 174.2 ± 3.08 mmol/L | | 176.4 ± 4.90 mmol/L | |
| | 8 | 161.4 ± 4.25 | −7.4 | 201.5 ± 7.93 | +14.2 |
| | | ($P < 0.05$) | | ($P < 0.02$) | |
| | $FR_{Na}$ = | 99.85 ± 0.06% | ($P < 0.001$) | 88.38 ± 2.01% | −11.49 |
| *Dipsosaurus dorsalis* | 0 | 169.1 ± 4.29 | | 163.4 ± 1.55 | |
| | 8 | 134.3 ± 3.04 | −20.6 | 193.5 ± 2.61 | +18.4 |
| | | ($P < 0.001$) | | ($P < 0.001$) | |
| | $FR_{Na}$ = | 98.86 ± 0.32% | ($P < 0.01$) | 95.47 ± 1.07% | −3.43 |

* Fluid loads were 10 mL $(100\ g.day)^{-1}$ of either distilled water or 150 mmol NaCl.

Two other species of desert lizards which have been studied intensively in the field are the North American chuckwalla, *Sauromalus obesus*, and the North African Fouette-Queue (whip-tail), *Uromastix acanthinurus*. Both species are large (300 g–600 g) herbivores and live in regions where rainfall is low and erratic and the only source of free-water is from the vegetation on which they subsist. Norris & Dawson (1964) identified salt-secreting nasal glands in *Sauromalus* which elaborate a secretion with a $K^+/Na^+$ ratio of about 2. They also described lateral lymph sacs which apparently function to expand the extracellular fluid volume during favourable periods. Nagy (1972, 1973) studied a marked population of chuckwallas on Black Mountain in the Mojave Desert between 1970 and 1971 and described changes in water and electrolyte balance during the summer period. He found that body weight declined markedly throughout the summer of 1970, to reach a mean of 63.4% of the initial (early April) weight by late October. From then onwards, activity declined but two individuals first captured in October had body weights of 53.5% and 55% of their initial values when recaptured in March, 1971, following the winter. Figure 2.23 shows this decline in condition of chuckwallas over a period of 12 months, together with data on rates of clearance of tritiated water (HTO). The half-life ($t_{1/2}$) of injected HTO increased from 14.5 to 28.4 days in spring to 475 to 887 days over summer, autumn and winter, indicating that the animals in the field were virtually starving with the only fluid input coming from metabolic water production, and even that at a reduced rate. Nagy (1972) also gives data on changes in fluid volumes of animals throughout this period and, when expressed as proportions of actual body weight, these show little variation, despite the large loss in condition. The manner in which such data are presented can be misleading, however, especially when dealing with cases of chronic and significant dehydration. Table XXVII recalculates some of these data and expresses them as a fraction of the original body weight, so that the relative contribution of the separate fluid departments to the water loss can be ascertained; this makes it very clear that chuckwallas respond to chronic dehydration in a manner very different from that of, for example, *Amphibolurus ornatus*. The extracellular fluid volume declines at the expense of cell water, and plasma and blood volumes, far from being protected, bear the brunt of the fluid loss, diminishing by 51.4% and 52% respectively. Plasma sodium concentrations were also significantly elevated in the sample of animals collected in October (to 188 mmol/L) but the high $K^+/Na^+$ ratio of nasal gland salt collected from these individuals (9.8) shows that sodium intake is low when compared with potassium and never likely to lead to persistent hypernatraemia as with *Amphibolurus* lizards.

The Fouette-Queue, *Uromastix acanthinurus*, has been studied by Lemire *et al.* (1979, 1982) in the region of Béni-Abbès in the Algerian Sahara. Preliminary studies by Grenot (1976) had suggested that *Uromastix* is an

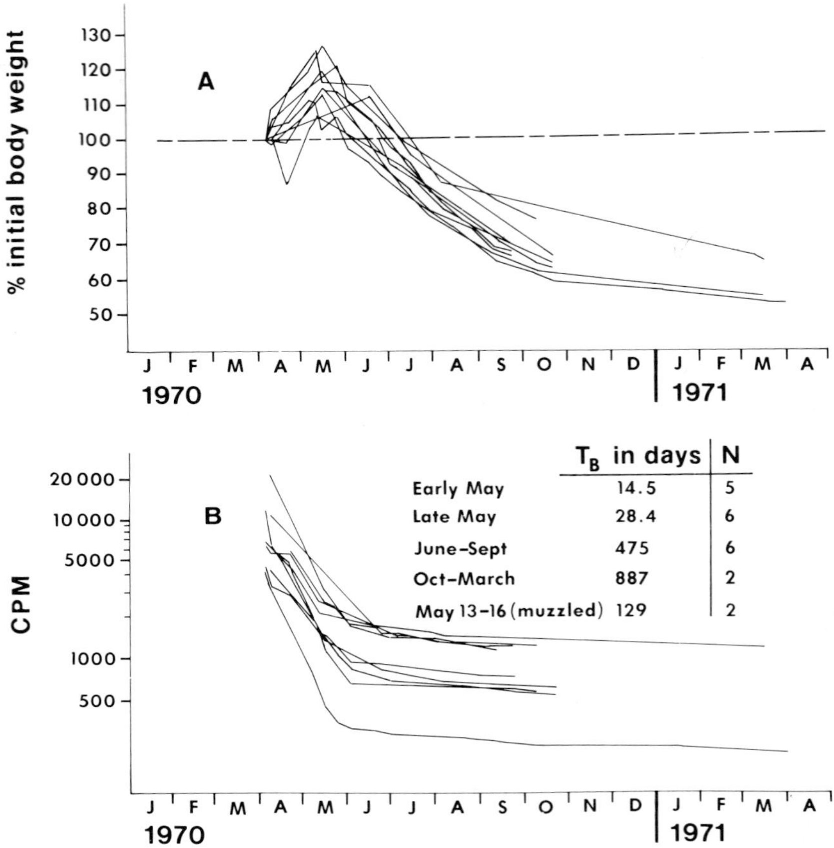


**Fig. 2.23** Seasonal changes in (A) body weight and (B) tritium disappearance from plasma of free-living chuckwallas (*Sauromalus obesus*) in the Mojave desert. Counts per minute (CPM) are from 10 $\mu$L plasma samples, and have been corrected for radioactive decay and adjusted to account for the curve shifts due to reinjection of HTO. $T_B$ is the mean biological half-life of HTO (from Nagy, 1972).

excellent regulator under field conditions (see also Bentley, 1971), but the recent results of Lemire *et al.* (1979, 1982) make it clear that this species experiences substantial changes in the composition of its body fluids throughout the year. Figure 2.24 shows the variation measured in plasma electrolyte concentrations in a population of *Uromastix* in the Ougarta mountains between 1970 and 1971. Plasma sodium levels varied from a minimum of 120 mmol/L in May 1971 to a mean of 219 mmol/L in September of the same year, the highest value recorded being 230 mmol/L.

**Table XXVII:** Seasonal change in fluid compartments of chuckwallas (*Sauromalus obesus*) in the Mojave Desert expressed both as a proportion of the dehydrated and the initial body weight (calculated from Nagy, 1972).

| Month | Mean body weight | Total body water content | Extra-cellular fluid | Intra-cellular fluid | Plasma volume | Blood volume |
|---|---|---|---|---|---|---|
| *April* | 100 g | 76.3 | 35.1 | 36.3 | 7.0 | 9.8 |
| *October* | | | | | | |
| (expressed as mL/100 g *actual* weight) | 55 g | 73.0 | 32.9 | 40.0 | 6.1 | 8.6 |
| (expressed as mL/100 g *initial* weight) | | 40.2 | 18.1 | 22.0 | 3.4 | 4.7 |
| Δ% | | −47.4% | −48.4% | −39.4% | −51.4% | −52.0% |

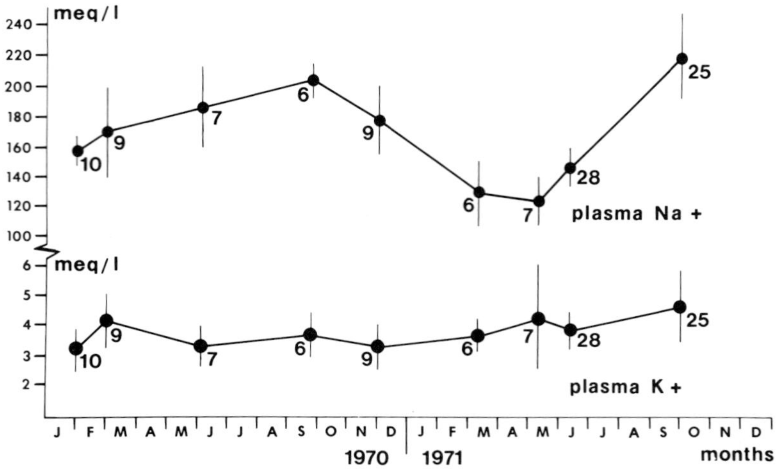


**Fig. 2.24** Seasonal variations in plasma concentrations of sodium ($Na^+$) and potassium ($K^+$) of the Saharan agamid lizard *Uromastix acanthinurus* from the Mountains of Ougarta (Algeria) during 1970 and 1971. Means ± S.D. with number of individuals in parentheses (from Lemire, Grenot & Vernet, 1982).

Plasma potassium concentrations showed less variation, remaining between 3–4.5 mmol/L over the same period. Changes in fluid volumes are shown in Figure 2.25, for a hypothetical 100 g individual, and it is evident that the lizards experience substantial fluid losses over the summer period (August–September). By the end of the summer of 1970, animals had fallen to approximately 60% of their initial body weight but much of this reduction reflected loss of solids and digestive-tract water (i.e., plant water) rather than loss of body fluids. The extracellular fluid volume was reduced from 29 mL/100 g to 15 mL/100 g over the 1970 summer and again from 33 mL/100 g to 20 mL/100 g in the succeeding summer. Plasma and blood volumes were unaffected by this fluid loss which was borne entirely by the interstitial fluid. Intracellular fluid volume declined slightly during these summer droughts, but to a much lesser extent. *Uromastix* is, in this respect, similar to the chuckwalla.

Water balance is clearly the key to survival in both *Uromastix* and *Sauromalus* and this depends almost entirely on the water content of the vegetation consumed. Typically, the water content of the vegetation falls below 50% in mid-summer (see Figure 2.26) and Nagy (1972) has calculated that chuckwallas cannot remain in water balance if their diet contains less than 1.7 mL/g $H_2O$ dry weight (63% $H_2O$). In mid-summer, for example, the

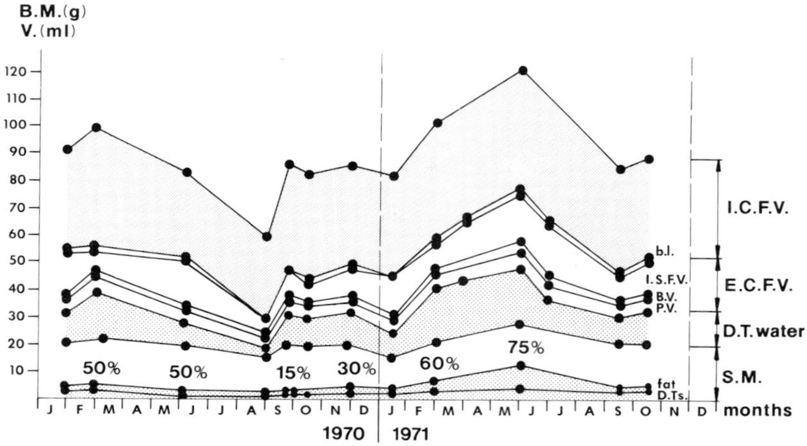


**Fig. 2.25** Seasonal changes in body composition of the Saharan agamid lizard *Uromastix acanthinurus*, adjusted for an individual weighing 100g (snout-vent length = 13.6 cm), from the Mountains of Ougarta during the years of 1970 and 1971. B.M. = body mass in g; ICFV = intra-cellular fluid volume; ECFV = extra-cellular fluid volume, = bladder fluid (b.l) + interstitial fluid volume (ISFV) + blood volume (BV) + plasma volume (PV). DT = water contained in the digestive tract; solid matter (SM) comprises fat and dry digestive tube content (D.Ts.). (%) = percentage of individuals captured with abdominal fat reserves (from Lemire, Grenot & Vernet, 1982).

chuckwallas in Nagy's study were losing weight because of a marked negative water balance, but they were not starving. Dry-matter intake was much the same as in spring and the energy intake was sufficient to ensure that the abdominal fat bodies increased in size right up until July, but the food they were eating was very dry and they were consequently becoming progressively dehydrated. Table XXVIII compares the water balance of chuckwallas and *Uromastix* lizards in their natural habitat at different periods of the year and highlights the prevalence of dehydration in both species. Both species are in water balance in spring (the difference between influx and efflux is not significantly different for *Uromastix*—Lemire *et al.*, 1979) but mid-summer is associated with a negative water balance in both species and, by autumn, all feeding and activity has virtually ceased in both species. By remaining inactive in rock crevices in the case of the chuckwalla, or at the bottom of their deep burrows in the case of *Uromastix*, both lizards are able to balance their water budget and survive on metabolic water production which equals roughly the daily water loss by evaporation. Grenot (1976) found that *Uromastix* could survive for long periods without food under these conditions and the same is

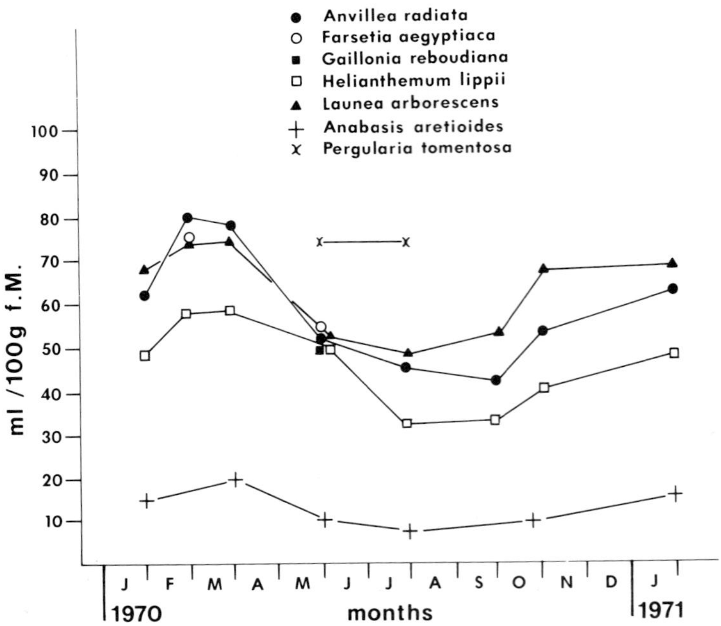


**Fig. 2.26** Seasonal changes in the water content (mL/100g fresh mass) of leafy and fleshy tips of flowers of plants consumed by the Saharan agamid lizard *Uromastix acanthinurus*. Each point represents the average value measured from 10 samples collected at various sites of the habitat (from Lemire, Grenot & Vernet, 1982).

probably true for *Sauromalus*, although Nagy's observations (Nagy, 1972) suggest that mortality may be elevated over the winter period. The nasal salt gland is obviously more important for the maintenance of long-term electrolyte balance in these two lizards than in the other herbivorous lizard, *Dipsosaurus dorsalis*. In the laboratory, Nagy (1972) found that 47% of the ingested potassium and 33% of dietary sodium was excreted via the nasal gland. Forty per cent of the dietary sodium was eliminated through the faeces (and probably not even assimilated in the body) and urate pellets accounted for a further 43% of the dietary potassium. Whether a similar situation pertains in the field is difficult to say, but the fact that plasma potassium concentrations vary little throughout the year suggests this and, given the rather limited capacity of the nasal salt gland of the chuckwalla to handle sodium ions (Templeton, 1966), it is not surprising that they suffer mild hypernatraemia in autumn and late winter. The activity of the nasal salt gland, which is well-developed in *Uromastix acanthinurus* (Lemire *et al.*, 1970, 1972) also appears critical for

**Table XXVIII:** Comparative water budgets during different seasons in the field of the two desert herbivorous lizards, the chuckwalla, *Sauromalus obesus*, and the Fouette-Queue, *Uromastix acanthinurus* (modified from Nagy, 1972; Lemire *et al.*, 1979, 1982; Lemire, 1983).

| | | Water Fluxes in mL $(kg.day)^{-1}$ | | | | | |
|---|---|---|---|---|---|---|---|
| | | Spring | | Summer | | Autumn* | |
| Species | | Gain | Loss | Gain | Loss | Gain | Loss |
| *Sauromalus obesus* | Dietary | 32.8 | $16.7^{E}$ | 13.7 | $11.1^{E}$ | 0 | $2.6^{E}$ |
| | Metabolic | 3.4 | $12.2^{E}$ | 3.4 | $8.1^{F}$ | 1.0 | |
| | | | $3.2^{U}$ | | $2.1^{U}$ | | |
| | Total: | 36.2 | 32.1 | 17.1 | 21.3 | 1.0 | 2.6 |
| | Balance: | +4.1 | | | −4.2 | | −1.6 |
| *Uromastix acanthinurus* | Dietary | 12.5 | $13.2^{E}$ | 7.9 | $8.7^{E}$ | 0 | $3.6^{E}$ |
| | Metabolic | 4.1 | $5.4^{F,U}$ | 3.5 | $5.4^{F,U}$ | 3.5 | $0.1^{F,U}$ |
| | | | $1.0^{N}$ | | $1.0^{N}$ | | |
| | Total: | 16.6 | 19.6 | 11.4 | 15.1 | 3.5 | 3.7 |
| | Balance: | | −3.0 | | −3.7 | | −0.2 |

E = Evaporative water loss
F = Faecal water loss
U = Urinary water loss, including urate pellets
N = Nasal gland fluid loss
} Estimated from laboratory experiments.

* Animals inactive in rock crevices or burrows.

potassium homeostasis in this species and large quantities of this ion are also eliminated through the formation of urate salts (Lemire, 1983). As with the chuckwalla, however, the salt gland of the Fouette-Queue has a very limited capacity to secrete sodium ions (Lemire *et al.*, 1980; Lemire & Vernet, 1982) and, consequently, this animal is forced to tolerate significant periods of hypernatraemia, especially in late autumn.

Rates of water turnover in free-ranging animals reflect the interaction between an individual's diet, and hence primary source of water, and the avenues by which it habitually loses water from the body. Potentially, water turnover could be reduced, and intake diminished, by decreasing the rate at which water is lost from the body. The principal avenue for this in reptiles is evaporation. Herbivorous lizards lose large amounts of water through the faeces and lizards are capable of excreting significant volumes of fluid urine when fully hydrated, but characteristically, survival in deserts is associated with reduced rates of feeding and activity, and pronounced anuria. Evaporation through the integument and from the respiratory tract thus remains the major avenue of water loss under these circumstances and might be expected to be lower in desert than in non-desert species of reptiles. Bentley & Schmidt-Nielsen (1966) confounded the conventional dogma of the day by showing that, almost invariably, the cutaneous component of total water loss in reptiles is quite large and, more often than not, exceeds the fraction lost from the respiratory tract. Bentley (1976) for example tabulates rates of evaporative water loss for 18 species of reptiles and in 14 of these the cutaneous component exceeds 50% of the total (see also Minnich, 1979, and his Table XIII). In general, species from more arid habitats show lower weight-specific rates of evaporative water loss, but this is not invariably the case. Water loss in the lizard *Uta stansburiana* for example was found to be similar in populations inhabiting either the arid Colorado desert or semi-arid montane, coastal and island communities in southern California (Roberts, 1968; Thorpe & Kontogiannis, 1977; Mautz, 1982) and Dmi'el (1972) found that two colubrid snakes from a semi-arid and an arid habitat both displayed higher rates of water loss than two viperid species from the same habitats. Interspecific comparisons of this sort are fraught with difficulties, however, not least of all because of the problems associated with the precise and reproducible measurement of rates of evaporative water loss. Zucker (1980) points out that most published data can be faulted on the grounds of the techniques employed and the lack of attention paid to factors such as activity of the animals, state of hydration, stage of moulting cycle etc., all of which profoundly modify rates of cutaneous water loss. Kobayashi *et al.* (1983) go further and, on the basis of acclimation responses to changes in humidity in *Anolis carolinensis*, speculate that all such differences may be acquired rather than genetic in origin. Mautz (1982) has plotted weight-specific rates of evaporative water loss of over 100 species of reptiles, classified according to habitat type (arid, semi-arid and mesic) and fitted three regression lines to the data; all the regression coefficients were significantly less than zero, but did not differ amongst themselves and a combined regression for all the data is thus given with:

$$\text{Log EWL} = 0.388 - 0.279 \log \text{W} \qquad r^2 = 0.80, N = 147$$

where: EWL = evaporative watr loss in mg $(\text{g.h})^{-1}$
W = body weight in g.

The exponent of $-0.279$ is not significantly different from 0.33' indicating that weight-specific rates are proportional to $\text{W}^{0.67}$.

This problem of scaling and its effect on rates of water loss has been appreciated by a number of workers but not satisfactorily resolved in the literature. On theoretical grounds the rate of cutaneous water loss would be expected to be a function of the surface area of the animal which should itself vary with the body weight to the power 0.67 (assuming that the animal be adequately represented as a simple cylinder). There are in fact only a few studies where surface area has been measured in a number of individuals of varying body weight and, as seen in Table XXIX, the actual exponent varies from 0.47 to 0.83. Henzell (1965) attempted to make interspecific comparisons based on the form of the equation relating rate of water loss and body weight in six species of lizards, from three different families occupying the one habitat. Total evaporative water loss (in $\text{mg.h}^{-1}$) was found to be a power function of body weight in all species (i.e. $\text{EWL} = a\text{W}^b$) but the exponent 'b' was not

**Table XXIX:** Mass–area scaling functions for reptiles of the form $A = a\text{W}^b$. n is sample size. (from Mautz, 1982).

| Taxon | a | b | n | Reference |
|---|---|---|---|---|
| Lizards | 10.0 | 0.67 | | Bennedict, 1932 |
| *Anolis carolinensis* | 11.0 | 0.83 | 4 | Claussen, 1967 |
| *Iguanidae* (9 species) | 12.7 | 0.58 | 17 | Calculated from Norris, 1967 |
| *Uromastix acanthinurus* | 8.0 | 0.66 | | Tercafs and Schoffeniels, 1965 |
| *Uta stansburiana* | 16.0 | 0.47 | 4 | Claussen, 1967 |
| *Varanus gouldii* | 12.0 | 0.67 | 30 | Green, 1969 |
| *Xantusiidae* (5 species) | 11.4 | 0.66 | 65 | Mautz, 1979 |
| Snakes | | | | |
| *Carphophis vermis* | 12.4 | 0.74 | 9 | Elick and Selander, 1972 |
| *Diadophis punctatus* | 10.5 | 0.76 | 9 | Elick and Selander, 1972 |
| *Elaphae climacophora* | 25.1 | 0.63 | 21 | Gans *et al.*, 1968 |
| *Virginia valeriae* | 9.8 | 0.76 | 10 | Elick and Selander, 1972 |

**Table XXX:** Constants from the power curves ($y = ax^b$) describing the relationships between evaporative water loss (y) and body weight (x) in 6 species of West Australian lizards (from Henzell, 1965).

| Family | Species | N | a | b |
|---|---|---|---|---|
| Scincidae: | | | | |
| | *Ctenotus labillardieri* | 8 | 5.87 ± 1.13 | 1.04 ± 0.10 |
| | *Ctenotus lesueuri* | 6 | 3.97 ± 1.42 | 0.65 ± 0.17 |
| Gekkonidae: | | | | |
| | *Underwoodisaurus milii* | 15 | 4.18 ± 1.11 | 0.97 ± 0.06 |
| | *Diplodactylus vittatus* | 9 | 7.89 ± 1.08 | 0.81 ± 0.10 |
| | *Gehyra variegata* | 12 | 8.71 ± 1.24 | 0.26 ± 0.07 |
| Agamidae: | | | | |
| | *Amphibolurus ornatus* | 12 | 3.16 ± 1.42 | 0.55 ± 0.18 |

Values given as Means ± S.E.

identical for each lizard. Table XXX tabulates the data from this study and shows that 'b' may vary from 0.26 to 1.00. The extent to which body weight may influence rates of water loss thus varies markedly between different species. In the gecko, *Gehyra variegata*, for example where $b = 0.26$, the weight-specific rate of water loss will decline dramatically as body size increases, but in another gecko, *Underwoodisaurus milii*, where $b = 0.97$, the rate will be constant, regardless of the size of the animal. The other constant in the power equation, 'a', represents essentially the rate of water loss of a 1 g animal (or 1 kg, depending upon units chosen for the abscissa) and this was also found to vary significantly between species. A plot of 'a' against 'b' for these sympatric species is shown in Figure 2.27 and it is evident that lizards such as *Ctenotus labillardieri* and *Diplodactylus vittatus* must, if they are to survive, restrict their movements to moist microhabitats, because of their high 'a' value and potential for desiccation. *Ctenotus lesueuri* and *Amphibolurus ornatus* on the other hand, by virtue of their low 'a' value, can exploit dryer microhabitats and also benefit from the fact that, as they increase in size, their weight-specific rate of water loss will fall substantially because of a 'b' value between 0.5 and 0.6. The other gecko, *Gehyra variegata*, is unusual in having a high 'a' value, and hence a high rate of water loss when small, coupled with a very low 'b' value which means that its water economy will improve as it grows in size. One would thus predict that this particular species would display a preference for humid microhabitats as a juvenile, but would be found in drier situations when adult.

The study by Henzell (1965) was very much a preliminary investigation and subsequent work by den Hollander (1970) has shown that temperature needs to be taken into account when estimating the constants 'a' and 'b' in the power

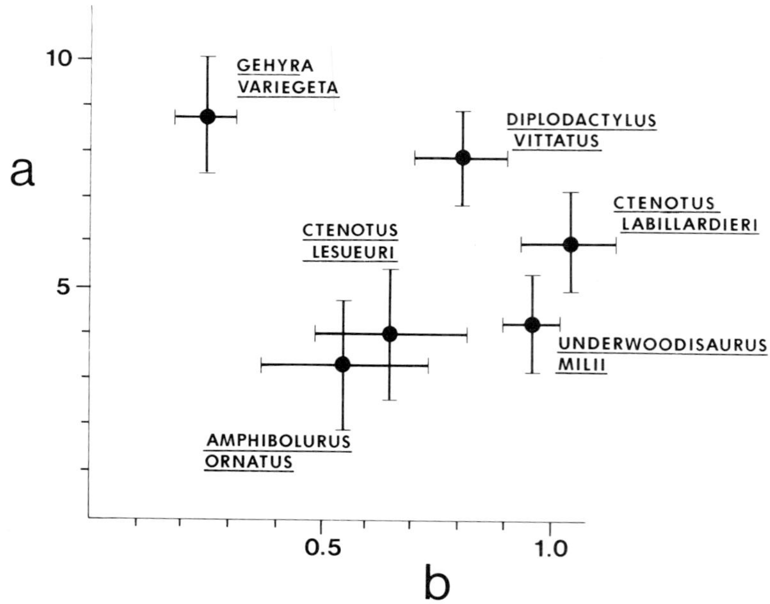


**Fig. 2.27** A plot of mean values for 'a' and 'b' from the respective allometric equations ($y = ax^b$) describing the relationship between body weight (x) and rate of evaporative water loss (y) for six species of West-Australian lizards from three families (modified from Henzell, 1965).

curve. As would be expected, 'a' was found to be highly temperature dependent in all species and there was some suggestion that 'b' may also vary with temperature, although the differences found were not statistically significant. What is apparent from the study of den Hollander (1970), however, is that, although weight-specific rates of water loss may vary markedly between species when measured at a given environmental temperature, they may be much more comparable when measured at the Preferred Body Temperature for each species (see Table XXXI) and, ecologically, this may be a more meaningful way to approach the problem of interspecific comparisons.

It is quite evident that attempts to measure and compare rates of evaporative water loss in reptiles leave much to be desired, and more thought needs to be given to the factors which may influence these rates, as well as to the theoretical bases of water exchange between an animal and its immediate environment. It is not surprising, therefore, that although one may cite a large number of studies in which desert-living reptiles have apparently been shown to possess reduced rates of evaporative water loss (e.g. Gans *et al.*, 1968;

**Table XXXI:** Rates of evaporative water loss of 5 species of lizards, estimated at the Preferred Body Temperature (PBT) for each species (from den Hollander, 1970).

| Species | PBT (°C) | Evaporative water loss [mg(g.hr)$^{-1}$] |
|---|---|---|
| *Gehyra variegata* | 33 | 3.5 |
| *Diplodactylus vittatus* | 27 | 2.3 |
| *Diplodactylus spinigerus* | 32 | 2.2 |
| *Underwoodisaurus milii* | 26 | 3.7 |
| *Amphibolurus ornatus* | 37 | 3.7 |

Claussen, 1967; Prange & Schmidt-Nielsen, 1969; Bradshaw, 1970; Heatwole & Veron, 1977; Elick & Selander, 1972; Cohen, 1975; Baeyens & Rountree, 1983), all of these record weight-specific rates from a very limited range of body weights, and invariably the measurements have been made at a single temperature, so that the conclusion must be accepted with some reservation. Efforts to identify some morphological basis for lower cutaneous rates of water loss have also met with little success and various studies have served to eliminate epidermal scales, the $\beta$-keratin layer (Licht & Bennett, 1972; Bennett & Licht, 1975) and the hinge-region between the scales (Maderson *et al.*, 1970) as the major barrier to water loss from the integument. Current speculations focus on the $\alpha$-keratin layer (Zucker, 1980) and epidermal lipids (Baeyens & Rountree, 1983; Maderson & Lillywhite, 1982) as factors restricting cutaneous evaporation.

Measurements of rates of water turnover in free-ranging animals thus emerge as critical data for any comparisons between the water economy of desert and non-desert reptiles and it is indeed fortunate that there has been such substantial progress in this field within the last 10 years (see Streit, 1982). Water influx rates have been estimated in a variety of reptiles (mostly lizards) following injections of either tritiated water (HTO) or doubly-labelled water ($^3HH^{18}0$). The assumptions on which these methods are predicated, their inherent errors, and practical limitations have all been discussed (see Nagy, 1975; Nagy & Costa, 1980; Nagy, 1980; Congdon *et al.*, 1978; Buscarlet *et al.*, 1978; Holleman *et al.*, 1980) and the methods have been well-validated for all but tropical species of lizards where water flux rates appear suspiciously high (see Nagy, 1982). The use of a stable oxygen isotope ($^{18}O$) in conjunction with a radioactive isotope of hydrogen ($^3H$), extends the technique and enables estimates to be made of $CO_2$ production rates (and hence field metabolic rate) because of the fact that the oxygen of water is in isotopic equilibrium with that of $CO_2$ in the body (Lifson & McClintock, 1966). Material and energy budgets have thus been established for a number of desert reptiles in the field, including the chuckwalla, *Sauromalus obesus* (Nagy & Shoemaker, 1975), the Western Fence Lizard, *Sceloporus occidentalis* (Bennett & Nagy, 1977) and the Australian agamid lizard *Amphibolurus nuchalis* (Bradshaw & Nagy, in prep., and see also Bradshaw *et al.*, 1976; Congdon *et al.*, 1982).

Minnich (1979) plotted published water turnover rates from six species of lizards and two tortoises (one desert and the other subtropical) and derived a power-curve relationship with body weight. The equation of the line was:

$$\text{mL water flux (day)}^{-1} = 21\ \text{kg}^{0.84}$$

which does not differ significantly from that calculated by Bradshaw (1981) for six species of desert lizards:

$$\text{mL water flux (day)}^{-1} = 16.7\ \text{kg}^{0.81} \quad r^2 = 0.99$$

Nagy (1982) has utilised a larger data set, much of it as yet unpublished, to compute equations for arid herbivores without drinking water and arid non-herbivores with access to drinking water, the equation for this regression being:

$$\text{mL water flux (day)}^{-1} = 20.5\ \text{kg}^{0.91} \quad r^2 = 0.98$$

The 95% confidence interval of the exponent was 0.85 to 0.97 and, again, this does not differ significantly from the exponents derived by Bradshaw (1981) and by Minnich (1979). Nagy (1982) records a second equation for tropical and subtropical reptiles, regardless of diet or drinking water availability as:

$$\text{mL water flux (day)}^{-1} = 45.0\ \text{kg}^{0.66} \quad r^2 = 0.99$$

The slope of this line differs significantly from that derived for arid reptiles and the two curves are shown in Figure 2.28. As mentioned previously, some caution is needed in interpreting the values for reptiles from tropical habitats as Nagy (1982) considers it possible that exchange of water across the skin and lungs may account for a relatively large fraction of total water flux measured with HTO in animals inhabiting moist and humid environments. Nonetheless, at this stage the data suggest a clear dichotomy between arid and humid-living species of reptiles which adds weight to the conclusions of the laboratory studies cited above indicating that a reduction in the rate of evaporative water loss is a common feature of the physiology of desert reptiles.

This short review of reptilian water and electrolyte balance has perhaps served to bring into sharper focus some of the central questions which need to

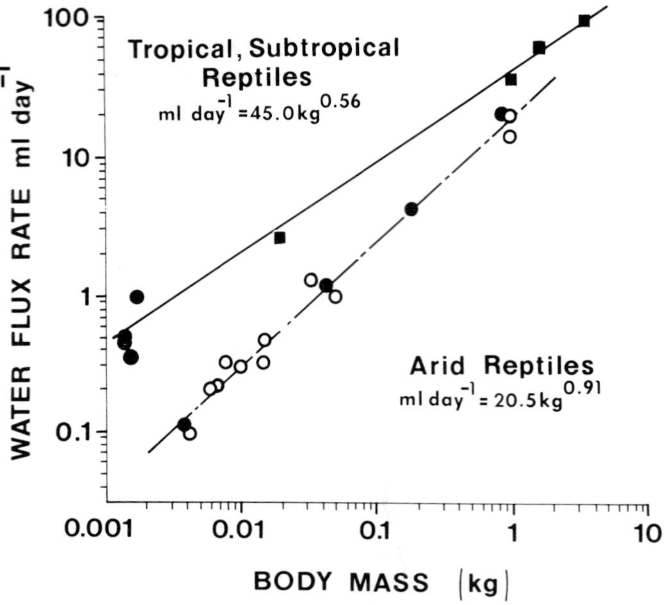


**Fig. 2.28** Relationship between water flux rate and body mass in free-living reptiles that are maintaining hydric balance in arid, tropical and subtropical habitats. Closed circles indicate insectivores and closed squares represent herbivores; open circles represent insectivorous, carnivorous, or omnivorous reptiles with access to drinking water (modified from Nagy, 1982).

be addressed in this field. Primary amongst these is whether a limited capacity for osmotic homeostasis is an adaptive response peculiar to desert reptiles, or instead, an inherent limitation of the Class as a whole, and whether desert species are subjected to periods of 'stress' as a consequence of these periodic perturbations. At this stage it seems clear that hypernatraemia in field populations is a simple consequence of the dietary régime adhered to by a particular species. Few reptiles seem to possess the ability to excrete sodium ions in great abundance (and then only via the nasal salt gland which is lacking in many groups) and, in the absence of other extra-renal avenues of electrolyte excretion, and lacking a counter-current osmotic multiplier in the kidney, they have no option but to store ions such as sodium in the body fluids at higher than normal concentrations. The careful analysis by Lemire *et al.* (1982) of the water and electrolyte balance of the Fouette-Queue has shown that this Saharan herbivore, initially thought to be an impeccable regulator, experiences significant hypernatraemia when sodium intake is high, in the same way as *Amphibolurus* lizards do in the Australian arid zone. Nagy (1982) comments that "the desert lizards *Dipsoraurus dorsalis* . . . and *Sauromalus obesus* . . . . do not show large changes in plasma solutes during prolonged droughts so this phenomenon [hypernatraemia] is not universal in reptiles" but this misses the point: any reptile will experience hypernatraemia if the sodium concentration of its diet exceeds that of its body fluids and it has no significant means for extra-renal elimination of the ions. Potassium is much less critical in this regard because most nasal salt glands secrete this ion by preference, and $K^+$ excretion appears to proceed via renal secretion and the formation of urate salts, even during periods of starvation and inactivity. The experiments on renal function of *Dipsosaurus dorsalis* carried out in the laboratory show very clearly that this lizard is in no way physiologically better able than *Amphibolurus ornatus* to handle and excrete sodium ions when salt loaded (in fact it is less able) and the fact that it enjoys a stable sodium balance in the field is only a reflection of the fact that its diet is virtually sodium-free. Having resolved this question, one must ask whether all reptiles will inevitably experience hypernatraemia when challenged with a hyperosmotic intake (as occurs in mammals if the osmolality of the fluid ingested exceeds the concentrating capacity of the species' kidney) and whether desert-living species are better able to tolerate this hypernatraemia? Data are again very limited on this point, but there are a number of observations which may be made. The large scincid lizard *Tiliqua rugosa*, studied by Bentley (1959) is by no means a desert reptile and much of its range occurs in semi-arid and mesic habitats, but this was the first species in which seasonal hypernatraemia was described. Similarly, within the genus *Amphibolurus, A. ornatus* is a semi-arid species evidencing pronounced hypernatraemia, whereas *A. nuchalis* is an arid species but experiences only mild hypernatraemia in summer due to the lower sodium content of its diet (Bradshaw, 1970). Significantly, Bradshaw (1970) described extremely high plasma sodium levels in another species, *A. minimus*, caught in mid-winter on the Abrolhos Islands off the coast of Western Australia when these lizards were subsisting on a diet of flowers which were particularly rich in sodium. In the absence of any significant evidence to the contrary, it must be concluded that the ability to tolerate extended periods of hypernatraemia appears to be a reptilian characteristic, somewhat astonishing when compared with the very limited ability of mammals to tolerate this state (Knowles, 1956), but apparently equally developed in all members of the Class.

The question of 'stress' is even more difficult to answer although, as we have seen when discussing thermoregulation, hypernatraemia is often associated with a general breakdown of normal bodily functions such as feeding, activity, behavioural temperature regulation, etc. Bradshaw (1975) found that plasma sodium levels were positively correlated with circulating levels of the adrenal corticosteroid, corticosterone, in the lizard *Amphibolurus ornatus* in the field, but it is far from clear that this particular hormone is responsible for the normal maintenance of sodium homeostasis in reptiles (see Chapter 2, Section IV). The other adrenal steroid which has been implicated in electrolyte regulation in reptiles, aldosterone (see Bradshaw & Rice, 1981), has yet to be measured in any field population of reptiles. By the definition of 'stress' advanced earlier (i.e., significant change of state of the *milieu intérieur* associated with elevated levels of regulatory hormones), the data presently available are not adequate to answer this question.

Identifiable 'adaptations' in the water and electrolyte physiology of desert reptiles are thus not obvious and perhaps restricted to reductions in cutaneous (and pulmonary?) rates of water loss, leading to generally lower rates of water turnover, and the protection of fluid volumes in chronic dehydration. In this latter case, however, the response of all species studied is by no means uniform. *Amphibolurus ornatus* and *Uromastix acanthinurus* preserve plasma and blood volumes at normal levels, even when weight losses exceed 40%, but they differ in the extent to which the extra-cellular compartment is also protected. *Sauromalus obesus*, which has also every claim to be considered as a 'desert species', differs markedly from these two species in that it is the intracellular compartment which is 'protected' in this animal at the expense of all other fluid volumes, including that circulating. From the study of just these three species there would appear to be considerable diversity in the responses of reptiles to chronic dehydration and it would be premature at this stage to characterise any of these as being of high adaptive significance.

## IV. HORMONAL CONTROL OF WATER AND ELECTROLYTE BALANCE IN REPTILES

Progress in this field has been rapid in the last decade, due to the development and deployment of highly sensitive and specific radioimmunoassays for neurohypophyseal and corticoid hormones, and to a resurgence of interest in reptiles as subjects for analysis. Much of this work has been undertaken by comparative endocrinologists whose interest in reptiles is as much phylogenetic as herpetologic. They see them as extant descendants of the progenitors of the birds and mammals, whose study may illuminate critical questions of endocrine evolution. But the study of hormonal control mechanisms also falls into the province of the ecophysiologist and some recent studies have been directed towards an analysis of changes in endocrine gland activity and hormone levels in field populations of reptiles.

There have been numerous recent reviews of pituitary–adrenal function in reptiles (see Bradshaw, 1975, 1978a,b, 1985; Callard & Callard 1978; Holmes & Pearce, 1979; Bradshaw & Rice, 1981). This section briefly documents recent advances and summarises our understanding, still unfortunately very meagre, of the extent to which water and electrolyte balance is subject to hormonal modulation. Typically, osmoregulation in reptiles and other lower vertebrates involves the cooperative interaction of a wider variety of effector organs than is the case in mammals, and post and extra-renal sites of water and electrolyte exchange assume a far greater importance. Little is known, however, of the endocrine control of the activity of these sites which, in reptiles, include the cloaca/colonic complex, the urinary bladder when present, and the various cephalic or nasal salt glands. As little as 15 years ago, opinion was divided as to whether reptiles evidence 'typical' vertebrate homeostatic responses, for example, at the level of the kidney when challenged with salt loading or deprivation, and there was doubt expressed regarding the presence of a pituitary adrenocorticotrophic hormone (ACTH) controlling the activity of the adrenal gland. This somewhat extraordinary state of affairs came about as a result of a series of laboratory investigations where adrenal tissue from reptiles was incubated *in vitro*, in the presence of radioactive steroid precursors plus mammalian ACTH or extracts of reptilian pituitary glands. Typically, when such preparations are made from amphibian or fish adrenocortical tissue, the rate of corticosteroidogenesis and the extent to which added precursors are incorporated into secretory products is enhanced by the addition of ACTH. Experiments with reptilian tissue were negative, however (see Phillips *et al.* 1962; Gist & DeRoos, 1966; Macchi & Phillips, 1966), and Macchi (1967) speculated that the reptilian adrenal may be autonomous and not subject to pituitary control. Classical histological and organ ablation studies (e.g. Schaefer, 1933; Miller, 1952; Wright & Chester Jones, 1957) did not support this conclusion, however, and the problem was resolved when Licht & Bradshaw (1969) and Huang *et al.* (1969) showed that the incubation temperature is much more critical for reptiles than for fish and amphibians. At that time it was customary for reptilian (or any 'lower vertebrate') tissue to be incubated at room temperature (around 20°C) as their thermoregulatory capacities were not at all appreciated by physiologists and comparative endocrinologists. Although this temperature is often conducive to effective steroidogenesis in animals such as fish and frogs which have lower activity temperatures, it is well below that needed in most reptiles. Licht & Bradshaw (1969) measured circulating corticosteroid levels in the small tropical lizard *Anolis carolinensis* and found that these fell rapidly to undetectable levels following hypophysectomy or dexamethasone blockade, suggesting the presence of a classical negative-feedback system via the pituitary. Injections of mammalian ACTH markedly elevated plasma corticosterone levels within 15 minutes, but only if the lizards were maintained at their preferred body temperature of 30°C. When held at 21°C, the response was barely detectable, highlighting once again the importance of temperature in studies of reptilian physiology. Shoemaker *et al.* (1967) found that rates of water and electrolyte excretion were temperature dependent in a number of species of Australian lizards and early failures to detect the presence of homeostatic renal responses in reptiles (e.g. Schmidt-Nielsen & Skadhauge, 1967), were due to a combination of two factors which are now obvious in retrospect: a failure to respect the thermal preferenda of the animals being studied, and a failure to wait long enough for the responses to develop. Bradshaw (1972) found, for example, when studying renal function in the lizard *Amphibolurus ornatus*, that homeostatic responses to salt loading and deprivation developed over a period of days, rather than hours, as is the case with mammals, and that it was necessary to work with spontaneously-voided urine, which has been subjected to post-renal modification, in order to gauge the full extent of the response. Why the physiological response of effector organs, such as the kidney, to relatively abrupt changes in circulating hormone levels should be so delayed in reptiles is not at all clear, and studies on the rate of receptor synthesis should prove interesting.

## A. Hormonal Control of Kidney Function

### 1. *Neurohypophyseal Hormones*

It has been known for some time that the reptilian nephron is sensitive to crude pituitary extracts containing the octapeptide arginine vasotocin (AVT) (see Dantzler & Schmidt-Nielsen, 1966; Dantzler, 1967, 1970; Green, 1972) displaying either a primarily glomerular or a tubular response, or a combination of both, depending upon the dosage injected and the species concerned (Dantzler, 1976, 1978; Sawyer & Pang, 1979; Dantzler & Braun, 1980). Characteristically, the antidiuresis which follows AVT injection is produced by a decrease in both glomerular filtration rate (GFR) and relative free-water clearance ($C_{H_2O}/C_{IN}$), the latter brought about, it is thought, by an increase in tubular permeability. In some species however, such as the lizard *Sceloporus cyanogenys* studied by Stolte *et al.* (1977), AVT has no effect on the permeability of any tubular segment, although it still reduces the GFR. Most workers have also found that, qualitatively and quantitatively, the antidiuresis produced by AVT injection in a given species parallels that induced by hyperosmotic saline loading, suggesting that this procedure normally leads to a physiological release of endogenous AVT.

In review a decade ago of water and mineral metabolism in Reptilia, Dantzler & Holmes (1974) commented that AVT is . . . "generally considered to be the natural antidiuretic hormone' . . . but that '. . . this has never been completely documented by the production of hormone deficiencies by ablation of the hypothalamic neurosecretory cells and correction of the defect by hormone replacement". Preliminary results of such a study in the agamid lizard

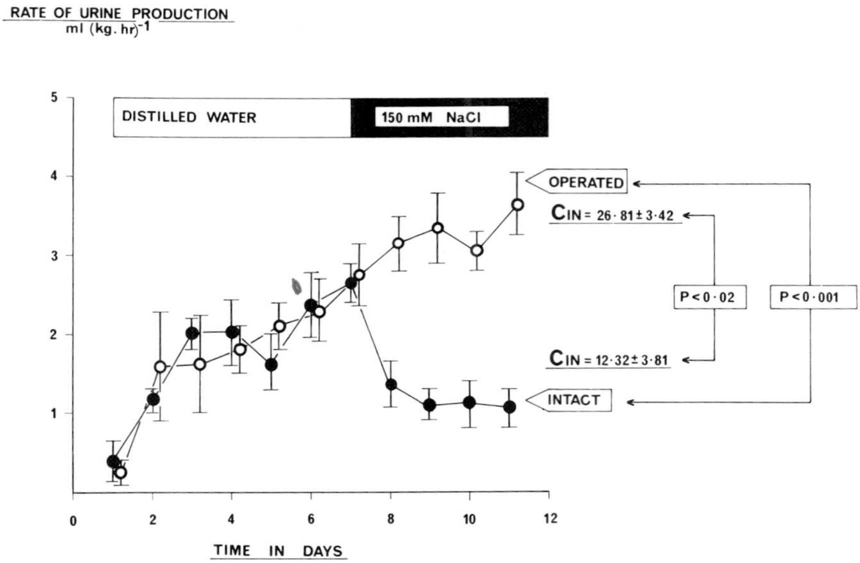


**Fig. 2.29** Rates of urine production with prolonged water and saline diuresis in intact and hypothalamic-lesioned (operated) *Amphibolurus ornatus* lizards. All fluid loads = 10mL (100g. day)$^{-1}$ and data presented as Mean ± S.E. (from Bradshaw, 1975).

*Amphibolurus ornatus* were reported by Bradshaw (1975) and full details of the effect of hypothalamic lesions on renal and post-renal function in this species were given later (Bradshaw, 1976, 1978a). Electrolyte lesions placed to interrupt the hypothalamo-hypophyseal tract were found to inhibit the normal antidiuretic response to saline loading, as shown in Fig. 2.29, which could be fully restored by injection of AVT as seen in Fig. 2.30. Fractional reabsorption of sodium ions was also found to be enhanced in tract-operated individuals when compared with intact controls during both water and saline diuresis (see Table XXXII), but it seems unlikely that this effect is due to the absence of AVT. Probenecid, which inhibits organic acid transport in mammals, has also been shown to be diuretic in reptiles (Dantzler *et al.*, 1970). It competitively inhibits the effects of exogenously administered AVT as seen in Fig. 2.31.

A highly sensitive radioimmunoassay for AVT was first reported by Rosenbloom & Fisher (1974) and circulating levels under various physiological conditions were reported soon after in frogs (Sawyer & Pang, 1975) and chickens (Koike *et al.*, 1977). To date, the only measurement of AVT levels in a reptile are those by Rice (1982) in the Australian varanid lizard, *Varanus gouldii*, but they amply confirm a rôle for this peptide as a physiological

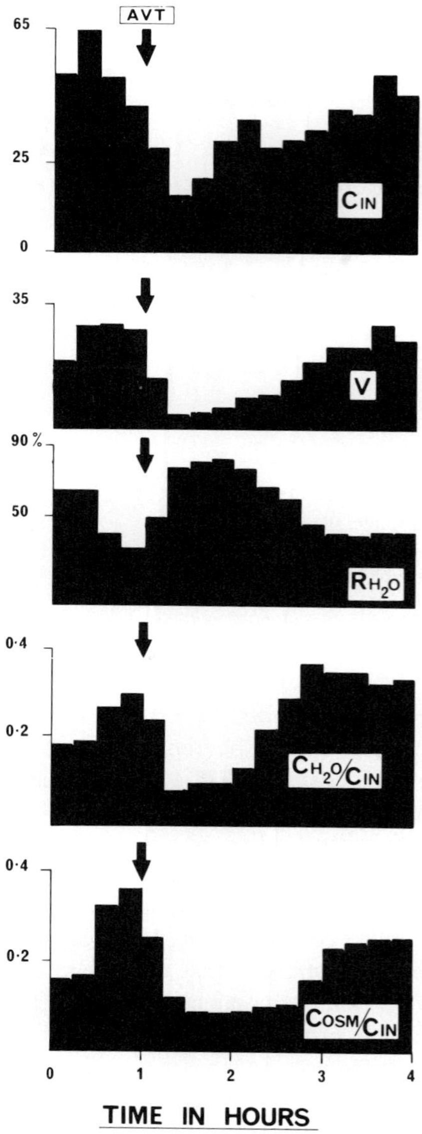


**Fig. 2.30** Representative renal response of hypothalamic-lesioned *Amphibolurus ornatus* to an intracardiac injection of 472ng/kg of arginine vasotocin (AVT). Clearance of inulin ($C_{IN}$) and urine volume (V) expressed in $mL(kg.hr)^{-1}$. The animal was maintained on a constant water load by stomach gavage and ureteral urine was collected through a cloacal cannula similar to that employed by Roberts & Schmidt-Nielsen (1966) (from Bradshaw, 1975).

**Table XXXII:** Renal function in intact and tract-operated *Amphibolurus ornatus* during water diuresis—ureteral urine (from Bradshaw, 1978a).

| | | | | | | | | Fractional reabsorption | | |
|---|---|---|---|---|---|---|---|---|---|---|
| Group | N | V | $C_{IN}$ | $C_{OSM}$ | $C_{H_2O}$ | $C_{OSM}/C_{IN}$ | $C_{H_2O}/C_{IN}$ | $H_2O$ | $Na^+$ | $K^+$ |
| Operated | 11 | 24.10 ± 1.09 | 47.13 ± 4.89 | 8.45 ± 1.37 | 15.47 ± 1.12 | 18.38% ±2.74 | 36.48% ±4.03 | 45.18% ±4.15 | 88.58% ±2.31 | 58.60% ±16.24 |
| Intact | 21 | 33.23 ± 1.07 | 51.30 ± 2.41 | 14.96 ± 1.03 | 18.10 ± 1.25 | 29.16 ±3.43 | 35.29 ±2.55 | 35.23 ±3.14 | 78.36 ±2.26 | −17.41 ±23.60 |
| Significance of difference between means | | $P < 0.001$ | NS | $P < 0.02$ | NS | $P < 0.05$ | NS | NS | $P < 0.02$ | NS |

Fluid load = 10 mL/100 g of distilled water administered intraperitoneally.
Means ± SE.

V, $C_{IN}$, $C_{OSM}$ and $C_{H_2O}$ in mL $(Kg.hr)^{-1}$

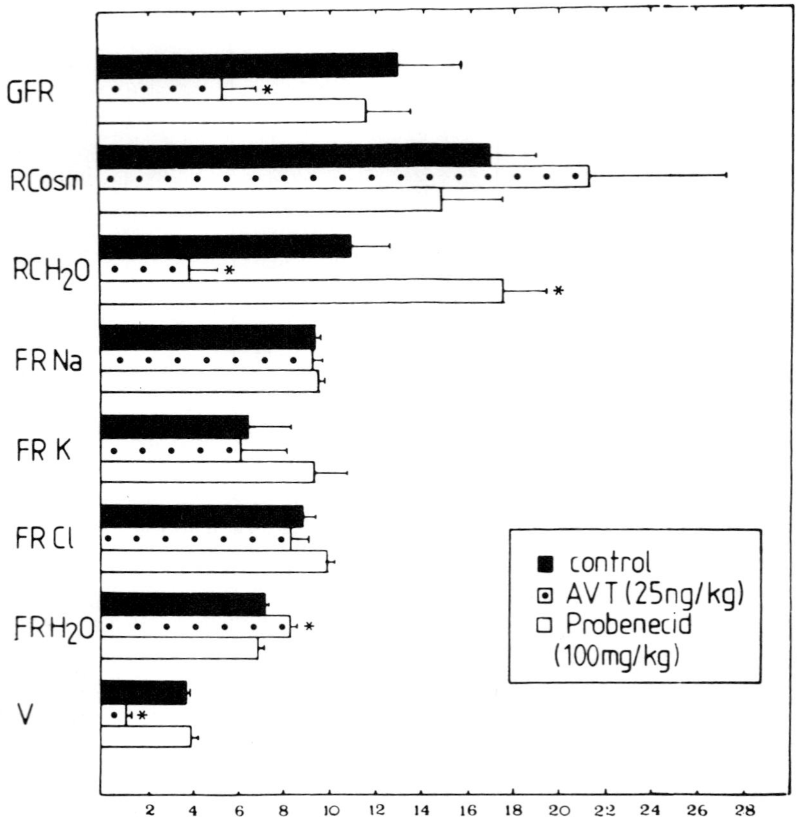


**Fig. 2.31** Renal function in the varanid lizard *Varanus gouldii*. Changes in glomerular filtration rates (GFR mL $(kg.hr)^{-1}$) relative osmolar clearance ($RC_{Osm}$, %) and free-water clearance ($RC_{H_2O}$ %), fractional sodium ($FR_{Na}$, %/10), potassium ($FR_K$, %/10), chloride ($FR_{Cl}$, %/10), and water reabsorption ($FR_{H_2O}$ %/10), and ureteral urine flow (V, mL$(kg.hr)^{-1}$), following the intra-cardiac injection of 25 ng/kg AVT followed 2.5 hr later by 100 mg/kg Probenecid. Statistical comparisons relate to control values with * = $P < 0.05$. Mean ± SE is reported (n = 6) (from Bradshaw & Rice, 1981).

antidiuretic hormone. Plasma levels in water-loaded individuals averaged 1.6 ± 0.4 pg/mL (around 0.2 $\mu$U) and were increased to 3.9 ± 0.3 pg/mL with 8%–10% dehydration and rose to an average of 7.1 ± 1.5 pg/mL in animals which had been salt-loaded for a period of one week by daily injections of 2.14 mol/L sodium chloride. Plasma AVT levels were also elevated above hydrated values in samples from animals caught in the field in summer, averaging 3.5 ± 0.4 pg/mL.

The renal responses of *Varanus gouldii* to dehydration and salt loading were studied by Bradshaw & Rice (1981) and the data are summarised in Table XXXIII. From this it is evident that effective antidiuresis results from a reduction in both GFR and relative free-water clearance ($C_{H_2O}/C_{IN}$) coupled with an increase in fractional reabsorption of filtrate. The antidiuresis associated with dehydration, which elevated plasma AVT levels to a lesser extent, was completely tubular in origin with no depression of the GFR. The tendency within the literature to characterise a given species' response to AVT as being either a glomerular or a tubular antidiuresis, may thus be in error, and the response may be simply a function of the circulating level of AVT. Rice (1982) found plasma AVT levels in *Varanus gouldii* to be positively correlated with plasma osmolality (see Fig. 2.32), as is the case with vasopressin in mammals (Robertson *et al.*, 1977) and as would be expected if AVT indeed functions as the physiological antidiuretic hormone in reptiles. It is to be hoped that AVT levels will be measured in other reptiles, and correlated with changes in renal function, in an effort to establish the generality of this hormone's rôle.

Neurohypophyseal peptides other than AVT have also been isolated from the reptilian pituitary gland, including oxytocin and 8-isoleucine oxytocin (mesotocin) (Sawyer *et al.*, 1959; Munsick, 1966) but their physiological significance is unknown. Oxytocin and mesotocin both produce a glomerular antidiuresis in water snakes (Dantzler, 1967) but only at what are certainly pharmacological dosages (20–40 mU/kg), and the effect seems to be secondary to a depression of the systemic arterial blood pressure. Sawyer & Sawyer (1952) reported that oxytocin produces a glomerular antidiuresis in the alligator *Alligator mississippiensis* at relatively low dosages (10 mU/kg) as is also the case in the scincid lizard *Tiliqua rugosa* (Bentley, 1959) but, again, these effects are probably due to a fall in renal perfusion pressure (Wooley, 1959). Galli-Gallardo *et al.* (1979) and Pang & Sawyer (1978) have reported that mesotocin induces diuresis in some amphibians but, as yet, renal effects of this peptide at 'physiological' doses have yet to be described in reptiles.

## 2. Adrenocortical Hormones

Aldosterone and corticosterone appear to be the major secretory products of the adrenal glands in the majority of reptiles (Sandor, 1972; Vinson *et al.*, 1979) although other steroids such as 18-OH-corticosterone have been reported, along with DOC (desoxycorticosterone) and, occasionally, cortisol. Their actions on the kidney have remained unclear until quite recently. Partly this is because of difficulties experienced by most workers in demonstrating any effect of exogenous aldosterone injections on renal function in lizards (Bradshaw, 1972; Bradshaw *et al.*, 1972; Templeton *et al.*, 1972b) and the failure of all experiments using spironolactone, which is a potent aldosterone antagonist in mammals (Elizondo & LeBrie, 1969; Butler & Knox, 1970;

**Table XXXIII:** Plasma hormone concentrations (AVT, aldosterone and corticosterone) and renal function in *Varanus gouldii* during prolonged water diuresis, saline diuresis and dehydration (from Bradshaw & Rice, 1981).

| Plasma hormone concentrations | Hydrated | | Dehydrated | | Salt-loaded |
|---|---|---|---|---|---|
| AVT (pg/mL) | 1.6 ± 0.4 | P < 0.01 | 3.9 ± 0.3 | P < 0.001 | 7.1 ± 1.5 |
| Aldosterone (ng/dL) | 45.9 ± 4.6 | NS | 42.0 ± 7.0 | P < 0.01 | 18.3 ± 6.0 |
| Corticosterone ($\mu$g/dL) | 9.1 ± 2.1 | NS | 9.3 ± 1.7 | P < 0.01 | 19.2 ± 2.6 |
| Renal paramaters | | | | | |
| V mL$(kg.hr)^{-1}$ | 7.8 ± 0.67 | P < 0.01 | 3.3 ± 0.40 | P < 0.01 | 1.2 ± 0.13 |
| $C_{IN}$ mL$(kg.hr)^{-1}$ | 20.8 ± 2.47 | NS | 20.6 ± 2.07 | P < 0.01 | 5.8 ± 0.78 |
| $R_{H_2O}$% | 53.0 ± 4.69 | P < 0.01 | 82.6 ± 1.34 | P < 0.01 | 73.4 ± 2.67 |
| $C_{OSM}/C_{IN}$ % | 20.0 ± 1.5 | P < 0.01 | 11.6 ± 0.7 | P < 0.01 | 22.5 ± 2.5 |
| $C_{H_2O}/C_{IN}$ % | 27.0 ± 4.1 | P < 0.01 | 5.8 ± 0.9 | NS | 4.1 ± 0.6 |
| $FR_{Na}$ % | 95.2 ± 1.19 | NS | 95.6 ± 0.64 | P < 0.01 | 88.9 ± 1.50 |
| $FR_K$ % | 29.5 ± 8.57 | NS | 11.4 ± 10.89 | P < 0.01 | −70.1 ± 30.16 |
| $FR_{Cl}$ % | 84.4 ± 2.64 | NS | 89.0 ± 0.65 | P < 0.01 | 76.65 ± 2.80 |

All data presented as Mean ± S.E.

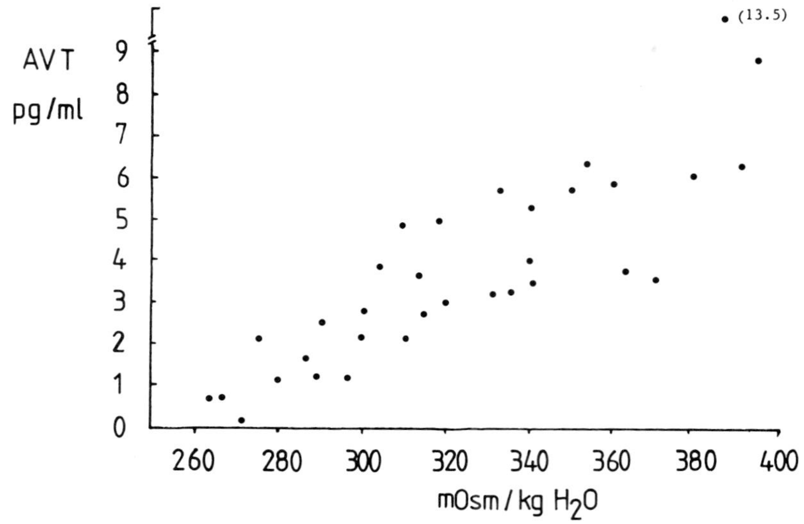


**Fig. 2.32** Relationship between circulating levels of arginine vasotocin (AVT) and plasma osmolality in the varanid lizard *Varanus gouldii* (from Rice, 1982).

Bradshaw *et al.*, 1972). Adrenalectomy studies have also been singularly uninformative (see Callard & Callard, 1978, for a review) due primarily to the technical problems associated with this operation in reptiles, where the adrenals closely adhere to the vena cavae.

LeBrie & Elizondo (1969) approached this problem somewhat differently and induced adrenal insufficiency in the water snake *Natrix cyclopion* by vascular occlusion rather than by attempting surgical adrenalectomy, and found that their 'operated' animals displayed an enhanced rate of renal loss of sodium and chloride ions when compared with intact controls. They were able to reduce this loss with injections of aldosterone which increased reabsorption of both sodium and water by approximately 24%, primarily, they thought, through an action on the proximal rather than the distal segment of the renal tubule. Circulating levels of aldosterone were first measured in reptiles by Bradshaw & Grenot (1976) and they described a significant negative correlation with plasma sodium concentrations in the North African lizard *Uromastix acanthinurus* and, as shown in Fig. 2.33. Bradshaw & Rice (1981) have confirmed this in *Varanus gouldii*. Interestingly, the relationship with corticosterone is also significant, but *positive* rather than negative in slope, as earlier found by Bradshaw (1972) in the lizard *Amphibolurus ornatus*. The fall

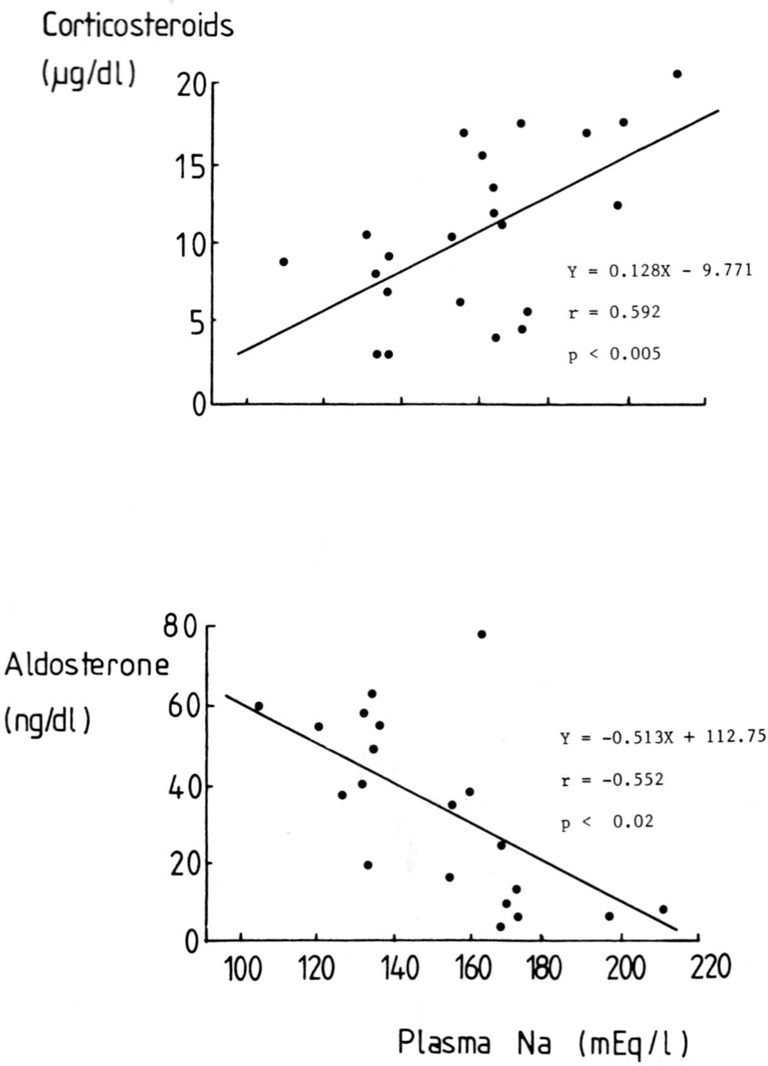


**Fig. 2.33** Correlation between circulating levels of the adrenal corticosteroids, corticosterone and aldosterone, and the plasma sodium concentration in the varanid lizard *Varanus gouldii* (from Bradshaw & Rice, 1981).

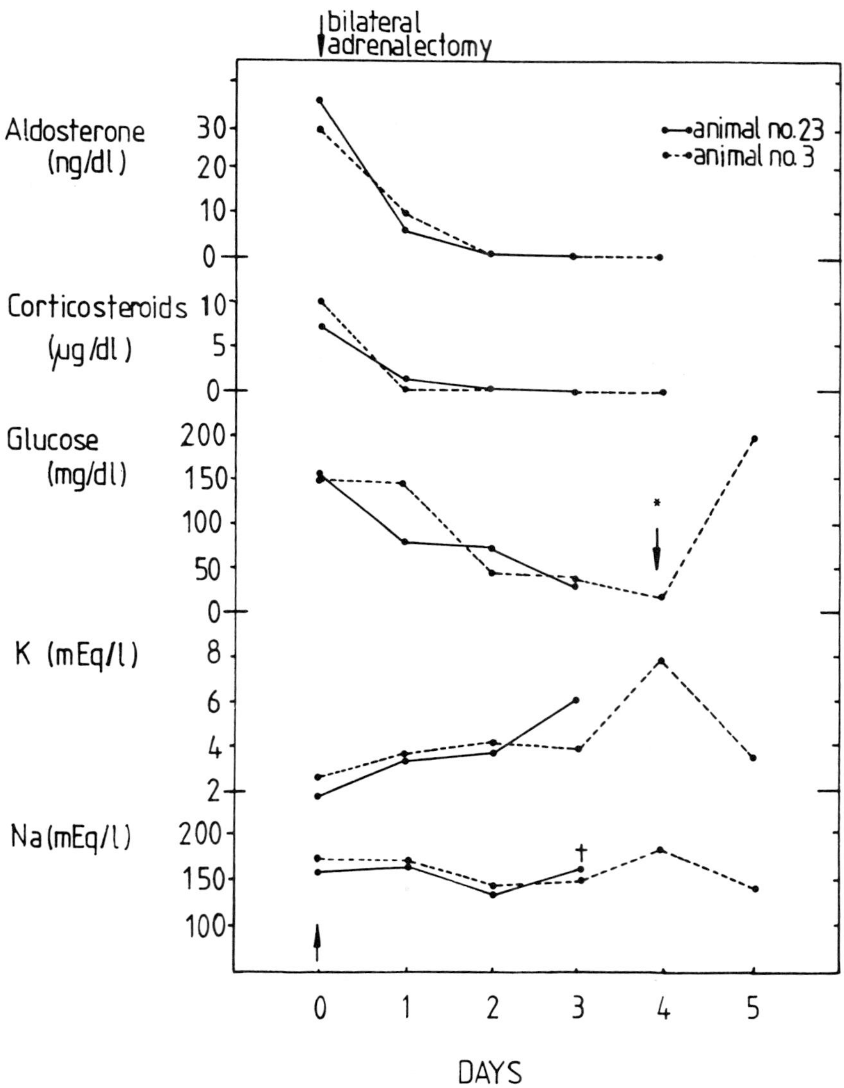


**Fig. 2.34** The effect of bilateral adrenalectomy on circulating levels of adrenal steroids, and plasma glucose and electrolyte concentrations in two individual varanid lizards, *Varanus gouldii*. Note that animal No. 23 died on day 3 and animal No. 3 was injected with dexamethasone (100μg) on day 4 (adapted from Rice, Bradshaw & Prendergast, 1982).

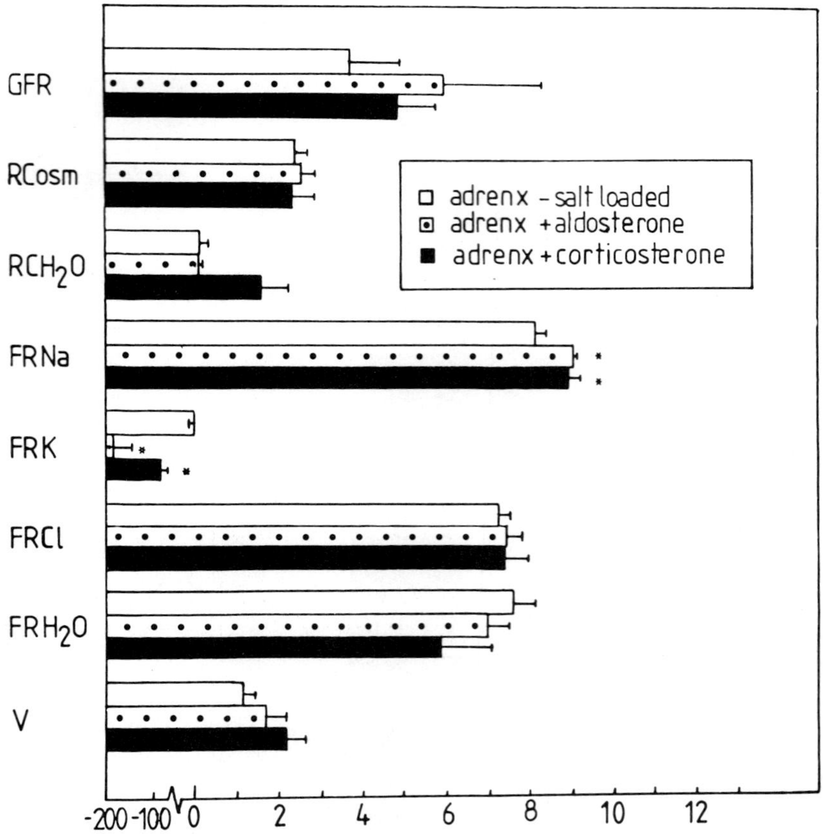


**Fig. 2.35** Renal response of salt-loaded, bilaterally-adrenalectomised *Varanus gouldii* to intravenous injection of either aldosterone or corticosterone at a dosage of 100 μg/kg. Columns represent Mean values ± SE with * = $P < 0.05$ (adapted from Rice, Bradshaw & Prendergast, 1982).

in plasma aldosterone levels in *Varanus gouldii* with saline loading is associated with a marked decrease in fractional reabsorption of sodium and chloride ions by the kidney, and an increase in the rate of secretion of potassium (see Table XXXIII), suggesting an obvious mineralocorticoid effect of aldosterone. Logically, however, this same effect could equally be attributed to the raised corticosterone levels, if this latter steroid were natriuretic in action.

Rice *et al.* (1982) developed a successful microsurgical technique for the adrenalectomy of *Varanus gouldii* and found that plasma sodium and glucose levels fell after the operation, and potassium levels rose, as in mammals (see Fig. 2.34). Renal function in adrenalectomised individuals was not altered by salt-loading (when aldosterone levels are normally low, and corticosterone high), but water-loaded animals showed a reduced GFR and enhanced relative-osmolar clearance ($C_{osm}/C_{IN}$) due to a significant fall in fractional reabsorption of both sodium and chloride ions. These changes could be repaired by the injection of 100 μg/kg of aldosterone in water-loaded individuals and this same dosage also significantly increased renal sodium reabsorption and potassium excretion in salt-loaded, adrenalectomised lizards, as shown in Fig. 2.35. Acute injections of the same dosage of corticosterone had an identical effect to aldosterone in salt-loaded animals and stimulated sodium reabsorption in water-loaded individuals. Both hormones thus function as mineralocorticoids in *Varanus gouldii* (i.e., natriferic and kaliuretic) and it is clear that the normal renal response to salt loading in this species, enhanced sodium and chloride excretion, is mediated by the fall in plasma aldosterone levels rather than by the concomitant increase in corticosterone concentrations.

Early studies on renal function in other lizards such as *Amphibolurus ornatus* and *Dipsosaurus dorsalis*, in which hypophysectomy and dexamethasone blockade were associated with an *increased* tubular reabsorption of sodium ions (see Bradshaw, 1972; Bradshaw *et al.*, 1972; Chan *et al.*, 1970), are difficult to interpret because these treatments would also be expected to decrease plasma aldosterone levels (Bradshaw & Grenot, 1976; Bradshaw, 1978a) and thus lead to enhanced rather than reduced rates of electrolyte excretion. At this stage there is clear evidence for aldosterone functioning as a mineralocorticoid hormone in at least two species of reptiles, the lizard *Varanus gouldii* and the water snake *Natrix cyclopion*, but not in the sea snake *Hydrophis cyanocinctus* (Duggan & Lofts, 1978, 1979). The rôle of corticosterone in reptilian water and electrolyte balance is much less clear and warrants further attention from researchers.

## B. Hormonal Control of Post-Renal and Extra-Renal Sites

### *1. Cloaca/Colon and Bladder*

It has been appreciated for many years that post-renal reabsorptive sites such as the cloaca and bladder are of importance for the water and electrolyte economy of reptiles (see Bentley & Schmidt-Nielsen, 1965; Junqueira *et al.*, 1966; Schmidt-Nielsen & Skadhauge, 1967), but their quantitative study has been difficult until relatively recently. *In vitro* measurements with isolated tissue have served to establish that water and salt are transported, but the actual rates measured are about one-twentieth of those estimated *in vivo* (Dantzler &

Schmidt-Nielsen 1966; Bentley & Bradshaw, 1972), due doubtless to the absence of vascularisation and adequate mixing. The 'difference technique' (i.e., between the composition of ureteral and spontaneously-voided urine) was used by Bradshaw (1975) to estimate transmural fluxes of water and electrolytes across the colon of the lizard *Amphibolurus ornatus*, which lacks a urinary bladder, and, as seen in Fig. 2.36, the colon in this species is responsible for reabsorbing about 90% of the fluid presented to it at all times, and a variable fraction of the electrolytes. This technique, however, is open to many objections and our understanding of the physiology of the cloaca/colon has been considerably enhanced by recent studies in which this segment is perfused, *in situ*, in conscious animals at known rates with solutions of known composition. This technique was first developed by Skadhauge (1967) to study the colon of the chicken, and has now been modified and applied to two lizard species, *Agama stellio* (Skadhauge & Duvdevani, 1977) and *Varanus gouldii* (Bradshaw & Rice, 1981).

Rates of water and electrolyte reabsorption from the perfused colon of *Varanus gouldii* are compared in Table XXXIV in water-loaded, salt-loaded and dehydrated individuals. Perfusion rates were adjusted with each treatment to approximate ureteral flow rates under these same conditions (see Table XXXIII). From Table XXXIV it is evident that saline loading is accompanied by a decrease in the absolute rates at which sodium, chloride and water are reclaimed from the colon, and potassium secretion is reduced. When the data are expressed fractionally, that is, relative to the presentation rates for each component, which are a function of the perfusion rate, it is seen that water reabsorption increases significantly from 23.7% to 40.6% with salt loading, and sodium reabsorption falls from 33.7% to 22.4% (both $P < 0.01$). Fractional reabsorption of chloride is unchanged and, as already noted, potassium secretion is reduced.

Surprisingly, colon function in dehydrated *Varanus gouldii* is essentially identical with that of saline-loaded individuals, with an increased rate of reabsorption of water and decreased fractional reabsorption of sodium ions. The electrolyte concentration of the reabsorbate in mmol/L of NaCl has been calculated under these three treatments and is also given in Table XXXIV. This is hypo-osmotic with both dehydration and salt loading but hyperosmotic in water-loaded lizards, suggesting in this latter case the presence of some solute-linked water flow as with the cloaca of the crocodile (Schmidt-Nielsen & Skadhauge, 1967), the intestine of the eel (Skadhauge, 1969) and the human rectum (Edmonds & Pilcher, 1972; see also Skadhauge, 1978). Skadhauge & Duvdevani (1977) reported similar findings when perfusing the colon of *Agama stellio* and both dehydration and saline loading were associated with increased water reabsorption and decreased rates of sodium transport (note however that the units in their Table IV should be $\mu$eq/kg.h and not meq/kg.h as listed).

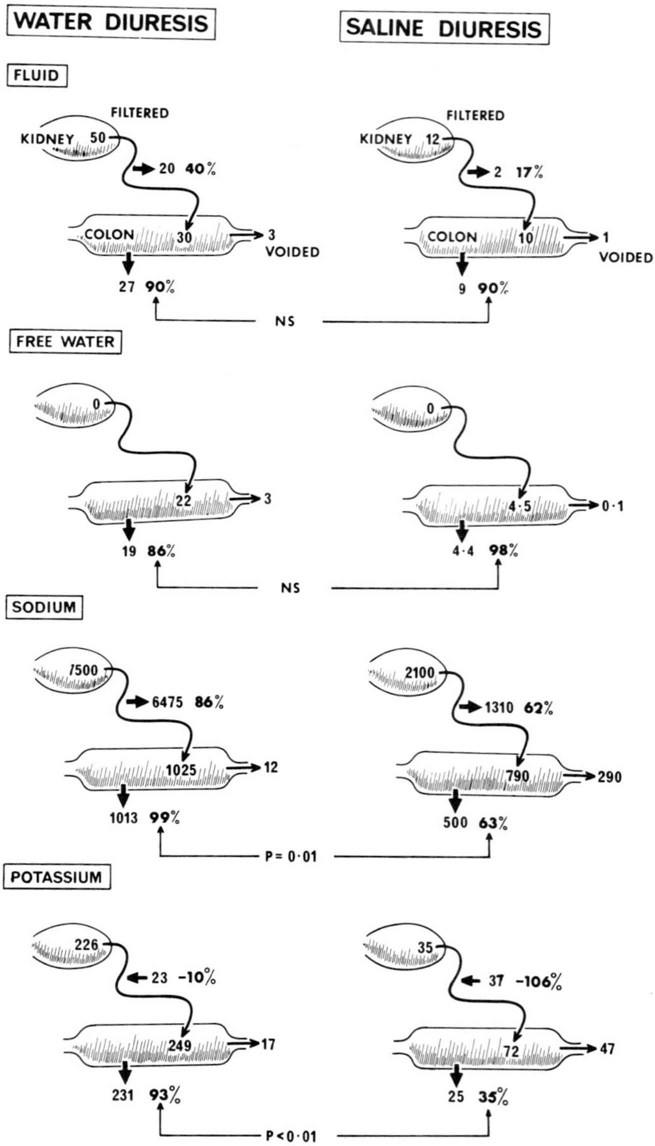


**Fig. 2.36** Diagrammatic representation of the parameters of colon function in the Australian agamid lizard *Amphibolurus ornatus* during prolonged water and saline diuresis. The animals were maintained with chronic injections of 10 mL $(100g.day)^{-1}$ of either distilled water or 150 mmol NaCl for a period of 10 days and transport parameters are expressed in $mL(kg.hr)^{-1}$ for volumes and $\mu eq(kg.hr)^{-1}$ for sodium and potassium. Filtered loads for sodium and potassium are also expressed in $\mu eq(kg.hr)^{-1}$ (from Bradshaw, 1975).

**Table XXXIV:** Cloacal parameters: Water and electrolyte reabsorption and secretion in *Varanus gouldii* (from Bradshaw & Rice, 1981).

| | Hydration (n = 6) | | Dehydration (n = 6) | | Salt-loaded (n = 6) |
|---|---|---|---|---|---|
| Perfusion rate [mL(kg.hr)$^{-1}$] | 6.52 | | 3.61 | | 2.16 |
| Reabsorbate [mL(kg.hr)$^{-1}$] | 1.46 ± 0.15* | NS | 1.67 ± 0.25 | P < 0.05 | 0.89 ± 0.11* |
| Na [$\mu$mol(kg.hr)$^{-1}$] | 123.9 ± 12.6* | P < 0.01 | 39.49 ± 7.35 | NS | 27.58 ± 4.08* |
| K [$\mu$mol(kg.hr)$^{-1}$] | −11.29 ± 2.95* | NS | −6.03 ± 0.94 | P < 0.05 | −2.19 ± 0.54* |
| Cl [$\mu$mol(kg.hr)$^{-1}$] | 75.27 ± 9.12* | NS | 50.71 ± 10.56 | P < 0.05 | 24.77 ± 5.50* |
| $FR_{Na}$ (%) | 33.72 ± 3.94* | P < 0.01 | 20.68 ± 1.39 | NS | 22.39 ± 3.34* |
| $FR_{Cl}$ (%) | 21.61 ± 1.62** | NS | 27.51 ± 5.45 | NS | 22.67 ± 5.34** |
| $FR_{H_2O}$ (%) | 23.70 ± 2.66* | P < 0.01 | 41.29 ± 4.31 | NS | 40.61 ± 1.66* |
| Electrolyte concentration of reabsorbate (NaCl mmol/L) | 140.3 ± 13.6* | P < 0.01 | 61.5 ± 5.3 | NS | 63.4 ± 13.9* |

Mean ± S.E. is reported; statistics, ANOVA–SNK; fractional values were arcsine transformed.
* p < 0.01, probability levels refer to within row comparisons only.
** NS.

These studies show that the colon functions to extend and amplify homeostatic renal responses to variations in water and electrolyte availability, and thus considerably enhance the overall effectiveness of the response. With either sodium loading or dehydration, reabsorption of water is further enhanced in the colon, and sodium reabsorption reduced, so that the osmolality of the urine rises during its sojourn in the colon, and is virtually isosmotic by the time it is voided. Conversely, electrolyte reabsorption is enhanced and water reabsorption reduced in water-loaded individuals, thus contributing further to the dilution of the urine and this must be an important mechanism in species such as the gecko *Hemidactylus* and the horned-toad lizard *Phrynosoma cornatum* in which ureteral urine is apparently always isosmotic (Roberts & Schmidt-Nielsen, 1966).

As yet there is no compelling evidence to implicate either pituitary or adrenal hormones in the control of post-renal water and electrolyte reabsorption in reptiles and Murrish & Schmidt-Nielsen (1970) have argued that water reabsorption is passive (but see Skadhauge, 1978 for a critique). The increased rate of water reabsorption from the colon of salt-loaded and dehydrated *Varanus gouldii* occurs, however, in the face of a decreasing osmotic gradient (the $U/P_{osm}$ ratio increases from 0.40 in hydrated animals to 0.98 with salt loading) and this suggests the involvement of hormonal mechanisms. Arginine vasotocin (AVT) is well-known for its effect of facilitating water reabsorption from the bladder of anuran amphibians (Bentley, 1974; Bakker & Bradshaw, 1977) and has been reported to stimulate water and sodium reabsorption from the isolated cloaca of *Varanus gouldii* by Braysher & Green (1970). Bradshaw & Rice (1981), however, have shown that dehydration and salt loading, which increase plasma AVT levels in this species, are both associated with *decreased* rates of sodium reabsorption from the colon. Bentley (1962) was unable to demonstrate an effect of AVT on the permeability of the colon of the terrestrial Greek tortoise (*Testudo graeca*) and AVT was similarly without effect on the rate of water transfer from the colon of two *Amphibolurus* lizards when measured *in vitro* by Bentley & Bradshaw (1972).

As seen in Table XXXV hypothalamic lesions which depress but do not completely abolish adrenal function in lizards (Daugherty & Callard, 1972; Bradshaw, 1976), also reduce the rates of sodium and potassium reabsorption from the colon of the lizard *Amphibolurus ornatus*, but the hormone(s) responsible for this effect have yet to be identified and this could well prove to be an effect due to a change in prolactin levels which have been implicated in both renal and post-renal function (see Mainoya *et al.*, 1974; Ensor, 1978). Bentley (1962) reported that aldosterone stimulates sodium transport from the isolated bladder of *Testudo graeca*, but it had no effect on the rate of transport from the colon of the same species. Although aldosterone is known to stimulate sodium reabsorption from the colon and coprodeum of the domestic fowl (Thomas *et al.*, 1979; Thomas & Skadhauge, 1979) and would seem an

**Table XXXV:** Estimated absolute and fractional* rates of water and electrolyte reabsorption from the colon of tract-operated and intact *Amphibolurus ornatus* during chronic saline diuresis (from Bradshaw, 1978a).

| Group | N | Fluid mL(kg.hr)$^{-1}$ | Free water mL(kg.hr)$^{-1}$ | Sodium μmol(kg.hr)$^{-1}$ | Potassium μmol(kg.hr)$^{-1}$ |
|---|---|---|---|---|---|
| Operated | 14 | 13.09 ± 1.31 (78.20%) | 9.35 ± 2.71 (87.38%) | 969.25 ± 59.34 (73.23%) | 20.26 ± 4.31 (33.68%) |
| Intact | 7 | 9.01 ± 1.15 (91.85%) | 4.24 ± 0.61 (93.42%) | 871.94 ± 33.41 (88.78%) | 62.89 ± 13.25 (68.34%) |
| Significance of difference between Means | Absolute | NS | NS | NS | $P = 0.001$ |
| | Fractional | NS | NS | $P < 0.05$ | $P = 0.025$ |

* Fractional rates expressed as a percentage in parentheses.
Fluid load = 10 mL(100 g.day)$^{-1}$ for a period of 10 days.

obvious candidate in reptiles, this is confounded by the data from *Varanus gouldii* where colon transport parameters are essentially identical in salt-loaded and dehydrated individuals, despite major differences in circulating levels of aldosterone (see Table XXXII).

Data on bladder function in reptiles are fragmentary and mainly restricted to *in vitro* studies of the bladder from the freshwater turtle *Pseudemys scripta* (see Dunson, 1979; Minnich, 1979). Investigations with this preparation, using both open and short-circuited bladders have demonstrated the active transport of both sodium and chloride ions from the mucosal to the serosal surface (Brodsky & Schilb, 1966; Gonzalez *et al.*, 1967a,b) and water transport is passive and secondary to both the existing osmotic gradient and the active transport of ions (Brodsky & Schilb, 1965). There is little information available on the effects of hormones on bladder function in reptiles. The reptilian bladder generally seems unresponsive to neurohypophyseal hormones, in contrast to its amphibian counterpart (see Bentley, 1962; Brodsky & Schilb, 1960; Gilles-Baillien, 1969). The only steroid to have been shown to stimulate sodium transport from the bladder is aldosterone in two species of turtle (Bentley, 1962; Le Fevre, 1973) and then only *in vitro*.

## 2. Cephalic Salt-Secreting Glands

Nasal salt-secreting glands were first identified as such in reptiles by Schmidt-Nielsen & Fänge in 1958 and much has now been written on their comparative anatomy and physiology (see Dunson, 1969, 1976, 1979; Peaker & Linzell, 1975; Minnich 1979; Templeton 1972; Dantzler & Holmes 1974; Lemire, 1983). These glands are of interest, not only because of their diverse embryological origins within different taxa, but also because they exhibit much greater secretory versatility than the analogous salt-secretory glands found in marine birds. Salt glands have now been identified in all major reptilian orders, including the Crocodilia which were long thought to lack extra-renal salt-excreting organs (Taplin & Grigg, 1981). Broad comparative data on reptilian salt-secreting glands are brought together in Table XXXVI from which it is clear that, in general, terrestrial species secrete salt solutions rich in potassium ions, whereas littoral and marine reptiles excrete primarily sodium.

Very little is known of the hormonal control of reptilian salt glands (see Sandor & Mendhi, 1980) although early studies established their sensitivity to adrenocortical steroids such as corticosterone (Holmes & McBean, 1964). Templeton *et al.*, first reported in 1968 that aldosterone inhibited sodium excretion by the nasal salt gland of the desert iguana, *Dipsosaurus dorsalis*, and Shoemaker *et al.*, (1972) further investigated the effect of adrenal corticosteroids on the secretory activity of this gland. As seen in Fig. 2.37, this species normally responds to salt loads by secreting within 24 hours a

**Table XXXVI:** Comparative data on reptilian salt glands.

| Taxon | Embryological origin of salt gland | Genus | Electrolyte concentration (mmol/L) $Na^+$ | $K^+$ | $\frac{K^+}{Na^+}$ | Reference |
|---|---|---|---|---|---|---|
| Chelonia → | Lachrymal | | | | | |
| Cheloniidae | | *Caretta* | 732–878 | 18–31 | 0.03 | Schmidt-Nielsen & Fänge (1958) |
| | | *Chelonia* | 685 | 21 | 0.03 | Holmes & McBean (1964) |
| | | *Lepidochelys* | 713 | 29 | 0.04 | Dunson (1969) |
| Emydidae | | *Malaclemys* | 682 | 32 | 0.05 | Dunson (1970) |
| Testudinidae | | *Geochelone* | ? | 233–260 | ? | Peaker (1978) |
| Sauria → | External nasal | | | | | |
| Iguanidae | | *Amblyrhynchus* | 1434 | 235 | 0.16 | Dunson (1969) |
| | | *Dipsosaurus* | 494–1032 | 640–1387 | 2.8–0.6 | Schmidt-Nielsen *et al.* (1963) |
| | | *Sauromalus* | 82–150 | 490–1102 | 6.0–7.4 | Templeton (1964), Nagy (1972) |
| | | *Ctenosaura* | 78–475 | 220–527 | 0.5–6.6 | Templeton (1964) |

| | | | | | | |
|---|---|---|---|---|---|---|
| | | *Iguana* | 507–728 | 497–290 | 0.4–1.0 | Schmidt-Nielsen *et al.* (1963) |
| | | *Conolophus* | 692 | 214 | 0.3 | Dunson (1969) |
| Agamidae | | *Uromastix* | 639 | 1398 | 2.2 | Schmidt-Nielsen *et al.* (1963) |
| Scincidae | | *Tiliqua* | 167–434 | 226–433 | 0.42–1.98 | Bradshaw, Tom & Bunn (1984) |
| Varanidae | | *Varanus* | 654 | 54 | 0.08 | Dunson (1974) |
| Ophidia | | | | | | |
| Hydrophiidae → | Posterior lingual | *Laticauda* | 686 | 57 | 0.08 | Dunson & Taub (1967) |
| | | *Pelamis* | 620 | 28 | 0.04 | Dunson (1968) |
| | | *Aipysurus* | 798 | 28 | 0.04 | Dunson & Dunson (1974) |
| | | *Lapemis* | 676 | 23 | 0.03 | *ibid.* |
| | | *Hydrophis* | 509 | 20 | 0.04 | *ibid.* |
| Acrochordidae → | Posterior lingual | *Acrochordus* | 483 | 15 | 0.03 | Dunson & Dunson (1973) |
| Homalopsidae → | Premaxillary | *Cerberus* | ? | ? | ? | |
| Crocodilia → | Lingual | | | | | |
| Crocodylidae | | *Crocodylus* | ? | ? | ? | Taplin & Grigg (1981) |

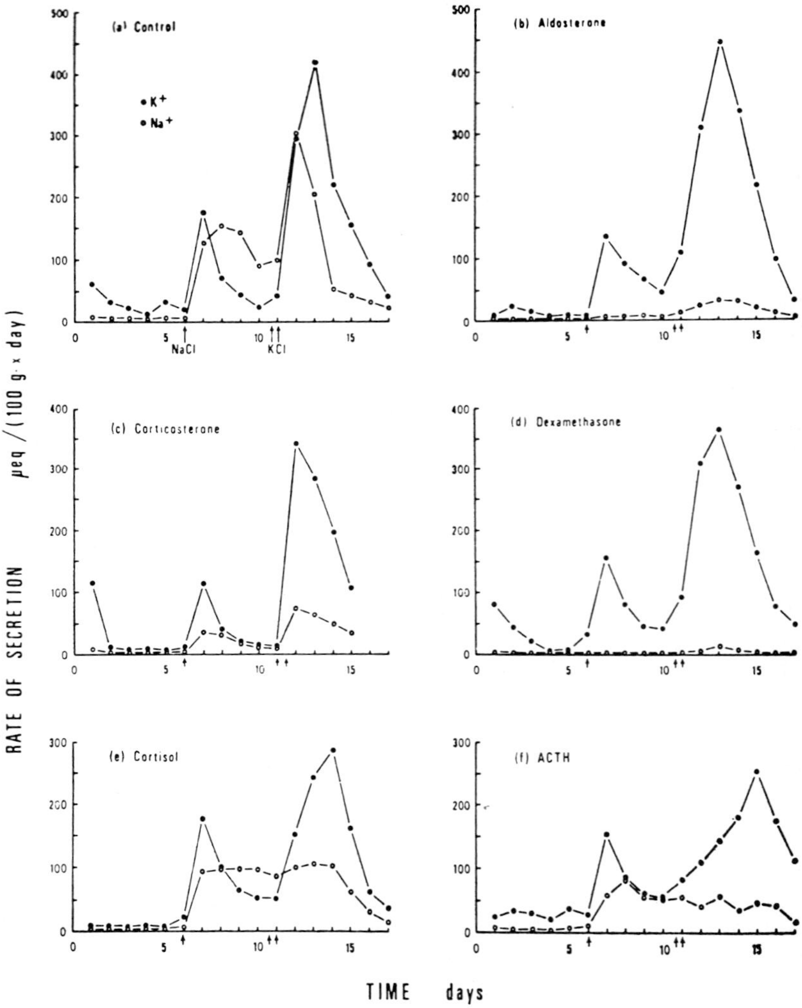


**Fig. 2.37** Rates of secretion of $K^+$ (closed circles) and $Na^+$ (open circles) by the nasal salt gland of the iguanid lizard *Dipsosaurus dorsalis*. Two m-moles of NaCl and KCl per 100 g were given on day 6 and day 11, respectively. Animals received daily injections of water (a) or various hormones (b-f) beginning on day 0. Points represent salt collected over previous 24 hr (from Shoemaker, Nagy & Bradshaw, 1972).

concentrated salt solution composed of a mixture of NaCl and KCl, regardless of which salt is injected. The 'fidelity' of the secreted solution improves with repeated injections of the one salt and $K^+/Na^+$ ratios can vary from about 0.3 to 4.0. Daily injections of 15 $\mu$g aldosterone per animal per day had the effect of virtually abolishing sodium excretion by the gland, whilst leaving the potassium excretion intact (see Fig. 2.37b). To a large extent the effect of other steroids, such as corticosterone, cortisol and dexamethasone as well as mammalian ACTH, was similar to that of aldosterone, although only dexamethasone was as potent. No effect of hormone treatments on the rate of potassium excretion was observed by Shoemaker *et al.*, (1972), even though $K^+$ is the primary ion excreted in this species, and, as in birds, hormones alone were ineffective in stimulating secretory activity. Injections of hyperosmotic solutions of sucrose and mannitol also did not trigger a secretory response in *Dipsosaurus dorsalis* and led these authors to speculate that the gland was stimulated by an increase in the plasma concentration of alkali metal ions.

Templeton *et al.* (1972b) confirmed the action of aldosterone on the nasal gland of the desert iguana and found that adrenalectomised individuals had greatly enhanced rates of sodium loss from the salt gland. A *prima facie* case thus exists for adrenocortical control of the salt gland in this species with aldosterone playing the rôle of a classical mineralocorticoid. The Fouette-Queue, *Uromastix acanthinurus*, also possesses a well-developed nasal salt gland, described by Grenot (1968a), and studied physiologically by Lemire *et al.* (1970, 1972, 1980) and Lemire & Vernet (1982, 1983). The endocrine rôle of aldosterone has recently been studied in this species by Bradshaw *et al.* (1984). The secretory response of this gland to two intraperitoneal injections of 1 mmol KCl is shown in Fig. 2.38, from which it is clear that this species is an almost 'pure' KCl excretor, elaborating a solution which is little-contaminated with NaCl, in contrast to *Dipsosaurus dorsalis*. Prolonged loading with NaCl has the effect of decreasing the $K^+/Na^+$ ratio from a maximum of about 81 to 2.5, but potassium ions are always secreted in excess of sodium in situations studied to date.

Changes in plasma levels of aldosterone and corticosterone of *Uromastix acanthinurus* with spaced injections of 1 mmol/100g of either 1 mol/L NaCl or KCl are shown in Fig. 2.39, from which it is evident that the endocrine response to these two treatments is quite different, with aldosterone levels increasing with KCl loading and falling with NaCl injections. Bradshaw *et al.* (1984) found that secretory rates for $K^+$ and $Na^+$ were significantly correlated with mean plasma aldosterone levels throughout the period of treatment, positively in the case of $K^+$ and negatively in the case of $Na^+$ as shown in Fig. 2.40. These data thus suggest a classical mineralocorticoid effect of aldosterone (i.e., natriferic and kaliuretic) on the nasal salt gland of *Uromastix acanthinurus*. It is also clear from this study why Shoemaker *et al.* (1972) and Templeton *et al.* (1972b) failed to find any effect of exogenous administration

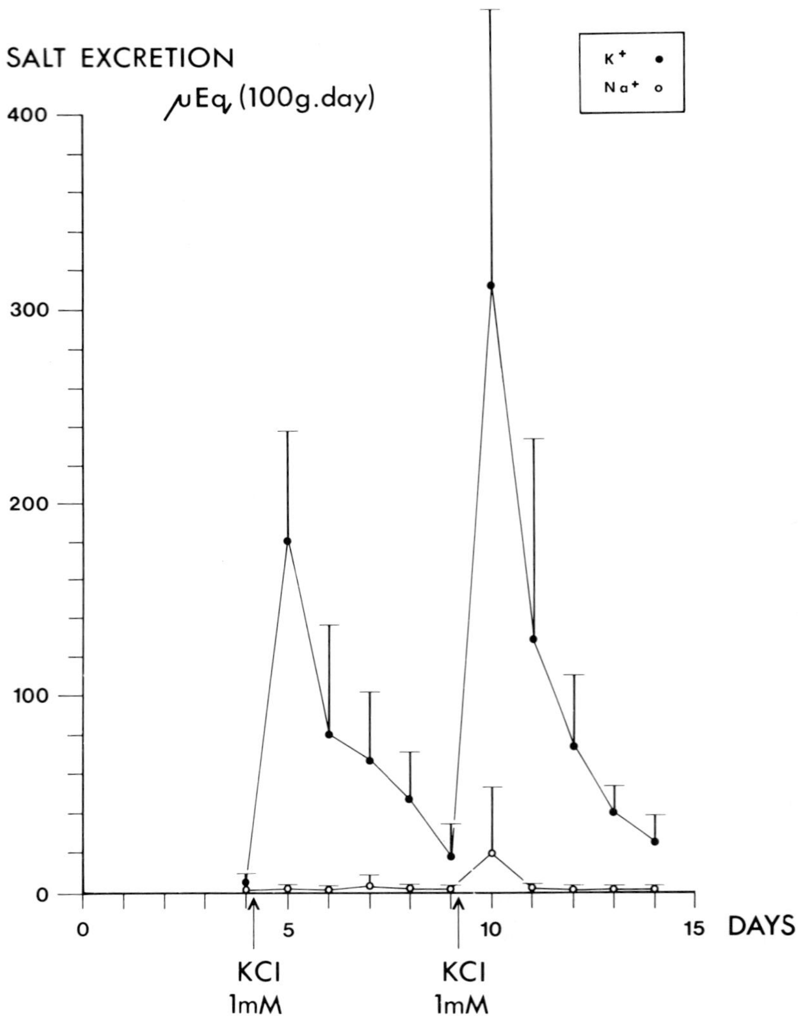


**Fig. 2.38** Excretion of salt (potassium = closed circles, sodium = open circles) by the external nasal gland of the Saharan agamid lizard *Uromastix acanthinurus*. Two injections of 1 mmol of KCl were given intraperitoneally on days 4 and 9 as indicated. Data presented as Means ± SE (from Lemire & Vernet, 1982).

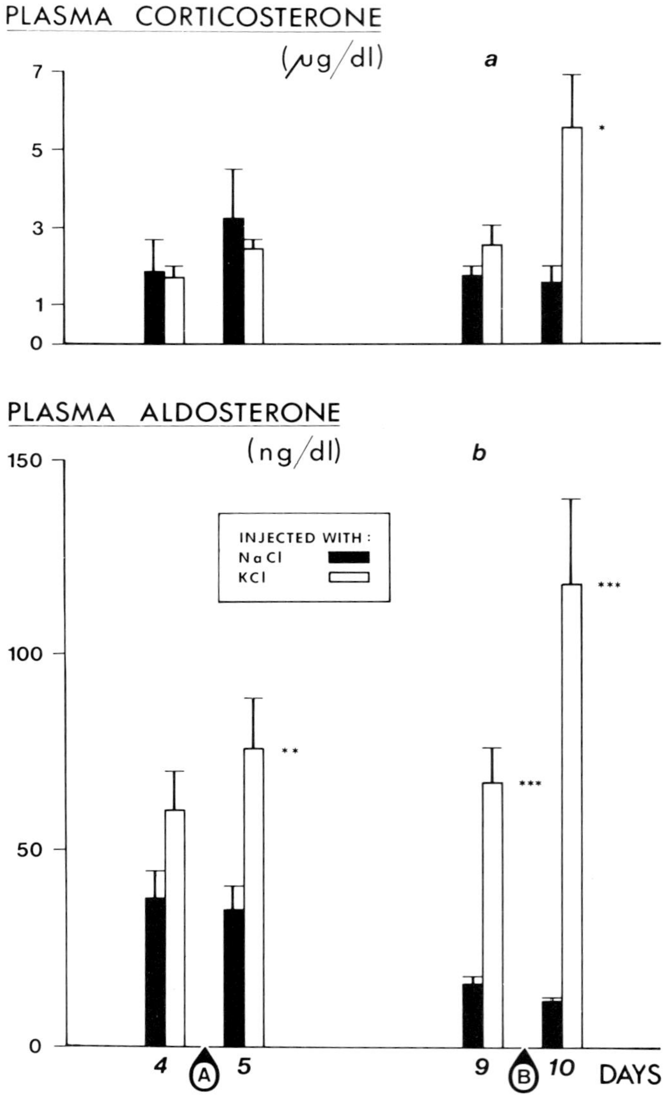


**Fig. 2.39** Progressive changes in plasma concentrations of (a) corticosterone (in µg/dL) and (b) aldosterone (in ng/dL) in the Saharan agamid lizard *Uromastix acanthinurus* following two intraperitoneal injections (A and B) of either 1 mol/L KCl (open columns) or 1 mol/L NaCl (closed columns). The significance of differences between hormone levels in KCl- or NaCl-injected animals on a given day is indicated with * $P < 0.05$, ** $P < 0.005$, and *** $P < 0.001$. Data shown as Mean ± SE (from Bradshaw, Lemire, Vernet & Grenot, 1984).

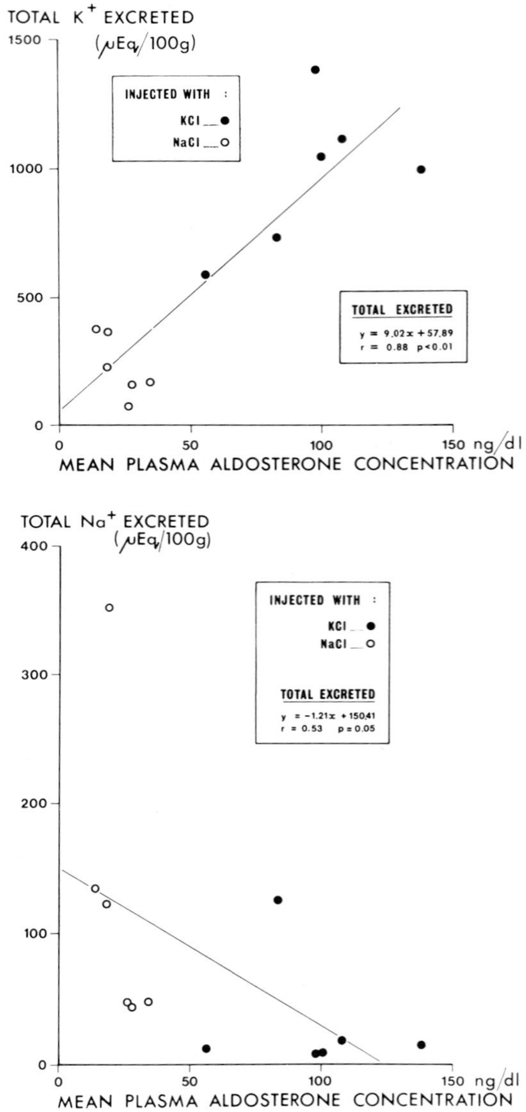


**Fig. 2.40** Correlations between total excretion of potassium and sodium ions from the external nasal gland of the Saharan agamid lizard *Uromastix acanthinurus* injected with either 1 mol/L KCl (closed circles) or 1 mol/L NaCl (open circles) and the mean plasma aldosterone concentration. The equations of the regressions and their statistical significance is given in each case (from Bradshaw, Lemire, Vernet & Grenot, 1984).

of aldosterone on rates of potassium excretion in KCl-injected animals. In all probability, plasma aldosterone levels would have been at a maximum as a result of the concomitant KCl loading, and further increases (to perhaps pharamacological levels) would be unlikely to influence the rate, which would have already been maximal, of secretion of $K^+$ ions by the gland.

One of the main reasons why work has been slow in elucidating control mechanisms of reptilian salt glands is their small size and the difficulty of access. It has become clear too that histological criteria alone are not adequate for determining whether a particular nasal gland plays an osmoregulatory rôle or not and purely morphological studies can prove deceptive. A case in point is the scincid lizard *Tiliqua rugosa*, first shown by Bentley (1959) to experience hypernatraemia in summer, and thought by most workers to be devoid of a functional salt gland. Braysher (1971) described a functional gland in this species but failed to describe the histology of the gland and did not show experimentally that a hyperosmotic fluid was elaborated by the gland, the *sine qua non* of a functional salt gland. Saint Girons *et al.* (1977) undertook to examine the light and electron microscopy of the gland and found that it was composed of two cell types: classical muco-serous cells and '*cellules striées*' (in the terminology of Gabe & Saint Girons, 1976) which are the cells producing the hyperosmotic secretion. The gland differed however from that of other species such as *Dipsosaurus dorsalis* and *Uromastix acanthinurus*, where the gland is composed entirely of '*cellules striées*' and it was clear that the electrolyte-excreting capacity of the gland must be reduced by comparison. Physiological studies by Bradshaw *et al.* (1984) have shown, nonetheless, that the external nasal gland of *Tiliqua rugosa* is capable of elaborating a hyperosmotic NaCl solution and this emphasises the importance of direct measurements on the fluid secreted by the gland (see also Saint Girons *et al.*, 1981). Bradshaw *et al.* (1984) also reported preliminary data on high-affinity binding by corticosteroids to cytosolic extracts of the nasal gland of *Tiliqua*. They found evidence of binding of both corticosterone and aldosterone. As shown in Fig. 2.41, the binding for corticosterone had a higher affinity than that for aldosterone ($K_d$ of $5.2 \times 10^{-9}$ mol/L for corticosterone as compared with a $K_d$ of $12.9 \times 10^{-9}$ mol/L for aldosterone) and, more importantly, as seen in Fig. 2.42, specific binding was found to be markedly seasonal and was only clearly demonstrable during the short breeding season of this species. The significance of these results is obscure at this stage, but they strengthen the supposition that nasal salt glands of lizards are target organs for adrenal corticosteroids.

## V. Conclusion

Reptilian endocrinology is still very much in its infancy and it is only in the last few years that it has been possible to identify common modalities in the

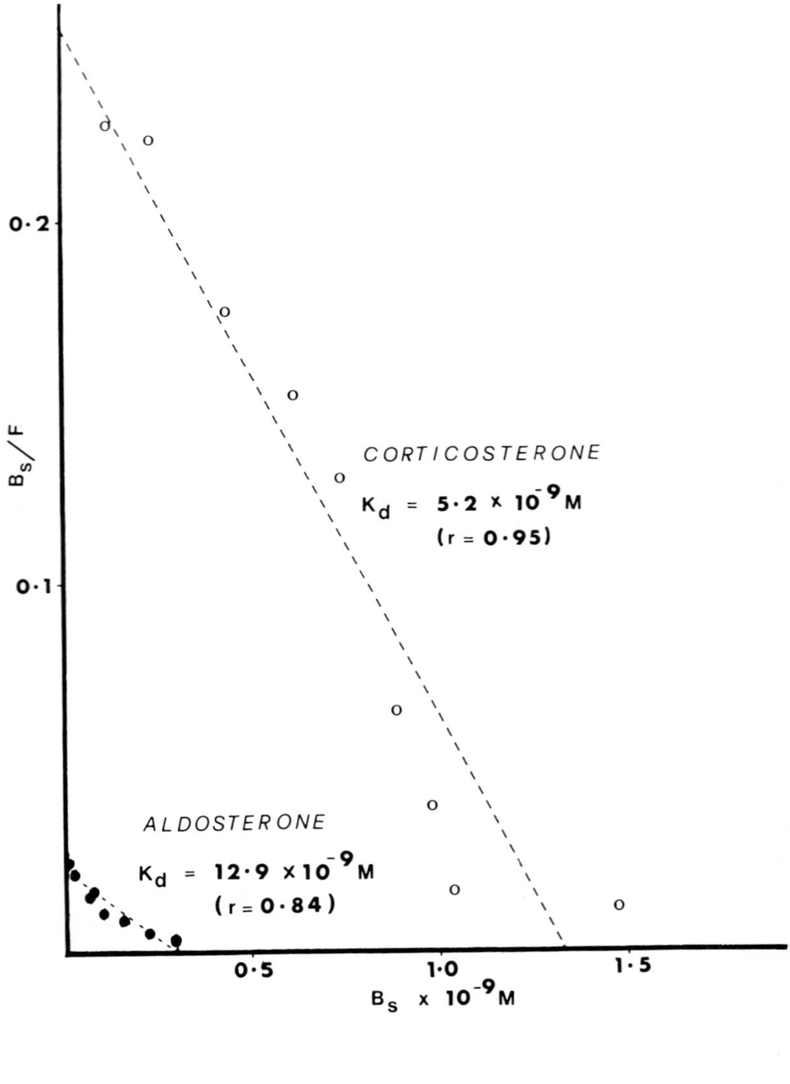


**Fig. 2.41** Scatchard analysis of high-specificity, low-capacity binding of the adrenal corticosteroids, corticosterone and aldosterone, by cytosol from the external nasal gland of the scincid lizard *Tiliqua rugosa* (from Bradshaw, Tom & Bunn, 1984).

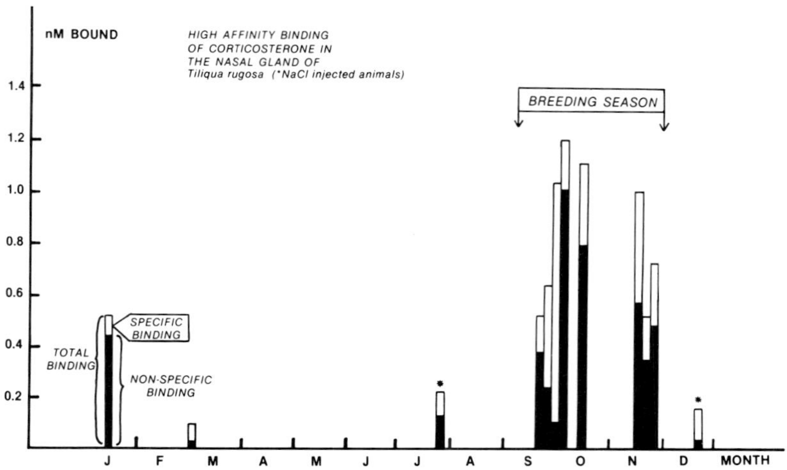


**Fig. 2.42** Seasonal variation in high-affinity binding of corticosterone by cytosol extracted from the external nasal gland of the scincid lizard *Tiliqua rugosa*. The breeding season for this species when binding, both specific and non-specific, appears highest is indicated (from Bradshaw, Tom & Bunn, 1984).

hormonal control of water and electrolyte balance of reptiles, including species which are not primarily associated with deserts. Our understanding in this area has advanced rapidly because of the development of highly sensitive and specific radioimmunoassays for neurohypophyseal and steroid hormones and there is now good evidence that AVT functions as a physiological antidiuretic hormone in at least two species of lizard. Work with *Varanus gouldii* has firmly established aldosterone as a classic mineralocorticoid hormone. Little is known however of the function of other peptides occurring in the reptilian neurohypophysis, such as mesotocin and oxytocin, and much more work is needed to clarify the renal effects of corticosterone in reptiles.

Post-renal sites of water and electrolyte exchange, such as the cloaca, colon and bladder, have emerged as vital adjuncts to the kidneys in reptiles and recent *in vivo* perfusion studies have underlined their pivotal rôle in extending and amplifying homeostatic responses initiated by the kidney to variations in water and electrolyte availability. As yet there is no compelling evidence to implicate either pituitary or adrenal hormones in the control of the activity of these sites, although there are many precedents for this in the Amphibia. Cephalic salt-secreting glands have evolved independently in representatives of all the major reptilian orders, attesting to their undoubted utility in osmoregulation. Little is known again of the factors which normally initiate secretion, nor of the underlying biochemical mechanism responsible for the elaboration of the hyperosmotic solution. Aldosterone appears to have a marked influence on the composition of the secretion and the rate at which both sodium and potassium ions are excreted, and again it appears to function as a classic mineralocorticoid. Hormonal receptors have only just been identified in reptilian nasal gland tissue and the fact that both aldosterone and dexamethasone are equally potent in inhibiting sodium excretion by the nasal gland is difficult to reconcile with current steroid-receptor theory. This may indicate that hormone receptors in reptiles have a lower specificity than those found in birds and mammals. Future studies should prove rewarding.

# 3

# Reproductive Strategies of Desert Reptiles

Reptiles are noteworthy amongst amniotes in displaying both oviparous and viviparous modes of reproduction along with a bewildering array of intergrades between these two conditions. They also have reproductive cycles which may be mono-oestrous or polyoestrous, seasonal or continuous and, although bisexual special predominate, a large number of parthenogenetic lizards has also been described and others are still being discovered (see, for example, Moritz, 1983). Despite a plethora of studies on the reproductive cycles of snakes and lizards (summarised in Fitch, 1970) the nature of the factors, both internal and external, which interact to control the secretory activity of the reptilian gonad is poorly understood and even less is known of the means by which associated behavioural attitudes essential for successful reproduction are cued at appropriate times. As with other areas of reptilian biology however, there is currently a great deal of interest in the resolution of these problems.

## I. HORMONAL CONTROL OF GAMETOGENESIS AND REPRODUCTIVE BEHAVIOUR IN REPTILES

In most reptiles the testes lie close to the vena cavae with the left testis often being somewhat smaller and located slightly caudad to the right. The seminiferous tubules and interstitial cells usually hypertrophy together during the breeding season and the sexual segment of the kidney is also androgen-dependent and undergoes a seasonal cycle of secretory activity corresponding to the spermatogenic cycle (Prasad & Reddy, 1972). The sexual segment is thought to play a similar rôle to the seminal vesicles and prostate gland of mammals, capacitating the sperm and providing nutrients during sperm storage (Ceuller *et al.*, 1972). Secretions from the sexual segment produce a copulatory plug in some snakes. In the garter snake *Thamnophis sirtalis* this has been shown to contain a pheromone which signals the mated status of the female (Ross & Crews, 1977). Licht & Gorman (1970) have categorised the stages in reptilian spermatogenesis according to the condition of the seminiferous tubules, the state of the epididymis and renal sexual segment, and the presence and location of spermatozoa. This system has been widely accepted and employed by others working with lizards.

The ovaries in most female species are bilaterally symmetrical, except in snakes, where the left ovary is less well-developed. The ovarian cycle typically consists of oogenesis (transformation of oogonia to oocytes), vitellogenesis, ovulation and formation of the corpus luteum and, depending upon the mode of reproduction, either shell deposition and oviposition or retention of the egg with the formation of a placenta. Follicle maturation is under pituitary control (Callard *et al.*, 1972a) and recruitment of follicles may reflect local secretion of steroid hormones from granulosa cells (Jones *et al.*, 1974). Vitellogenesis occurs as the primary oocyte achieves a characteristic diameter and the corpus luteum forms at the site of ovulation, as in mammals. The precise function of the corpus luteum in reptiles is still a matter of conjecture although there seems little doubt that it is the major source of progesterone secreted during pregnancy in most viviparous species (Veith, 1974; Callard *et al.*, 1972b; Chan *et al.,* 1973; Highfill & Mead, 1975a). Yaron (1972, 1985), in reviewing published work on the effects of ovariectomy or lutectomy on the course of gestation in viviparous species, concluded that embryonic development up to the normal time of parturition is not dependent on any secretory product of the copora lutea, nor of the entire ovary. Highfill & Mead (1975b) found, however, that the duration of pregnancy was significantly lengthened in lutectomised garter snakes and concluded that the corpus luteum secretes some substances, as yet unidentified, which is responsible for ensuring that parturition occurs at the correct time. Callard *et al.* (1972b) similarly found the course of gestation in the lizard *Sceloporus cyanogenys* to be unaffected by gonadectomy, but noted that parturition was delayed or prevented. Badir (1968) working with *Chalcides ocellatus* could detect no effect of deluteinisation on the length of gestation and recorded normal parturition. Evidence that the corpus luteum may not be the sole source of progesterone comes from a recent study by Guillette *et al.* (1981) on the lizard *Sceloporus jarrovi* where they found a clear dissociation between plasma progesterone levels and the size and secretory activity of the corpus luteum, which degenerates well before parturition in this species (see Fig. 3.1). They

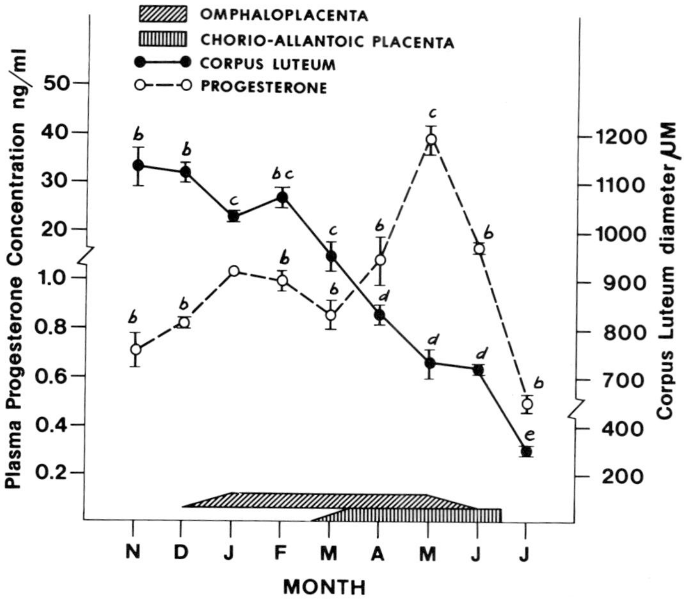


**Fig. 3.1** Monthly averages (± SE) for luteal diameter and plasma progesterone concentration in relation to the presence of the omphalo- or chorioallantoic placenta. Different superscripts indicate significant differences among monthly values (Duncan's multiple range test, $P < 0.05$) (from Guillette, Spielvogel & Moore, 1981).

speculated that the high plasma progesterone levels seen in this lizard in late gestation result from the secretory activity of either the chorioallantoic placenta or the corpora atretica, but there is as yet no direct evidence to support this and the adrenal gland is another possible source.

Licht (1979) has pointed out that the superficial resemblances in reproductive endocrinology amongst tetrapod vertebrates have led many workers to assume that the hormones in lower vertebrates merely reflect, albeit less perfectly, the functions of their counterparts in the mammals. In fact this has not been found to be the case in reptiles and substantial differences have emerged, especially in the extent to which the reptilian gonad is controlled by pituitary gonadotrophins. Although two separate gonadotrophins have been isolated in chelonians and crocodilians (as well as in amphibians), only a single hormone has been demonstrated in squamate reptiles (snakes and lizards). This fact alone enhances the variability between the various taxa (see Licht *et al.*, 1977). Early studies *in vivo* on the actions of mammalian gonadotrophins follicle-stimulating hormone [FSH] and luteinising hormone [LH] in lizards, snakes and turtles indicated that testis growth and spermatogenesis were highly FSH-specific and the small effects discerned with LH preparations were attributed to contamination with FSH. Licht & Papkoff (1973) were able to show using immunodiffusion techniques that both mammalian FSH and LH have intrinsic spermatogenic-stimulating activity in lizards and, although potency ratios for the two hormones may vary widely in different species, both are always active. Licht (1979) suggests that variations in FSH/LH potency ratios may be explained by differences in the rate of metabolic clearance of the two hormones from the circulation rather than as a reflection of innate differences in receptor specificities. This highlights the dangers of extrapolating from results of experiments utilising heterologous rather than homologous hormones. As Licht (1979) observes: "the relative potencies of heterologous preparations of FSH and LH in squamates may have little relevance to the endocrinology of these reptiles if indeed they have only a single gonadotrophin". The major conclusion to be drawn from these studies at this stage is that a number of important components of gonadal physiology, including growth, gametogenesis and oestrogen secretion, exhibit little FSH/LH specificity in the Reptilia (or in the Amphibia) whereas ovulation, testicular androgen secretion and ovarian progesterone secretion are either FSH-sensitive or non-specific.

Patterns of gametogenesis and steroidogenesis in both male and female reptiles are beginning to emerge as more species are studied in detail. Information on both sexes has recently been reviewed by Lofts (1977) and Callard & Vance (1977). These latter authors categorised follicular cycles into three types as shown in Fig. 3.2: females which produce a single large clutch of eggs (e.g. the turtles *Pseudemys* and *Chrysemys* which inhabit northern latitudes), those producing several large clutches (such as the green sea turtle *Chelonia mydas*), and the third category with multiple, small clutches. This last is common in many lizards from tropical and temperate regions as well as tropical turtles such as *Kinosternum leucostomum* which produce small clutches throughout the year. Seasonal changes in plasma levels of immunoreactive LH, progesterone and 17$\beta$-oestradiol ($E_2$) have been reported in *Chrysemys picta* (Type I) by Callard & Vance (1977) and are shown in Fig. 3.3. The concentration of all these hormones was maximal at the time of ovulation in mid-summer with a similar secondary increase in $E_2$ levels associated with ovarian recrudescence in the autumn. They also reported that plasma testosterone (T) levels show a minor increase to 420 $\pm$ 75 pg/mL in spring (May–June), then fall to a low of 65 $\pm$ 7 pg/mL in September and increase markedly to levels of 1838 $\pm$ 678 pg/mL during the autumn period of ovarian recrudescence. In *Chelonia mydas* (Type II) peak progesterone levels again coincide with ovulation but in contrast $E_2$ levels were found to be very

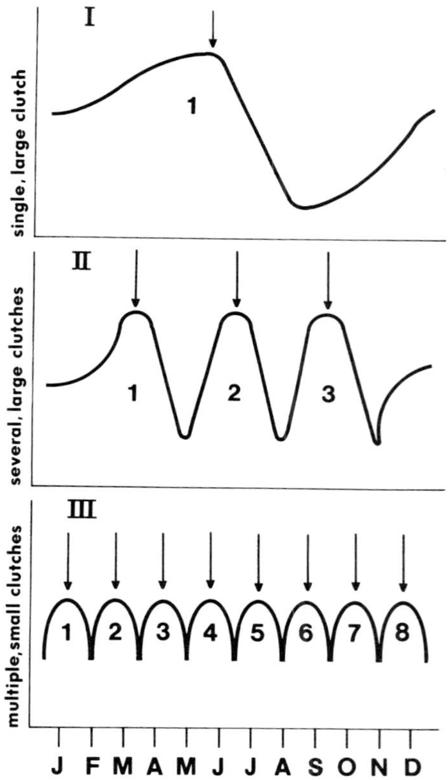


**Fig. 3.2** Principal types of reproductive cycle seen in reptiles. The major variables are clutch size and clutch interval. The arrow indicates the approximate time of ovulation (from Callard & Lance, 1977).

low. As this turtle is polyoestrous there is a series of peaks of progesterone in the plasma, coinciding with the maturation and ovulation of each batch of eggs as seen in Fig. 3.4 (see also Licht *et al.*, 1979a,b). Licht *et al.* (1982) have also described a similar 'ovulatory surge' of progesterone secretion in the Olive Ridley sea turtle, *Lepidochelys olivacea*, and demonstrated a significant positive correlation between plasma levels of progesterone and turtle LH.

Callard & Vance (1977) suggest that there are two basic patterns of progesterone secretion seen in reptiles—a preovulatory rise of progesterone as seen with the oviparous *Chelonia*, followed by a declining level of secretion associated with the formation and regression of the short-lived corpus luteum

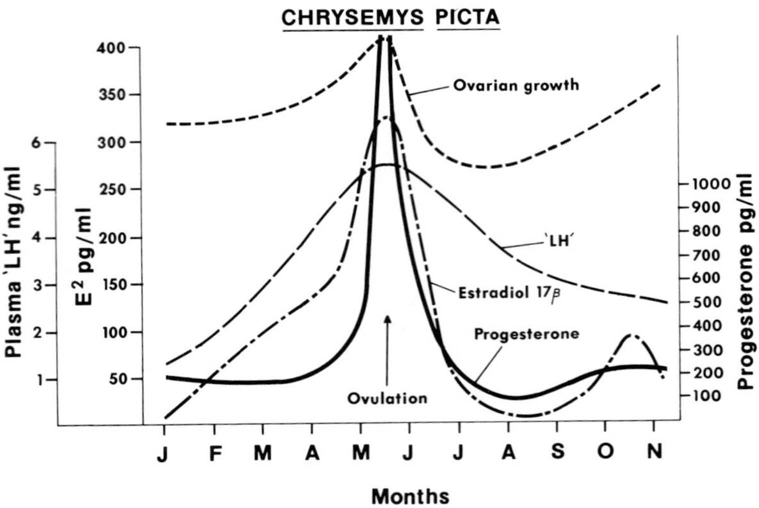


**Fig. 3.3** Temporal changes in plasma IR-LH, Progesterone and $E_2$ associated with the ovarian cycle in the painted turtle, *Chrysemys picta*. The dashed line for ovarian growth indicates changes in ovarian weight. For the plasma steroids, values were obtained from monthly bleedings in all months except January and February, and 'n' in all groups was not less than 10. For plasma IR-LH, values were obtained for the months of March, May, June, July, August, September and November (n = 5/group) (from Callard & Lance, 1977).

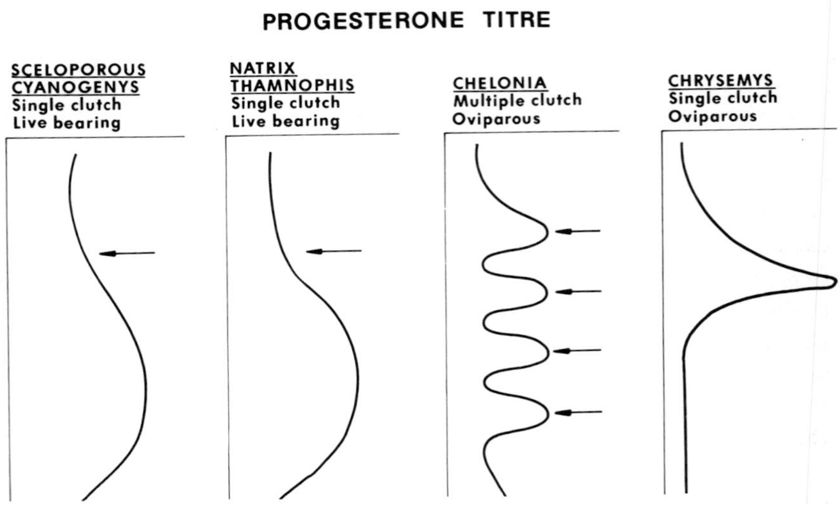


**Fig. 3.4** Patterns of progesterone secretion in reptiles (from Callard & Lance, 1977).

which only persists for the time that the eggs are contained within the oviduct (see also Joss (1983, 1985)). The second pattern is that seen in viviparous species of lizards and snakes such as *Sceloporus cyanogenys* (Callard & Doolittle, 1973), *Chameleo pumilis* (Veith, 1974), *Natrix sipedon* (Chan *et al.*, 1973) and *Thamnophis elegans* (Highfill & Mead, 1975a) where the major peak of progesterone secretion is postovulatory rather than preovulatory and coincides generally with the development of the corpus luteum. Interestingly, the first of these two patterns is found in birds, whereas eutherian mammals such as the rat combine features of both cycles with a preovulatory surge associated with oestrus followed by a second phase of progesterone secretion from the corpus luteum and placenta.

Licht & Gorman's (1970) classification of testicular cycles is shown in Fig. 3.5 and five patterns are recognised: in Type I the testes undergo rapid spermatocytogenesis and spermiogenesis after the animals emerge from hibernation in the spring. After a short breeding season of one to two months, the testes regress and remain quiescent throughout the winter hibernation period. Type II is similar to Type I, differing only in that the breeding season is longer and testicular activity continues throughout summer. In Type III cycles, testicular recrudescence commences during the winter and is completed in spring, being followed by a decline in summer. Type IV differs only in that the breeding season is more protracted and testicular regression occurs in late rather than early summer. Types I–IV are found in temperate-zone species and Type V describes sub-tropical species such as the lizard *Anolis carolinensis* which have a prolonged breeding season during spring and summer followed by a regressive or refractory phase in late autumn and throughout the winter months.

Although this schema has been very useful in helping to group the diverse data available on male sexual cycles in reptiles, it is clear that it is nevertheless an oversimplification and this is more obvious in the case of snakes than with lizards. Even so, Crews (1979a) points out that the testicular cycles of some lizards do not correspond to any of Licht & Gorman's Types—in the North-American viviparous lizards *Sceloporus cyanogenys* and *S. jarrovi*, for example, maximum testicular volume, weight and activity occur in mid-winter (see Crisp, 1964). In the tropical Australian skink *Leiolopisma rhomboidalis*, studied by Wilhoft (1963), spermatogenesis is continuous except for a brief quiescent phase during winter hibernation. Small desert species of lizards such as *Acanthodactylus pardalis* and *Eremias olivieri* from the northern Sahara also show continuous spermatogenesis and may breed throughout the whole year (Saint Girons, 1985a).

Saint Girons (1963, 1982b, 1985b) and Bons & Saint Girons (1982) have developed an alternative classificatory scheme which may be compared with that of Licht & Gorman (1970). They divide spermatogenic cycles into four major types: Aestival, Mixed, Vernal-Prenuptial and Continuous which, to

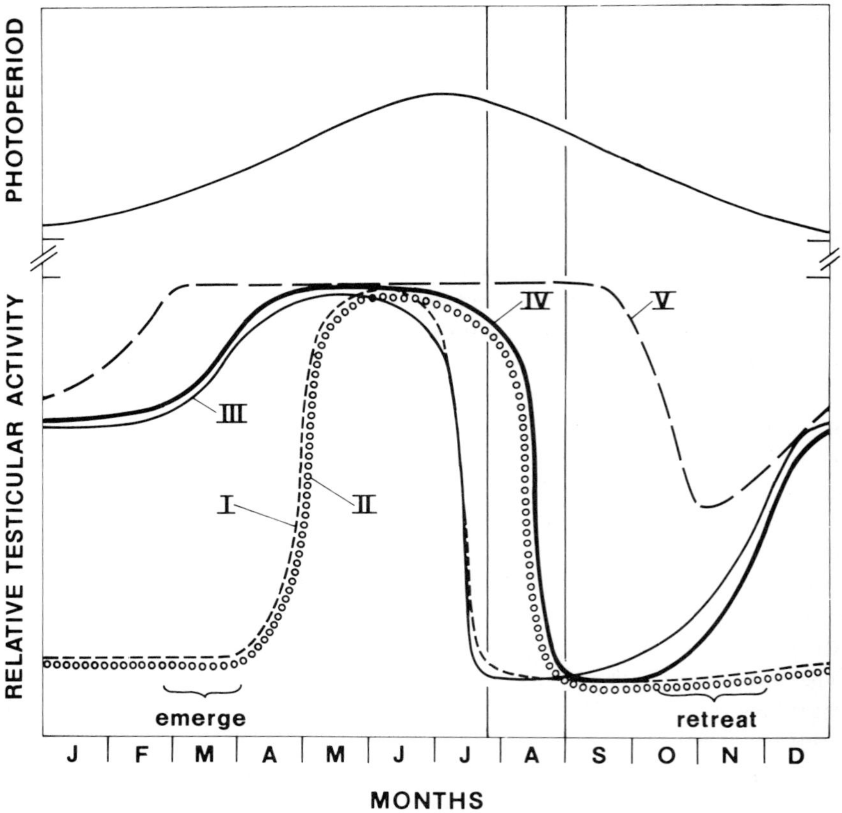


**Fig. 3.5** Schematic representation of the diverse patterns of annual testicular activities in lizards. Patterns I to IV represent temperate-zone species. The testis cycles of the most temperate species of *Anolis* (*A. carolinensis*) is shown by type IV. Type V summarises the 'typical' testis cycle of the Caribbean *Anolis*. The temperate species differ in the time when testes regress, when recrudescence commences, and in the rate of recrudescence. For example, recrudescence may occur before or after hibernation (times indicated by 'retreat' and 'emerge'). Fine vertical lines demarcate the times when most temperate species show a cessation of all spermatogenic activity (from Licht & Gorman, 1970).

some extent, may be equated with Licht & Gorman's Types. In cool, temperate zones and elevated habitats in the subtropics, the warm season is too short to enable reptiles to complete the entire reproductive cycle in a single year and spermatogenesis either occurs in the year preceding reproduction (Aestival) or begins in the summer, or is checked during winter and completed at the beginning of spring (Mixed). The Mixed category corresponds to Licht & Gorman's Types III and IV and does not include any species from arid regions.

In the subtropical deserts, reproductive cycles are almost exclusively of the Vernal-Prenuptial type. This corresponds to Licht & Gorman's Types I and II. Fig. 3.6 shows diagrammatically the spermatogenic cycles of a number of desert species of reptiles from North Africa which illustrates this pattern. In the northern hemisphere, development of the spermatogonia usually commences in September in these desert lizards with some spermatocytogenesis in autumn and at the beginning of winter. Spermiogenesis may commence in March with mating in May and ovulation in June (e.g. *Xantusia vigilis*, Miller, 1948); or it may run from the end of April to the end of June when mating occurs with ovulation in early July (e.g. *Scincus scincus*, Kehl, 1944; Badir, 1958; and *Varanus griseus*, Vernet, 1977). In some lizards such as *Agama bibroni* and *Acanthodactylus erythrurus* in Morocco (Saint Girons, 1967; Bons, 1969) and *Uma notata* and *Uma inornata* in California (Mayhew & Wright, 1970), spermiogenesis may extend from February–March until July–August with mating from April to July. Up to three clutches of eggs may be produced by these species between early spring and late summer. In lizards which are perhaps the most characteristic of the desert regions of both the Sahara and North-America such as *Uromastix acanthinurus* (Courrier, 1929; Kehl, 1944), *Dipsosaurus dorsalis* (Mayhew, 1971), *Urosaurus graciosus* (Vitt *et al.*, 1978), *Urosaurus ornatus* (Asplund & Lowe, 1964) and *Phrynosoma solare* (Blount, 1929), the spermatogonia do not commence development prior to winter, or even spring, and this is followed by a very rapid spermatogenesis in spring. In all these desert lizards, mating and ovulation occur later than in species occupying cool, temperate regions and the young emerge towards the end of summer. It appears that the survival of the eggs in the soil over this extremely hot, dry period is not a major problem and the juveniles thus avoid being exposed to the most unfavourable period of the year. Adult animals in the Sahara typically also show reduced activity during the hottest part of the summer months and two environmental constraints appear to be critical in moulding reproductive cycles: the availability of sufficient quantities of food to sustain vitellogenesis in the females and the need to ensure that the young emerge at the time of the year most propitious for their survival. In general, it seems clear that the cycle of the male is not so much adapted to the environment in which it lives as being attuned to that of the female.

Although one may identify the Vernal-Prenuptial cycle as being characteristic and predominant in reptiles living in arid sub-tropical regions, this does not mean that all animals inhabiting these regions share this pattern exclusively. We have already mentioned that some small lizards such as the *Acanthodactylus spp.* may breed continuously in the North African Sahara and the case of the chameleon *Chamaeleo chamaeleon* is particular. This lizard is the only member of this genus to be found in arid regions and is quite common for example in Morocco (Bons & Bons, 1960). This species is oviparous and characterised by an extremely long incubation period of from 250–290 days,

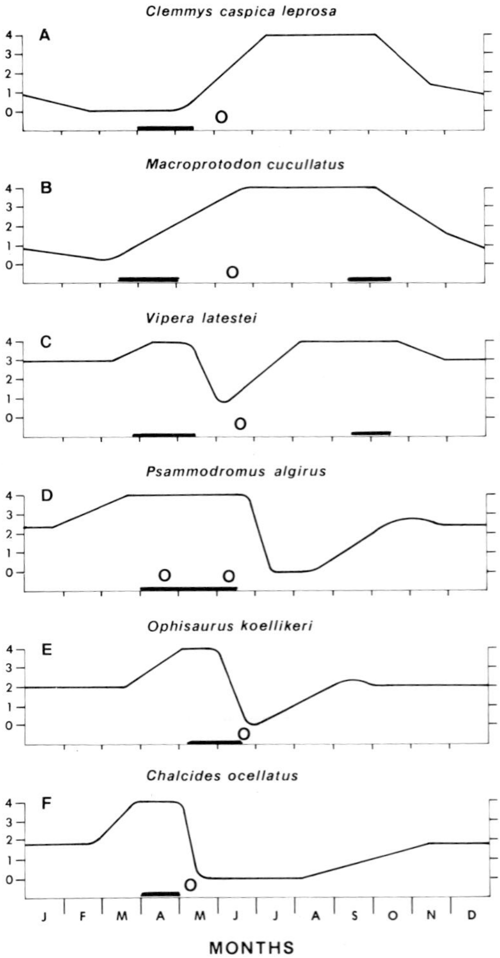


**Fig. 3.6** Schematic representation of *aestival* (A and B) and *mixed* (C to F) reptilian spermatogenic cycles. Abscissa—time in months; ordinate—stage of spermatogenesis.

0–1: Multiplication of spermatogonia; first normal primary spermatocytes appear at 1.
1–2: Primary spermatocytes, of different stages, more numerous than spermatogonia. First young spermatids appear at 2.
2–3: Young spermatids as common as spermatocytes and beginning to develop.
3–4: Process of spermatogenesis; first spermatozoids appear at 4; continuous spermatogenesis with all elements of seminal lineage present.
4–0: End of spermiogenesis with elimination of spermatozoids from the seminiferous tubules.

Frequently, multiplication of spermatogonia begins before the total elimination of the spermatozoids, so that stage 0 is never really attained.

0 = mean ovulation date (in the case of polyovulators, the date is that of the females with the greatest clutch size); mating period indicated by solid bar (from Bons & Saint Girons, 1982).

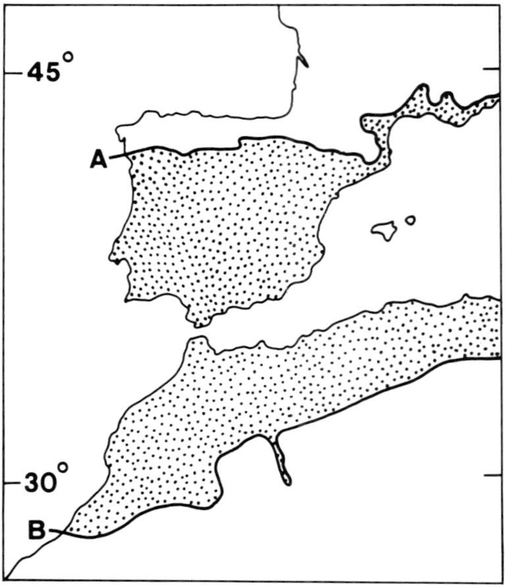


**Fig. 3.7** Distribution of the different types of spermatogenic cycles among snakes in southern Europe and northwestern Africa. A = northern limit of species having vernal-type cycles; B = southern limit of species having aestival or mixed cycle; stippling = area where the three types of cycles can occur. Species having vernal cycles are absent from mountains above 500–2500 m, depending on latitude (from St. Girons, 1982b).

most of which occurs during the winter period, but the length of gestation is due to embryonic diapause at an early stage of development rather than low environmental temperatures throughout this period. Vitellogenesis and spermatogenesis are aestival (occur in the summer) but mating occurs in August–September and ovulation in October so that the cycle is in reality *prenuptial* with birth of the young in the summer of the following year. Similar cycles are found in a number of species of the North-American genus *Sceloporus* inhibiting mountainous regions of the intertropics (e.g. *Sceloporus grammicus*, Guillette & Casas-Andreu, 1980; Ortega & Barbault, 1984; *Sceloporus jarrovi*, Goldberg, 1971). Arid and desert regions may thus harbour species showing a variety of reproductive patterns. Fig. 3.7 shows the description by Saint Girons (1982) of the geographical limits of reptiles displaying all three types of reproductive cycles in Morocco. As may be seen, species with typically desert, Vernal cycles extend as far north as northern Spain and reptiles with Aestival or Mixed cycles may be found as far south as the Haut Atlas. In between, there is a broad semi-arid region where all three cycles may be found.

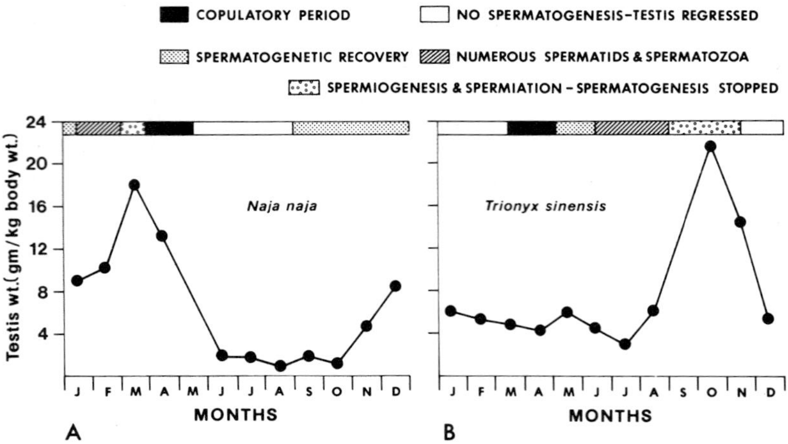


**Fig. 3.8** (A) Seasonal changes in the testicular weight and spermatogenic condition of the cobra *Naja naja*. The main copulatory period is in April. The peak of spermatogenetic development is reached in the period immediately preceding the breeding phase. (B) Seasonal changes in the testicular weight and spermatogenetic condition of the soft-shelled turtle, *Trionyx sinensis*. Unlike the cobra, the seminiferous tubules are fully regressed during the April copulatory period. Spermatogenesis begins after this period and is completed several months before the following breeding phase (from Lofts, 1977).

Despite the obvious variety of reptilian reproductive patterns, Lofts (1977) has suggested that all be accommodated in two basic hormonal categories based on whether or not the species retains spermatozoa in the epididymides for a prolonged period after the cessation of the spermatogenic cycle. Lofts (1977) considers the Chinese cobra *Naja naja* and the soft-shelled turtle *Trionyx sinensis* as examples of these two categories and, although the reproductive cycles of the females of these two reptiles are almost identical with mating in spring and eggs being laid in late May–June, the spermatogenic patterns of the males are entirely different, (Fig. 3.8). *Naja* has a *prenuptial* pattern of spermatogenesis and after copulation the testes rapidly regress into a quiescent phase for three to four months. Spermatogenesis resumes in September and a few spermatids may appear before the winter hibernation retards development, with the final phases of spermatogenesis occurring in the following spring. In *Trionyx*, on the other hand, the spermatogenic cycle is a *postnuptial* phenomenon. Spermatogenesis begins immediately after the breeding phase and by mid-July, when the testes of the cobra are fully regressed, the seminiferous elements of the soft-shelled turtle are greatly enlarged and filled with a complete spectrum of developing germ cells, including spermatozoa. Spermatogenesis declines in August to be replaced by spermiogenesis and spermiation and the spermatozoa so produced are stored in the epididymal canals. With the termination of spermatogenesis and the evacuation of the spermatozoa, the seminiferous tubules regress and remain quiescent for a further six months. Thus, at the height of the breeding season, the testes of *Trionyx sinensis* are completely regressed and spermatozoa are ejected from the epididymal reservoir (Lofts & Tsui, 1977). Lofts (1977) points out that this phenomenon of extratubular sperm storage is common in turtles and is widespread in colubrid and crotaline snakes (e.g. *Thamnophis*), but is less frequent in lizards.

Crews (1984) has termed these two patterns *'Associated'* and *'Dissociated'* reproductive tactics in a recent review article in which he questions the conventional wisdom that steroid sex hormones activate mating behaviour in all animals. Associated reproductive tactics are universal in laboratory and domesticated mammals as well as being usual in birds, temperate and tropical lizards and crocodilians, and this may explain the durability of the paradigm. He documents instances of dissociation between gonadal activity and mating behaviour in both males and females and highlights a problem which was long ago identified (e.g. Courrier, 1927, in bats; Herlant, 1933; Kehl & Combescot, 1955, in squamates). Lofts (1977) developed an index for testosterone production in *Naja* and *Trionyx* which compensates for differences in testicular mass (testicular weight multiplied by the mean percentage conversion of labelled precursor into testosterone) and this is shown for the two species throughout the year in Fig. 3.9. As would be expected, the testosterone peak occurs in spring with the prenuptial type and in autumn with the postnuptial cycle. Licht (1982) has recently measured circulating levels of testosterone in *Trionyx sinensis* and they confirm the measurements by Lofts (1977) *in vitro* with a peak that coincides with testicular recrudescence. Seasonal changes in plasma testosterone levels have now been measured in a number of lizard species and these all conform to the prenuptial type. Arslan *et al.* (1978) found that plasma testosterone levels peaked during the spring mating season (late March to early April) in *Uromastix hardwicki* in Pakistan and fell rapidly in mid-May along with the regression of the testes (Fig. 3.10). A second peak in testicular testosterone concentration, but not plasma levels, was also found in late August and early September and the authors speculate that this may reflect intratubular androgen synthesis by the Sertoli cells. This theory has also been invoked by Callard *et al.* (1976) to account for the recrudescence of spermatogenesis in the turtle *Chrysemys picta* when plasma testosterone levels are very low. A similar pattern of testosterone levels was reported in the viviparous lizard *Lacerta vivipara* studied in France by Courty & Dufaure (1979) and, as may be seen in Fig. 3.11, the secondary peak of testosterone in the testis which follows the regression of the testes and the secondary sex

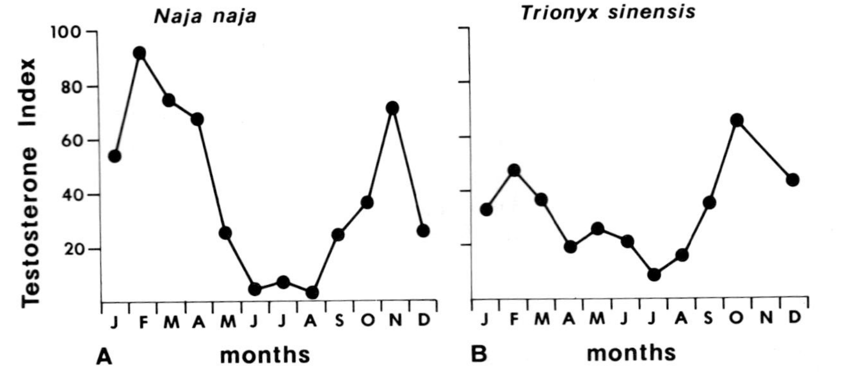


**Fig. 3.9** Seasonal fluctuations of *in vitro* testosterone synthesis from $\Delta^5$-pregnenalone-16-$^3$H. Testosterone levels are expressed as an index from the mean percentage conversion of a 250 mg sample of testicular tissue multiplied by the mean total testicular weight for each month. Each point is the average from six animals. (A) The annual pattern in the cobra shows a main peak in spring and a second smaller peak in autumn, but the soft-shelled turtle (B) has its main peak in autumn and then declines to a lower level of productivity. In both animals there is a decline during the winter hibernation period (from Lofts, 1977).

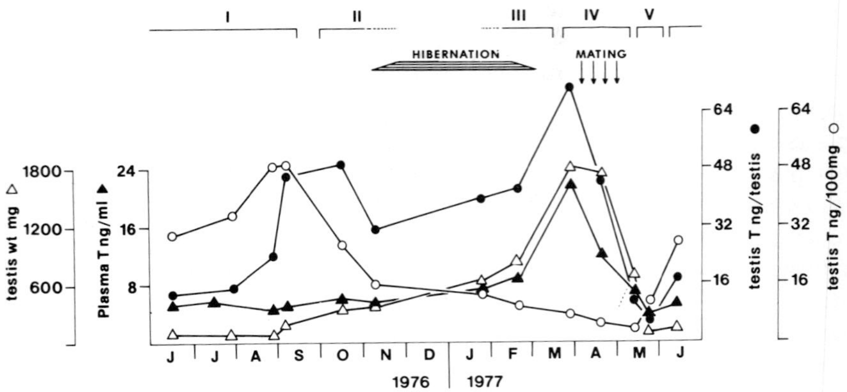


**Fig. 3.10** Monthly changes in testicular weight and plasma and testicular androgen concentrations in *Uromastix hardwicki*. Roman numerals at the top of the diagram indicate changes in the spermatogenic activity: I, germinal quiescence; II, germinal recrudescence; III, germinal maturation, early spermatids; IV, late spermatids and mature sperm; V, germinal regression (from Arslan, Lobo, Zaidi, Jalali & Qazi, 1978).

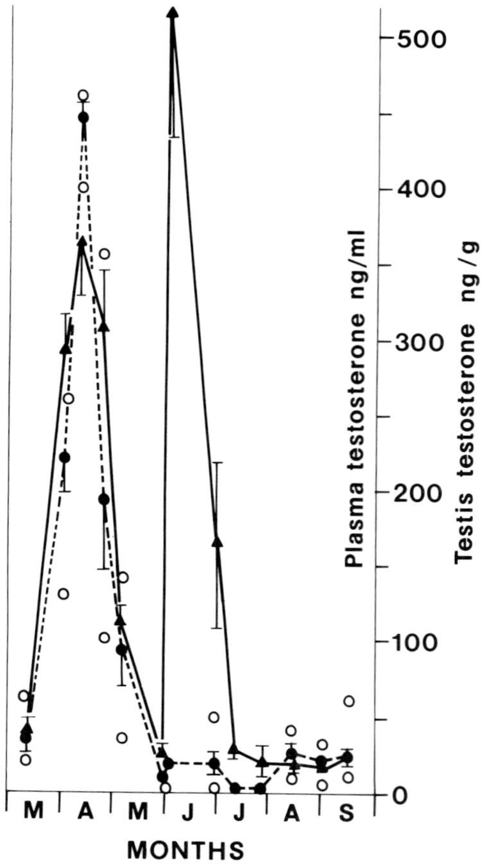


**Fig. 3.11** Seasonal changes in testis and plasma testosterone concentration in *Lacerta vivipara*. Testis testosterone is expressed as ng/g wet tissue. Data (mean ± SEM). (▲) Testis testosterone; (●) plasma testosterone; (o) highly dispersed individual values in plasma testosterone concentration (from Courty & Dufaure, 1979).

characters is even more marked than in *Uromastix hardwicki*. These data are consistent with earlier work in *Lacerta vivipara* by Dufaure & Chambon (1978) in which they found that $^{3}$H-testosterone was selectively retained by cell nuclei of the epididymis in spring but not in summer and autumn.

Seasonal changes in plasma levels of 17$\beta$-hydroxysteroids (subsequently identified as epi-testosterone, Bourne & Seamark, 1978) were reported by Bourne & Seamark (1975) in the viviparous Australian skink *Tiliqua rugosa*. These also conform to the prenuptial pattern with a peak in September–October (spring in the southern hemisphere). Perhaps the most detailed study of hormonal profiles in a reptile is that of Bona-Gallo *et al.* (1980) on the

Chinese cobra *Naja naja*. In this species plasma testosterone levels are clearly associated with increasing testis size and secretory activity but it is unfortunate that the time of mating is not known. Weil & Aldridge (1981) studied seasonal androgenesis in the water snake *Nerodia* (= *Natrix*) *sipedon*, which exhibits postnuptial spermatogenesis, and found two peaks of testosterone in the plasma, one in the spring when the animals mate soon after emerging from their winter hibernation, and the second in autumn coinciding with the time of maximum spermiogenesis and spermiation in this species. Testosterone levels peak in October when testicular volume is also greatest in the multiclutched fresh-water turtle *Sternotherus odoratus* studied by McPherson *et al.* (1982) and animals were observed copulating in autumn. Thus the turtle *Chrysemys picta* studied by Callard *et al.* (1976) appears to be the only well-established case to date where there is a marked temporal dissociation between interstitial cell activity and spermatogenesis and this may thus not be entirely characteristic of postnuptial cycles. Lofts & Chui (1968) found testosterone to be important for the completion of spermatogenesis in the lizard *Gekko gecko*, but there are unsufficient data to say with certainty whether the Sertoli cells are responsible for the production of testosterone in autumn. Interestingly, plasma vitellogenin levels in *Chrysemys picta* also show two peaks during the year, one in spring and another towards the end of autumn (Gapp *et al.*, 1979).

The occurrence of 'dissociated reproductive tactics' in temperate chelonians and some snakes (see Licht, 1984) highlights the important question of the hormonal control of reproduction and, especially, mating behaviour which hitherto has been thought, by analogy with domestic animals, to be sex-steroid dependent. Postnuptial gametogenesis is extremely rare in lizards however (some examples may be *Hemiergis peroni* studied by Smyth & Smith, 1968; *Phyllodactylus marmoratus* studied by King, 1977, and some of the geckos and skinks of New Zealand—Saint Girons, personal communication) and it is also important to note that, even though plasma levels of sex steroids may be low at the actual time of mating, the secondary sex characters are always hypertrophied. As gonadectomy in reptiles is known to be followed by a marked decrease in the secretory activity of the epithelial cells lining the renal sexual segment, epididymal tubules, and oviduct; disappearance of sexual coloration; regression of the femoral and preanal pores in the male and a rapid decline in sexual behaviour (see Licht, 1974), this suggests that the relationship between the maintenance of secondary sex characters and circulating levels of sex steroids needs further study. The femoral pores in male *Gekko gecko* for example atrophy following castration (Chui *et al.*, 1970) and Fergusson *et al.* (1985), working with the agamid lizard *Amphibolurus ornatus* found that their secretory activity could be restored by injections of either testosterone-propionate or 5$\alpha$-dihydrotestosterone (DHT), with the latter steroid being the more potent of the two. These two species both exhibit prenuptial reproductive patterns, however, and it would be of interest to study the rôle of androgens in maintaining secondary sex characters in species with a postnuptial pattern of spermatogenesis.

The study of reproductive behaviour in reptiles and its possible hormonal control was pioneered by Noble & Bradley (1933) and Noble & Greenberg (1941). After a long gap there has been a significant revival of interest in this field (see Crews, 1975, 1977, 1979a, 1979b, 1980, 1983). Much of this work has been done with the iguanid green anole, *Anolis carolinensis*, which in many respects cannot be considered typical of most lizards. This lizard is common in the south-eastern parts of North America. Males emerge from winter dormancy in January and establish breeding territories and engage in aggressive bouts. As breeding territories are established and females begin to emerge, territorial males court the newly arrived females and, as shown in Fig. 3.12, the presence of intact males significantly increases the rate of environmentally-induced ovarian recrudescence. Receptivity to the attentions of the male is associated with follicular development and ovulation and female *Anolis carolinensis* undergo successive cycles of receptivity as alternate follicules mature and ovulate. As seen in Fig. 3.13, the ovaries in this species alternate in producing follicles, due to disproportionate supply of gonadotrophin to the largest ovarian follicle (Jones, 1975), and a single egg is laid every 10–14 days for the duration of the breeding season. Receptivity is restricted to the week just prior to ovulation and the female is nonreceptive during the three to four day period when the egg is being shelled in the oviduct.

Crews *et al.* (1978) investigated the rôle of testosterone, DHT and 17$\beta$-oestradiol in controlling reproductive behaviour in male *Anolis carolinensis* and found testosterone to be the only hormone capable of reliably restoring full behaviour in castrated individuals (see Fig. 3.14). Both testosterone and DHT, but not oestradiol, also stimulated hypertrophy and protein production by the renal sex segment, which is a secondary sex characteristic in males. The authors concluded that DHT is peripherally, but not centrally, active whereas oestradiol was ineffective in stimulating either behaviour or secondary accessory structures in this lizard. Intracranial implants of oestradiol benzoate, however, have been shown to induce sexual behaviour in castrated rats if DHT is administered systemically at the same time (Davis & Barfield, 1979). Fig. 3.15 shows that this treatment is also effective in some but not all male *Anolis carolinensis*. These data are of relevance to the 'androgen aromatisation hypothesis' of Naftolin *et al.* (1975) which proposes that the action of testosterone in inducing sexual behaviour is secondary to its conversion in the brain to oestrogen by aromatase enzymes. Callard *et al.* (1977, 1978) have identified aromatase enzymes in the brain of turtles and snakes and suggest that androgen aromatisation is a phylogenetically ancient property of the vertebrate nervous system.

The control of female receptivity by 17$\beta$-oestradiol of ovarian origin seems well-established in *Anolis carolinensis*. Ovariectomy abolishes sexual receptivity

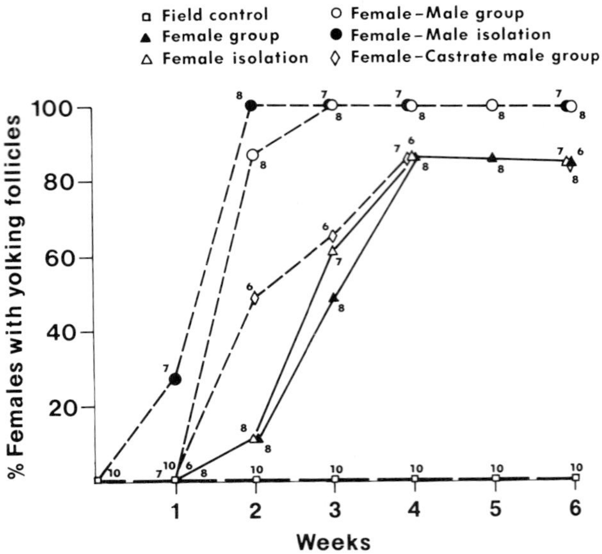


**Fig. 3.12** Patterns of ovarian growth in female *Anolis carolinensis* exposed to a stimulatory environmental régime alone or in different social situations. The sample size is shown next to each point (from Crews, 1975).

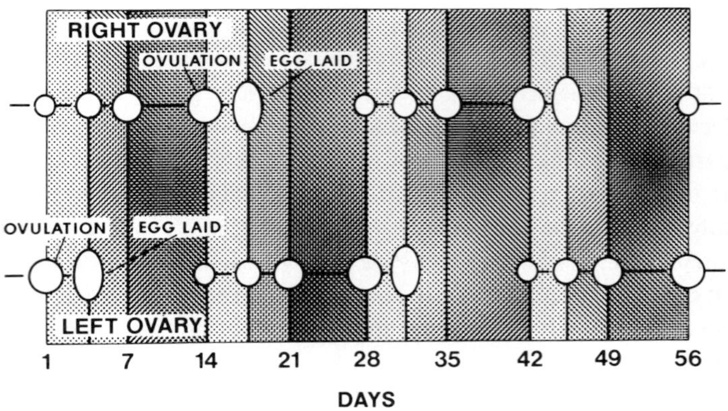


**Fig. 3.13** Sexual cycle of the female green anole lizard (*Anolis carolinensis*) is represented schematically. The light regions indicate when the female is sexually unreceptive, the next grade of stippling shows the period of onset of receptivity and the darkest colour represents the receptive period. The ovaries alternate in producing the follicles; its state of maturity is indicated by the size at which it is represented (from Crews, 1977).

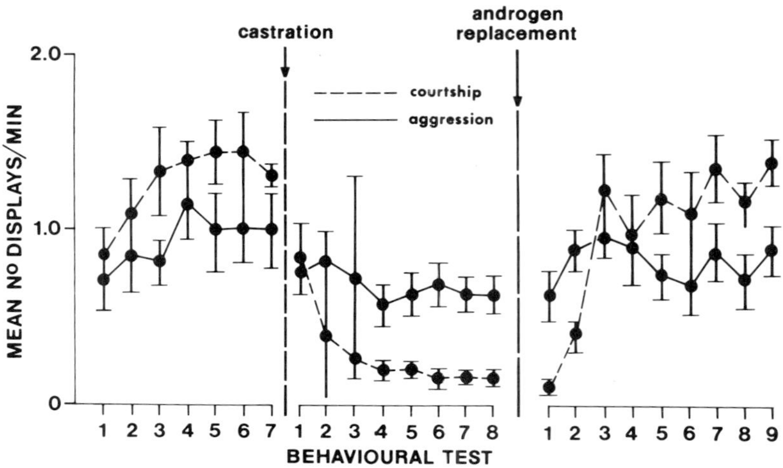


**Fig. 3.14** Effect of castration and testosterone replacement therapy on the sexual and aggressive behaviour of the male lizard, *Anolis carolinensis*. In these experiments, sexually-active males were castrated after baseline levels of sexual and aggressive behaviour were determined. Two weeks following castration, males were given silastic implants (0.06 cm i.d. × 0.12 cm o.d.) containing testosterone s.c. Males were tested daily first with a male intruder and then with a female intruder for 15 min. each. Mean and SEM shown (n = 12) (from Crews, 1979b).

and oestrogen therapy reinstates this behaviour in a dose-dependent manner (Fig. 3.16). Injections of progesterone alone do not activate sexual receptivity, but McNicol & Crews (1979) found that receptivity was enhanced if progesterone was given 24 hours after oestrogen priming. Tokarz *et al.* (1981) have also recently characterised high-affinity cytosolic binding sites for the synthetic progestin $^{3}$H-R5020 in the oviduct and hypothalamus of *Anolis carolinensis* and binding was induced by oestrogen pre-treatment, as is the case with classical progesterone receptors in mammals. Plasma progresterone levels are high in *Anolis* immediately before and after ovulation, and Crews (1979b) has suggested that progesterone of luteal origin may be responsible for the inhibition of receptivity in the female which occurs after mating and persists until the next follicle ovulates in the contralateral ovary (Crews, 1973). Experiments by Crews (1973) have shown that the presence of the ovaries is critical to coition-induced inhibition of female receptivity. He divides this abbreviation of oestrus behaviour into an *initiation phase*, induced by neurally mediated stimuli arising from intromission, and a *maintenance phase* in which the non-receptive state is presumably maintained by a change in circulating levels of gonadal hormones such as progesterone. Experiments by Tokarz & Crews (1981) and Crews (1983) suggest that prostaglandins such as $PGF_{2\alpha}$ may be responsible for the rapid *initiation phase* as suggested by the results shown in

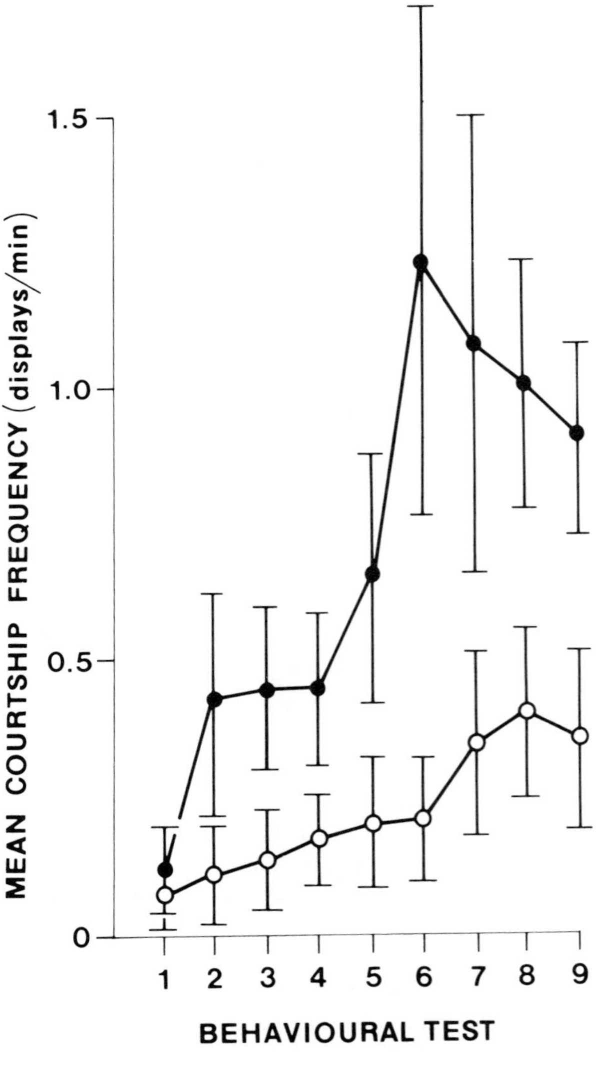


**Fig. 3.15** Effect of dihydrotestosterone and oestradiol in combination on male courtship behaviour of castrated, sexually-inactive *Anolis carolinensis*. Animals were assigned to one of two groups on the basis of their courtship behaviour following hormone administration. Group I (responders, filled circles, n = 10 included castrates exhibiting levels equivalent to or greater than that shown while intact. Group II (nonresponders, open circles, n = 10) included castrates exhibiting no significant change in courtship frequency (from Crews, 1980).

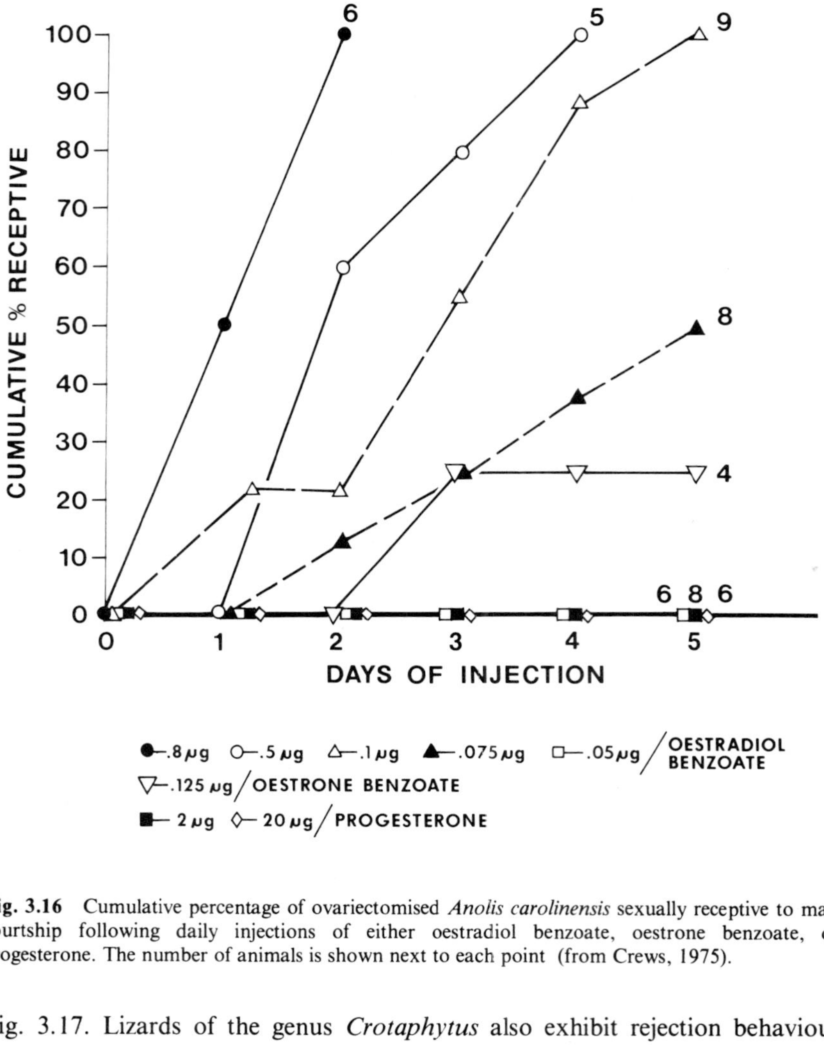


**Fig. 3.16** Cumulative percentage of ovariectomised *Anolis carolinensis* sexually receptive to male courtship following daily injections of either oestradiol benzoate, oestrone benzoate, or progesterone. The number of animals is shown next to each point (from Crews, 1975).

Fig. 3.17. Lizards of the genus *Crotaphytus* also exhibit rejection behaviour following ovulation which discourages males from mating and, coincident with this change in behaviour, gravid females rapidly acquire bright orange spots on the sides of the body and the flanks which they display to courting males. Cooper & Ferguson (1972a,b) found that progesterone and testosterone were both able to induce this coloration in non-gravid females and further studies have shown that 17$\beta$-oestradiol is ineffective (see Ferguson, 1976; Medica *et al.*, 1973).

The extensive work of Crews and his colleagues with *Anolis carolinensis* has done much to emphasise the critical importance of psychosocial as well as

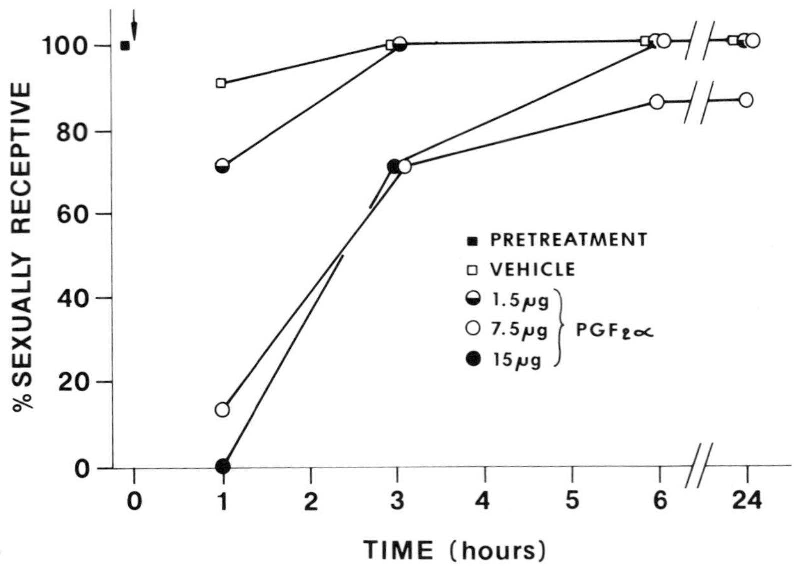


**Fig. 3.17** Effect of prostaglandin $F_{2\alpha}$ (PGF) on sexually-receptive green anole lizards, *Anolis carolinensis*. Ovariectomised, sexually-receptive females (square) which had been pretreated with 1.4μg oestradiol benzoate were injected (arrow) with varying doses of $PGF_{2\alpha}$ or vehicle and tested for sexual receptivity at 1, 3, 6 and 24 hours respectively. (from Crews, 1983).

hormonal cues in regulating reproduction in this species. The courtship behaviour of the male, for example, has an important influence on the reproductive physiology of the female and, as seen in Fig. 3.18, high levels of courtship frequency accelerate ovarian development in winter-dormant females, presumably through enhanced gonadotrophin secretion. The rôle of the ovary in directing the mating effort of the male has also emerged in recent studies where it is clear that males mount more frequently on the side opposite to the ovary containing the pre-ovulatory follicle and hence insert the appropriate hemipenis (see Fig. 3.19 and 3.20). The corpus luteum in *Anolis carolinensis* has also been shown to exert a local inhibitory effect on the ipsilateral oviduct, rendering it insensitive to the contractile effects of the neurohypophyseal peptide, arginine vasotocin (AVT) which is thought to be the hormone responsible for oviposition or parturition in reptiles (Jones *et al.*, 1982; La Pointe, 1969, 1977). Jones *et al.* (1983) have also shown that short-term release of adrenalin can inhibit uterine contractions and oviposition in *Anolis* and prostaglandin $F_{2\alpha}$ (PGF) has been identified by Guillette *et al.* (1984) as a luteolysin in this species. Hormonal control of the post-reproductive *refractory*

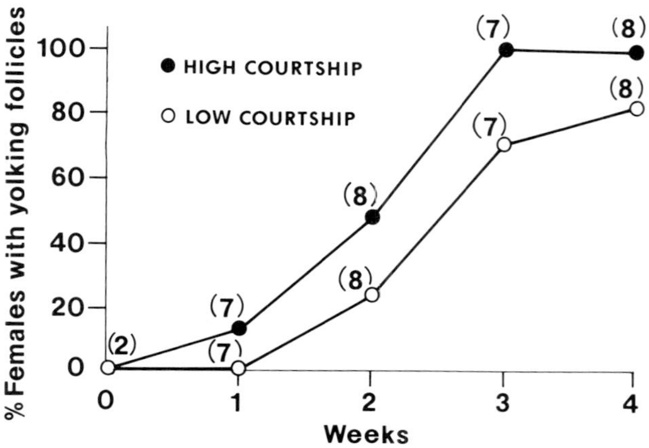


**Fig. 3.18** Ovarian response of winter-dormant female *Anolis carolinensis* exposed to high *versus* low frequencies of male courtship behaviour. All females were exposed to the same stimulatory environmental régime. Sample sizes are shown in parentheses (from Crews, 1980).

INTACT, REPRODUCTIVELY-ACTIVE FEMALES

a

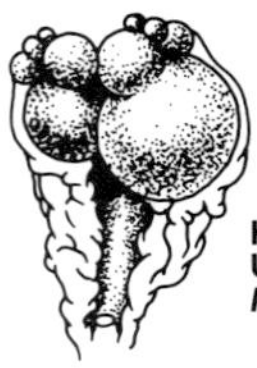

| HEMIPENIS USED DURING MATING | OVARY WITH LARGEST FOLLICLE: right | OVARY WITH LARGEST FOLLICLE: left |
|---|---|---|
| right | 5 | 0 |
| left | 1 | 8 |

p=.003, C=.65

INTACT, FSH-STIMULATED FEMALES

b

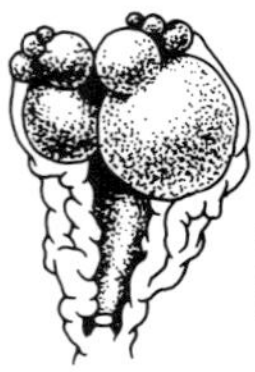

| HEMIPENIS USED DURING MATING | OVARY WITH LARGEST FOLLICLE: right | OVARY WITH LARGEST FOLLICLE: left |
|---|---|---|
| right | 8 | 4 |
| left | 3 | 9 |

p=.044, C=.39

**Fig. 3.19** Dissection of copulating female *Anolis carolinensis* captured during the breeding season reveals a significant correlation between the side on which the male mounts the female, and hence the hemipenis used, and the ovary containing the preovulatory follicle (a). Winter-dormant females that have received daily injections of 10μg mammalian follicle-stimulating hormone (FSH) for 10–14 days mate in a similar manner (b). Fisher's exact probability test and contingency coefficient (C) are shown (from Crews, 1980).

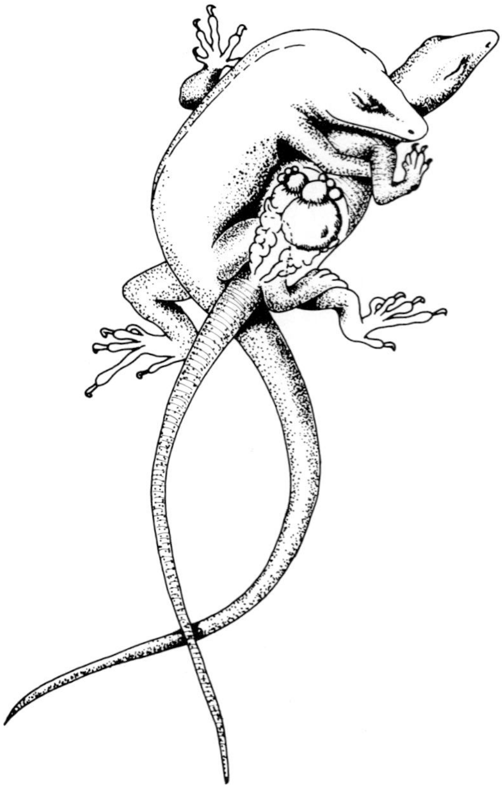

**Fig. 3.20** Schematic illustration of the relationship between the side on which male *Anolis carolinensis* mount females and the female's ovarian condition (from Crews, 1980).

*period* in reptiles has also revealed the importance of the ovary and, although alterations of the photo-period will overcome refractoriness in *Anolis carolinensis* (Licht, 1967, 1973) in contrast to lizards such as *Podarcis sicula* (Botte & Angelini, 1980), *Xantusia vigilis* and *Dipsosaurus dorsalis* (Licht, 1973), experiments by Crews & Licht (1974) suggest that the corpora atretica of *Anolis* secrete a substance which markedly reduces ovarian sensitivity to environmentally-induced gonadotrophin secretion.

The extent to which the large body of information on one tropical lizard species, *Anolis carolinensis*, can be generalised to other reptiles is debatable. Recent work by Crews (1983) with the garter snake, *Thamnophis sirtalis parietalis*, which has a *postnuptial* rather than a *prenuptial* pattern of

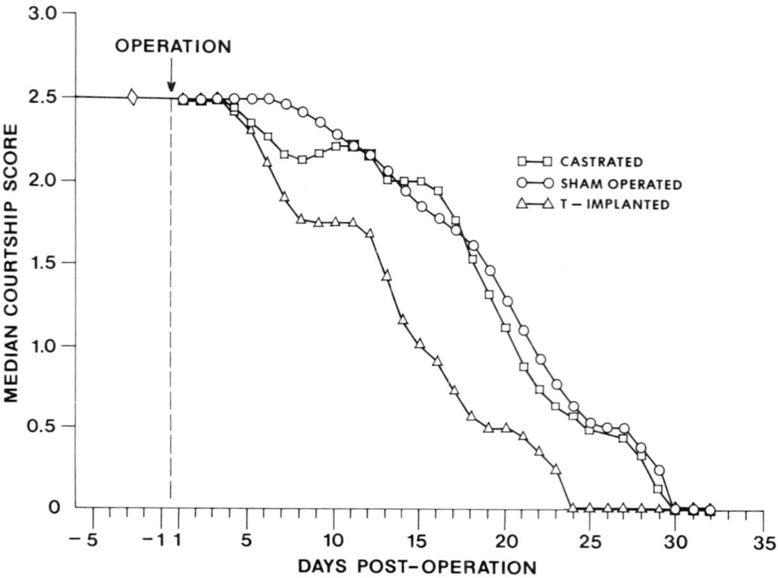


**Fig. 3.21** Effects of castration and androgen-replacement therapy on male courtship behaviour in the Canadian red-sided garter snake, *Thamnophis sirtalis parietalis*. Following castration (arrow), there is no difference in the decline in courtship of oestrogen-primed stimulus females by castrated, intact (sham-operated), or castrated males that have received testosterone replacement therapy. Male courtship score is a quantitative measure of chin-rubbing activity. Each dot represents the average of tests of all males on that day. Horizontal bar represents the mean test score for all tests during testing period. (from Crews, 1983).

reproduction suggests that such extrapolation would be premature. Spring mating occurs in this species following a long winter hibernation. At this stage the testes of the male are completely regressed and the sperm necessary for mating are stored in the epididymides. Males discriminate females on the basis of a pheromone released from the dorsal skin region. Production of the pheromone is stimulated by injections of oestrogen and can be induced in either males or females. Garstka & Crews (1981) have studied the production of this pheromone, which is carried in the blood of the female rather than being produced in an exocrine gland, and further studies by Garstka *et al.* (1982, 1983) suggest that the pheromone is produced in the liver and is identical with vitellogenin, the circulating precursor of yolk. Mating behaviour in the female may thus, ultimately, have an hormonal basis but the same cannot be said for the male of this species. In fact, as would be expected from the fact that mating males have regressed testes, castration and injections of testosterone had no effect on the course of courtship behaviour and mating in *Thamnophis* when tested by Camazine *et al.* (1980) as seen in Fig. 3.21. Crews & Garstka (1982) suggested that the pituitary gland might act directly on the brain to elicit sexual behaviour but hypophysectomised individuals continued to exhibit normal mating behaviour following hibernation. Injections of a great variety of hormonal substances, including AVT, dopamine, melatonin, LHRH, noradrenalin and adrenalin, thyroxine and various ions have all failed to elicit any behavioural responses in sexually inactive garter snakes.

Crews (1983) suggests that the most conservative conclusion to draw from these unusual findings in *Thamnophis* is that the neural systems subserving male sexual behaviour in this snake are activated either during, or prior to, the onset of hibernation when spermatogenic activity is at its maximum. (See also Licht, 1982, who has similarly suggested that mating behaviour in turtles may be 'primed' by elevated androgen levels in the summer preceding the mating season.) Further studies by Crews (1983) with parthenogenetic lizards of the genus *Cnemidophorus*, in which typical male mating behaviour can be induced by treatment with either testosterone or DHT, and male-like secondary sex characters can be induced by early androgen treatment, suggest that much more study of a greater variety of reptiles is needed before any attempt is made to generalise the data currently available.

## II. CLIMATIC CORRELATES OF REPRODUCTIVE PATTERNS IN REPTILES

The publication of a seminal article by Lamont Cole in 1954 on the "population consequences of life history phenomena" and the subsequent difficulty experienced in explaining his famous 'paradox' concerning the ubiquity of the iteroparous compared with the theoretically more-advantageous semelparous mode of reproduction, have led to the emergence of a body of theory concerned with the evolution of different 'tactics' promoting successful reproduction and, ultimately, species survival in diverse habitats. Reptiles have figured largely in this debate, due primarily to the important early studies of Tinkle (1969) and Tinkle *et al.* (1970) in which they attempted to discern the existence of discrete reproductive 'strategies' in lizards and account for their evolution. The important work of Williams (1966) on "adaptation and natural selection" also stimulated greatly the development of Fisher's early ideas on "reproductive effort" (Fisher, 1930) and a great deal of effort has been expended in recent years in an attempt to quantify these concepts. They are obviously of relevance to the question of how desert species of reptiles exploit the exigencies of environments which, at first sight, would appear inimical for their survival and long-term persistence. The utility of these concepts will be discussed in this section.

Tinkle *et al.* (1970) analysed literature records of some 88 species of lizards, 37 of which were relatively complete in terms of the data sought, and these were used to generate a Prim Network of clustered characters. Two major groups emerged: small, early-maturing species with relatively short life spans which produce a number of clutches of eggs each year; and relatively larger species, maturing later and producing only one clutch per year. Viviparity was seen as one extreme form of this latter category. If we re-examine their data it is evident that, although late-maturing single-brooded species predominate in temperature environments, many desert species occur in both categories (e.g. *Uma notata* and *Uma inornata* fall into the early-maturing, multiple-brooded category whereas *Xantusia vigilis* and many of the desert geckoes are characteristic of the second). These authors were unable to separate tropical from temperate species, but Barbault (1975) was able to show that fecundity is generally higher in tropical species and that they may be divided into two subgroups: hygrophilic forest dwellers, which produce small numbers of eggs throughout the year; and those living in the savanna which breed seasonally. This last group lays substantially larger clutches, so that overall fecundity of the two groups is much the same.

Weekes (1935), in an extensive review of modes of placentation in the Reptilia, first speculated on the possible ecological advantages of viviparity in different environments and noted that viviparous forms were common in both mountainous regions in south-east Australia and also in the dry interior of New South Wales. She commented that: " . . . the inland plain is very hot and dry, and species with parchment-shelled eggs exposed to the danger of desiccation would have supposedly fewer chances of survival than viviparous species with eggs safely carried within the female". Discussing lizards living in elevated situations, she noted " . . . Cold is suggested as the most likely external factor associated with high altitudes that may influence either directly or indirectly the adoption of viviparity". Thus began the debate on the evolution of the viviparous mode of reproduction in reptiles and, until recently, the commonly accepted view was that viviparity was an adaptation to the colder conditions prevalent at high altitudes and had in fact evolved in such environments (see Packard, 1966; Packard *et al.*, 1977; Shine & Bull, 1979) although aridity has also occasionally been invoked in some cases (e.g. Greer, 1977, when discussing the evolution of viviparity in the scincid genus *Lygosoma*, and Saint Girons & Saint Girons (1956) had already clearly resolved the problem as seen from the following quotation: "Le froid ne favorise pas spécialement l'évolution des espèces vers la viviparité, puisqu'elles sont tout aussi abondantes en plaine et dans les régions chaudes. Il élimine seulement les espèces ovipares en empêchant leurs oeufs de se développer". [p. 215]). Tinkle & Gibbons (1977) first questioned the cold-climate theory in an extensive analysis of the distribution of viviparous forms of reptiles throughout the world and pointed out that, although the *proportion* of viviparous species in the total reptilian fauna does increase as one moves towards the poles or high elevations, such faunae are in general impoverished and in fact in both North America and Australia, the *absolute numbers* of viviparous species increases monotonically as one proceeds south, towards warmer latitudes, rather than towards the north. As might be expected from their ectothermy, Tinkle & Gibbons (*op cit*) emphasise that reptiles, regardless of their mode of reproduction, decrease in both numeric and specific abundance as one progresses towards the poles or to higher elevations. The fact that, at extremes of latitude and altitude, the fauna is often composed almost entirely of only a few viviparous species, only suggests that, once having reached such localities, they are better able to survive there than oviparous species. These authors briefly examined the prevalence of viviparous forms in hot and xeric regions and found that examples of both oviparous and viviparous species were common. Similarly Guillette *et al.* (1980) in examining the question of the evolution of viviparity in the genus *Sceloporus* in North America found that xeric habitats were not exclusively occupied by oviparous forms, although they were significantly more abundant (Table XXXVII). Shine & Bull (1979) also found that viviparous forms were found about as frequently in dry habitats in China, Africa, Australia and North America as one would expect from the overall proportion of squamates occurring in such a habitat (Table XXXVIII), thus emphasising that this mode of reproduction is not a prerogative of species living in cold climates.

Shine & Berry (1978) utilised a stepwise multiple-regression procedure in order to investigate the strength of the correlation between a large number of climatic variables and geographic patterns in the distribution of viviparous species of reptiles in both Australia and North America. The geographical distribution of the proportion of viviparous species in the respective squamate faunae on the two continents is shown in Fig. 3.22 and the tendency for this proportion to increase in cool and northerly latitudes is immediately obvious. However, when the authors assessed the extent of the correlation with specific climatic, rather than geographic variables, they found that low temperature *per se* was of little significance. In fact, precipitation and evaporation measures

**Table XXXVII:** Contingency table comparing mode of reproduction of species of *Sceloporus* with environmental moisture (from Guillette *et al.* [1980]).

| Habitat | Oviparous | Viviparous |
|---|---|---|
| Mesic | 15 | 22 |
| Xeric | 30 | 10 |

Chi-square = 9.39 with 1 df ($P < 0.005$)

**Table XXXVIII:** Distributions of squamate species with respect to habitat and occurrence of parental care: Proportions in each category (from Shine & Bull, 1979).

| | All squamates | Live-bearers in genera with both egg-laying and live-bearing members (n = 38) | Species with egg retention (n = 60) |
|---|---|---|---|
| Cold climate | 0.44* (0.13–0.61) | ⩾0.74 | 0.72 |
| Wet soil | 0.12* (0.04–0.22) | 0.11 | 0.13 |
| Dry soil | 0.18* (0.03–0.35) | 0.24 | 0.05 |
| Aquatic | 0.05* (0–0.14) | 0.03 | 0.02 |
| Arboreal | 0.10* (0.02–0.28) | 0.08 | 0.08 |
| Maternal care (at generic level) | 0.11† | 0.24** | 0.33 |

* Based on date for 1,007 species from Africa, China, Australia, and North America, from Pitman (1974), Pope (1935), Cogger (1975), Conant (1958), and Stebbins (1966); brackets gives ranges of the four continental means. Categories were determined by the respective authors' criteria and their personal knowledge of the species and habitats (North America and Australia).

† Based on data for 700 egg-laying species (231 genera) in Fitch (1970).

** In egg-laying forms.

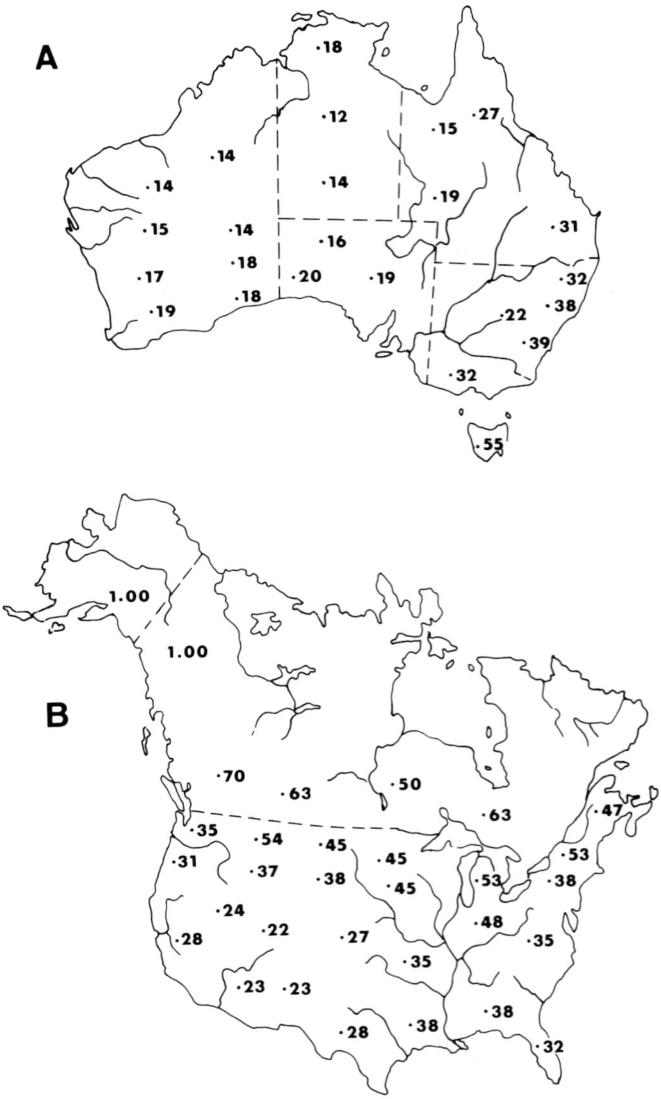


**Fig. 3.22** The geographical distribution of the proportion of live-bearing species in the squamate reptile faunas of Australia (A) and North America (B). Proportions for 17 of 48 North American areas are omitted for simplification (from Shine & Berry, 1978).

were often better predictors of the prevalence of species with a viviparous mode of reproduction than were temperature variables and they concluded that: " . . . over the broad range of environmental conditions encountered in Australia and America, the geographic distribution of live-bearing is only poorly correlated with environmental temperature". The results of a more detailed analysis are shown in Fig. 3.23 and reveal that the proportion of live-bearing squamates drops markedly with increasing temperature in colder climates, but this trend is much weaker or even absent in warmer climates. Few Australian environments are cold enough to reveal this trend and it is only in northern America and Europe that one finds typically impoverished faunae composed entirely of viviparous forms. Shine & Berry (1978) conclude that their data effectively refute the hypothesis that reptilian live-bearing is primarily important as an adaptation to cold climates and maintain that it should be viewed instead as a strategy of high-parental investment which permits occupancy of a wide variety of different habitats. The potential advantages of viviparity are obvious—more stable conditions for incubation, avoidance of egg predation and competition for suitable nest sites etc.—and add weight to their conclusion. In a more recent ecophysiological study Shine (1983a) provides evidence that the eggs of oviparous species would not be able to develop successfully if laid in the ground in high-elevation areas of south-eastern Australia, and suggests that viviparity probably *first evolved* in response to such environmental restraints but that its subsequent radiation into animals living in vastly different environments had little to do with climatic factors.

Shine (1983b) has further analysed available data on lizards and measured the proportion of the total development period for which eggs are retained *in utero* in a variety of species of squamates and found that most are in effect mid-way between 'true oviparity' and 'true viviparity', retaining eggs *in utero* for about half of the total period of embryonic development. This underlines the extent to which egg retention is an extremely plastic character and subject to selection. Egg retention has always been viewed as a necessary intermediate stage in the evolution of viviparity (e.g. Packard *et al.*, 1977) and both viviparous and oviparous forms have even been reported to occur in the one species, such as the Mexican lizard *Sceloporus aeneus* (Guillette, 1981, 1982). Blackburn (1982) notes that viviparity appears to have arisen on at least 45 separate occasions in the Sauria, with 22 in the Scincidae, 10 in the Iguanidae, 5 in the Anguidae, 2 each in the Lacertidae and Gekkonidae and one each in the families Chamaeleontidae, Xantusiidae, Agamidae and Cordylidae. It is notable that viviparity has never evolved in a number of families (e.g. Varanidae, Pygopodidae, Teiidae, Dibamidae and Helodermatidae). Blackburn (1982) suggests that phylogenetic constraints which operate in these cases include temperature-dependent sex determination rather than the male-heterogametic type, and the formation of highly-calcified eggshells which militates against any form of egg retention. The question of the nature of the selective forces which

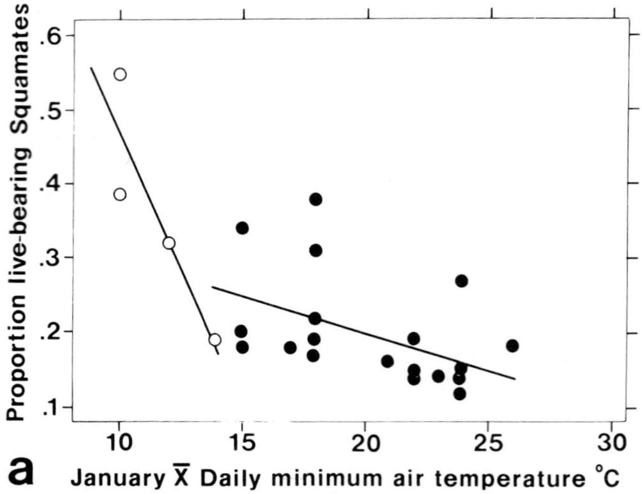


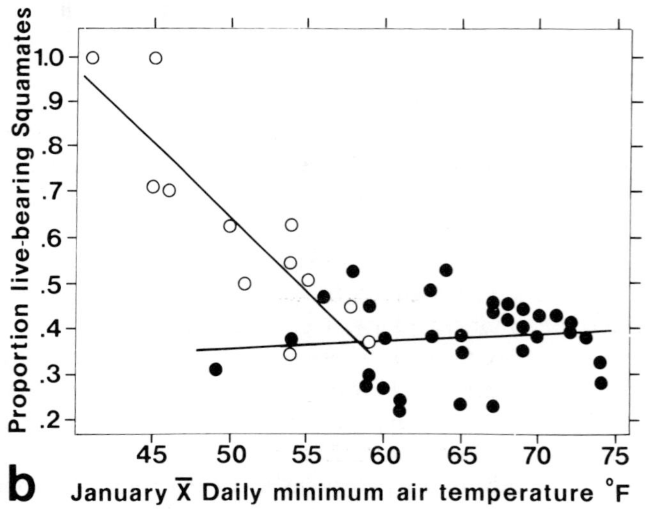


**Fig. 3.23** The relationship between midsummer mean daily minimum air temperature, and the percent of live-bearing species in the squamate fauna, for areas in Australia (a) and North America (b). Australia: open circles = areas above 33° S latitude: closed circles = all other areas. North America: open circles = areas above 45° N latitude: closed circles = all other areas (from Shine & Berry, 1978).

**Table XXXIX:** Mode of reproduction and aridity in Australian Scincidae*.

| Genus | No. of species | No. arid | Mode of reproduction of arid species |
|---|---|---|---|
| *Anomalopus* | 10 | 0 | — |
| *Carlia* | 15 | 2 | oviparous (O) |
| *Cryptoblepharus* | 2 | 1 | O |
| *Ctenotus* | 41 | 26 | O |
| *Egernia* | 22 | 9 | viviparous (V) |
| *Emoia* | 2 | 0 | — |
| *Eugongylus* | 2 | 0 | — |
| *Hemiergis* | 8 | 1 | V |
| *Leiolopisma* | 11 | 0 | — |
| *Lerista* | 35 | 21 | O |
| *Menetia* | 1 | 1 | O |
| *Morethia* | 6 | 6 | O |
| *Notoscincus* | 4 | 3 | O |
| *Proablepharus* | 2 | 1 | O |
| *Pseudomoia* | 2 | 0 | — |
| *Saiphos* | 1 | 0 | — |
| *Sphenomorphus* | 19 | 3 | O |
| *Tiliqua* | 8 | 4 | V |
| *Tropidophorus* | 1 | 0 | — |

* Data set provided by courtesy of Dr R. Shine, University of Sydney.

have led to the repeated evolution of the viviparous mode of reproduction in diverse groups of reptiles is still very much under debate and recent reviews by Shine (1985) and Blackburn (1985) revive to some extent the 'cold-climate hypothesis' but, as Shine admits: ". . . these results are based on correlations, . . . (and) . . . causation is inferred but cannot be demonstrated."

The family Scincidae is the group in which the viviparous mode of reproduction reaches its highest development and it is of interest to examine the occurrence of both oviparous and viviparous forms in relation to aridity of habitat. In Australia, only the scincid lizards exhibit viviparity and species occupying arid regions are grouped by their characteristic mode of reproduction in Table XXXIX.

It is evident from Table XXXIX that both oviparous and viviparous skinks are common in arid regions of Australia and the mode of reproduction characteristic of a given genus would appear to have little to do with the aridity of the environment. Twenty-six of the 41 species of *Ctenotus*, for example, occur in arid habitats and all are oviparous, as are all the species in this genus. All of the 22 *Egernia* species on the other hand are viviparous with 9 occurring in arid regions. This compares with 21 of the 35 *Lerista* species, all of which, however, are oviparous.

## III. MEASUREMENTS OF REPRODUCTIVE EFFORT IN DESERT REPTILES

Although there has been considerable theoretical interest in the concept of reproductive effort (RE) in recent years, empirical data are still too few to test many of the predictions from the various models that have been proposed to account for differences in reproductive tactics (see Congdon *et al.*, 1982, for a review of this subject in reptiles and Table XL). Difficulties have also been experienced, too, in finding adequate means for measuring RE which was loosely conceived by Fisher (1930) as the extent of an organism's investment in any current act of reproduction. Pianka (1976) suggests that an '*optimal reproduction tactic*' would maximise an individual's *reproductive value* (the sum of all present plus the expected probable number of all future offspring) at every age, and that RE should therefore vary inversely with residual reproductive value. Although there is some evidence to support this in lizards (e.g. Tinkle, 1969; Pianka & Parker, 1975; but see Congdon *et al.*, 1982), Clutton-Brock (1984) argues that RE often does not increase towards the end of the lifespan in many birds and mammals where reproductive value declines with age, and suggests that there has been an overemphasis placed on measuring the *energetic costs* of reproduction (usually expressed in terms of the number, weight or kilojoule content of eggs or neonates) rather than the *survival costs* of reproduction, measured in terms of the effects of breeding on the individual's future probability of survival and breeding potential. These 'costs' may be very high as shown by Shine (1980) who found that gravid females in a number of species of Australian skinks were more vulnerable to predation than non-gravid females, due to their propensity to bask more frequently and their significantly reduced running speeds.

Thus, although most workers agree that RE is defined and measured as "that proportion of the total energy budget of an organism that is devoted to reproductive processes" (Hirshfield & Tinkle, 1975), Shine 1980 argues that such energy-allocation measures are unlikely to offer useful estimates of RE, because the 'trade-off' they assess (fecundity *versus* growth) is probably of minor importance in the evolution of reptilian life-history tactics. Pianka (1976) similarly stresses that 'risks' associated with reproduction are not assessed with energetic estimates of RE. Nevertheless, these remain the only quantitative estimates that we have of RE at the present time and it is of interest to examine these, especially as some have been concerned with desert species of lizards. Tinkle & Hadley (1975) presented data for total energy budgets of three species of iguanid lizards, which are shown in Table XLI, together with the fraction of energy devoted to reproduction. This varied from 10% in *Sceloporus jarrovi* to 24% in two-year-old *Uta stansburiana*, the latter species being one which

**Table XL:** Selected predictions of major life history theories (from Congdon, Dunham & Tinkle, 1982).

| Theory | Selective factors | Relative RE | Lipid storage |
|---|---|---|---|
| *r* and *K* selection | Resource supply/demand ratio | | |
| | Ratio high → *r* selection | High | Low |
| | Ratio low → *K* selection | Low | Low |
| | Demographic environment | | |
| | High extrinsic adult mortality | High | Low |
| Bet-hedging | Juvenile mortality | | |
| | Relatively constant | High | Low |
| | High and unpredictable | Low | High |
| | Variable but predictable | Low in unfavourable periods | High |
| | | High in favourable periods | Low |

**Table XLI:** Total energy budgets for different age classes of reproductives in three species of iguanid lizards. Figures are total calories except for proportions shown in parentheses (from Tinkle & Hadley, 1975).

| | First breeders | Second breeders | Third breeders |
|---|---|---|---|
| *Sceloporus jarrovi* | | | |
| Maintenance | 21,720 | 47,040 | 72,210 |
| Growth | 1,877 | 11,768 | 7,138 |
| Reproduction | 2,530 (0.10) | 7,261 (0.11) | 11,958 (0.13) |
| **Total:** | 26,127 | 66,069 | 91,306 |
| *Uta stansburiana* | | | |
| Maintenance | 31,020 | 34,680 | — |
| Growth | 1,156 | 694 | — |
| Reproduction | 7,747 (0.19) | 8,394 (0.19) | — |
| **Total:** | 39,923 | 43,768 | — |
| *Sceloporus graciosus* | | | |
| Maintenance | 23,040 | 26,640 | 30,540 |
| Growth | 96 | 710 | 480 |
| Reproduction | 6,812 (0.23) | 8,630 (0.24) | 9,517 (0.23) |
| **Total:** | 29,948 | 35,980 | 40,537 |

**Table XLII:** Reproductive effort of free-living *Uta stansburiana* in southern Nevada during 1975–1976 (from Nagy (1983)).

| | First clutch | Second clutch | Breeding season | Annual |
|---|---|---|---|---|
| 1-year females: | | | | |
| Reproduction | | | | |
| Chemical energy to eggs ($R_c$) | 8.3 | 11.2 | 19.5 | 19.5 |
| Respiration cost ($R_R$) | 9.0 | 16.3 | 25.3 | 25.3 |
| Total ($R_r$) | 17.3 | 27.5 | 44.8 | 44.8 |
| Ingestion (I) | 54.3 | 41.4 | 95.7 | 170.5 |
| Metabolisable (M) | 40.1 | 32.6 | 72.7 | 132.5 |
| Storage (S) | 2.7 | $-2.7^b$ | 0.0 | 6.9 |
| $R_c\,(M-S)^{-1}$ | 0.222 | 0.317 | 0.268 | 0.155 |
| $R_R\,(M-S)^{-1}$ | 0.241 | 0.462 | 0.348 | 0.201 |
| $R_r\,I^{-1}$ | 0.319 | 0.664 | 0.468 | 0.263 |
| $R_r\,(M-S)^{-1}$ | 0.463 | 0.779 | 0.616 | 0.357 |
| 2-year+ females: | | | | |
| $R_c$ | 8.3 | 7.9 | 16.2 | 16.2 |
| $R_R$ | 18.1 | 11.0 | 29.2 | 29.2 |
| $R_r$ | 26.4 | 18.9 | 45.4 | 45.4 |

| | | | | |
|---|---|---|---|---|
| I | 38.4 | 54.6 | 93.0 | 168.7 |
| M | 29.0 | 42.8 | 71.8 | 132.3 |
| S | −2.6 | 0.6 | −2.0 | 2.4 |
| $R_c\,(M-S)^{-1}$ | 0.263 | 0.187 | 0.220 | 0.125 |
| $R_R\,(M-S)^{-1}$ | 0.573 | 0.261 | 0.396 | 0.225 |
| $R_r\,I^{-1}$ | 0.688 | 0.346 | 0.488 | 0.269 |
| $R_r\,(M-S)^{-1}$ | 0.835 | 0.448 | 0.615 | 0.349 |
| Males: | | | | |
| $R_R \cong R_r$ | — | — | 21.5 | 26.6 |
| I | — | — | 95.4 | 198.7 |
| M | — | — | 69.5 | 148.8 |
| S | — | — | 1.7 | 6.8 |
| $R_r\,I^{-1}$ | — | — | 0.255 | 0.134 |
| $R_r\,(M-S)^{-1}$ | — | — | 0.317 | 0.187 |

Numerical values shown are either rates of energy flow, in kJ per animal per total time period[a] or ratios of various flow rates.

a. Time periods are first clutch, 1-year females, February 22 to May 3 (71 days); first clutch, 2-year+ females, February 22 to April 22 (60 days); second clutch, 1-year females, May 4 to June 18 (46 days); second clutch, 2-year+ females, April 23 to June 18 (57 days); breeding season, February 22 to June 18 (117 days); and annual, January 1 to December 31 (365 days).

b. Negative signs for storage values indicate that the sum of chemical energy in somatic lipid and other somatic tissues, and fat bodies declined during that time interval.

occupies a more arid habitat. In a more exhaustive study of *Uta stansburiana* in southern Nevada, Nagy (1983) has used the stable isotope $^{18}O$ to measure rates of respiration associated with reproduction alone, and derived estimates of RE expressed as the fraction of available energy flow used for reproductive purposes (chemical potential energy of the eggs plus associated respiratory costs). Data from this study, which are the most extensive available for any reptile, are shown in Table XLII. Estimates of RE are higher, ranging from 45% to 85% of available energy being channelled into reproduction in females *during the breeding season*. The RE expressed as a fraction of the annual energy budget (which brings it into line with other estimates of RE in the literature) is 34% for *Uta stansburiana*. Nagy (1983) concedes that, although it is difficult to assess the risk of mortality as an additional energetic cost of reproduction, the values presented should include any additional metabolic costs due to predator avoidance but that they probably still exclude some risks, such as those due to disease and old age, which are not measurable in energetic terms.

There are few other data available from which to assess the RE of desert reptiles compared with that of species living in more equitable environments, but Andrews (1979) has estimated RE in the tropical lizard *Anolis limifrons* in Panama and suggests that the *absolute* allocation of energy to reproduction in this species is higher than that of temperate-living species, but not the *fraction* of its total energy budget so diverted. At this stage it seems prudent to conclude that, although the concept of RE is a powerful one, its precise measurement at this time is still problematic and there are insufficient data available from which one may speculate reliably. It would seem reasonable to anticipate, however, that, to the extent that desert habitats represent hazardous environments for reptiles (i.e. through the combination of both significant abiotic and biotic restraints to survival) one might expect inhabitants of such regions to devote more of their available energy to reproduction than species occupying less rigorous situations. This assumes, however, that rates of mortality are intrinsically higher in desert than in non-desert environments. This has yet to be shown to be the case and it may well prove to be that only the *modalities* of mortality differ, rather than its overall severity.

## IV. CONCLUSION

Reproductive patterns in reptiles appear to conform to two basic types—'*prenuptial*', where gametogenesis immediately precedes the mating period, and '*postnuptial*' where gonadal recrudescence occurs after the breeding season and sperm are stored in the epididymides for extended periods. These two patterns correspond to Crews' (1984) '*associated*' and '*dissociated reproductive tactics*', the latter being quite common in snakes and chelonians, but relatively rate in lizards. Hormonally, the interesting difference between these two strategies concerns the induction of mating behaviour which, in the case of the postnuptial cycle, occurs when circulating sex steroids are typically at their lowest level during the year. In snakes such as *Thamnophis*, for example, vigorous sexual behaviour can proceed in the absence of either the gonads or the pituitary gland. This is a challenge to conventional paradigms.

Vernal-Prenuptial cycles are characteristic of reptiles inhabiting desert areas of the world, but these do not appear to be a *specific adaptation* to this type of habitat, being instead a result of the longer and more intense period of high temperature available to these animals which assure that the complete gonadal cycle can be accommodated within the one season (spring). 'Aestival' and 'Mixed' cycles also occur to some extent in arid and semi-arid regions, but they are essentially characteristic of species living in more temperate areas which, because of the limitations imposed by shorter warm periods, are forced to apportion parts of the gonadal developmental cycle to different seasons, and often different years. This emphasises the important rôle played by temperature rather than photoperiod in regulating gonadal cycles in reptiles (see Licht, 1972; Duvall *et al.*, 1982) and underlines a major difference between environmental cueing of reproduction in birds and reptiles.

Despite the prevalence of the Vernal-Prenuptial cycle in desert reptiles, some small lizards such as, for example, *Acanthodactylus spp.* appear capable of breeding almost continuously throughout the year and it is only the viviparous mode of reproduction which is rarely encountered in arid and hyper-arid regions. Almost without exception, snakes and lizards that are most commonly associated with desert environments in North Africa, North America and Australia are all oviparous. Efficient means of ensuring that the eggs are protected from desiccation during the period of incubation in the soil must be widespread in these animals. Recent reviews of the occurrence and distribution of the viviparous mode of reproduction in diverse species of reptiles suggest that it is not constrained so much by environmental temperature *per se*, although it is a typically successful strategy in cold and mountainous regions, and should be viewed instead as a high-parental investment strategy which permits occupancy of a wide variety of habitats. The only family of lizards to exhibit viviparity in Australia is the skinks, and both viviparous and oviparous forms occur commonly in arid situations, and are often found living sympatrically. Australia however lacks hyper-arid regions such as occur throughout much of the Sahara desert and in which it would appear that low-parental investment strategies are more appropriate for long-term persistence of reptiles. The scientific study of the 'costs' and 'benefits' associated with reproduction is still in its infancy, but reptiles have already proven of value in developing theoretical insights and refining measures of 'reproductive effort'. Future investigations, particularly of desert-living species, are likely to prove rewarding.

# 4

# Growth and Population Dynamics of Desert Reptiles

It is a commonplace that many species of desert reptiles are able to maintain large and apparently stable populations of individuals, despite the obvious rigours of their environment. The quantitative study of growth and recruitment is essential for any understanding of the dynamics of such populations. Just as natural populations only increase through natality and immigration, so also they only decline through the processes of death and emigration of their component individuals, and a knowledge of the *causes* and *incidence* of mortality in the field is equally important in interpreting population phenomena. Little is known of the aetiology of death in reptiles, as is the case with most natural populations of wild animals, but this problem will be addressed in Chapter 5 through the study of individual case histories. In this chapter, attention will be focused upon the interaction between growth and life-history phenomena.

## I. GROWTH

One of the most extensive studies of growth in a reptile is that of Tinkle (1967) on the side-blotched lizard *Uta stansburiana* in western Texas. The study extended from 1960 until 1966. He found that hatchlings grow initially at quite a slow rate, but accelerate as they begin to feed. Growth rates for males and females until sexual maturity varied between 0.28–0.32 mm/day and 0.17–0.27 mm/day respectively. From a review of the available literature, Tinkle (1967) concluded that growth rates of 0.2–0.3 mm/day could be taken as the 'average' growth rates for lizards, although he noted that considerable variability exists. Individual variability in rates of growth was also a feature of the *Uta stansburiana* data and Tinkle commented that "some animals show practically no growth and others grow at more than twice the mean rate". He speculated that this may result from differences in food supply. There were also significant differences between the mean rates of growth of *Uta* in different years, with higher growth rates occurring in years of high population density. This effect was again attributed to variations in available food. Blair's (1960) study of the rusty lizard, *Sceloporus olivaceus*, in central Texas also gave evidence of considerable individual variation in growth rates as yearling females ranged in size from 63 mm to 107 mm and in weight from 16 g to 26 g and similar differences were found with males. Unfortunately, Blair (1960) did not pursue the source of this variation other than to suggest that it was probably due to dietary factors.

Variations in environmental temperature and food availability are the factors most commonly invoked in the literature to account for individual variability in growth rates of reptiles (e.g. Cagle, 1946) and Gibbons (1967) suggested that differences in rate of growth of different populations of the painted turtle *Chrysemys picta* could be explained by variations in the quality of the food available. Andrews (1976) found that insular populations of *Anolis* lizards grow at slower rates than mainland populations and suggested that food is more limited in the island situations, although no quantitative data were presented to support this hypothesis. Schoener & Schoener (1978) compared the efficacy of three different growth models (Von Bertalanffy, logistic-by-length, and logistic-by-weight) in describing growth data derived from four species of *Anolis* from the West Indian islands of Abaco and Bimini and concluded that the logistic-by-weight model gave the best fit to the data and that among-population variations in *individual* growth rate reflected differences in food availability. A similar analysis of growth in the iguanid lizard *Sceloporus merriami* was carried out by Dunham (1978) in an effort to test the hypothesis that variation in food availability is a proximate factor influencing individual growth rates in this lizard. Abundance of arthropod prey was estimated by two techniques and found to be serially correlated with seasonal precipitation and, when taken over a three-year period, variations in food availability resulted in predictable variation in individual growth rates which were positively correlated with active season and annual precipitation, as well as estimates of food availability. Krekorian (1984) studied a marked population of the desert iguana (*Dipsosaurus dorsalis*) at Thousand Palms in California over a period of seven years and found that growth during high-rainfall years was twice that observed in low rainfall years. He attributed this to the nutritional effects of spring and summer-flowering plants stimulated by the rainfall. These data are consistent with earlier studies in which a causal relationship was inferred between growth and food abundance in the field (see Ballinger, 1977; Smith, 1977; Medica *et al.*, 1975), but a more recent study by Stamps & Tanaka (1981) suggests that the nature of the interaction may be more complex. They studied the influence of water and food on growth rates in the tropical lizard *Anolis aeneus* and found that water deprivation may limit juvenile growth, even when food is abundant. Such an effect is of course well-documented in mammals where water deprivation leads inevitably to reduced food intake and, if prolonged, retarded growth. Hormonal factors could also be involved as Licht (1967) found that injections of exogenous prolactin stimulated somatic growth in *Anolis carolinensis* and corticosteroids such as corticosterone are known to have marked effects on appetite in mammals.

Andrews (1982) has reviewed the rôle of both extrinsic and intrinsic factors as determinants of rates of growth in reptiles and concluded that the effects of annual variation in food availability on reptilian growth are best documented for arid regions in which community productivity is highly variable from year to year. There is a common assumption in the literature that herbivorous species grow less rapidly than carnivores (see MacCulloch & Secoy, 1983), and this is supported by the observation that growth rates in many primarily herbivorous turtles can be stimulated by the addition of protein-rich foods to their diet (e.g. Hulse, 1976; Gibbons, 1967; Parmenter, 1980). Andrews (1982) points out however that strictly herbivorous lizards such as *Iguana iguana* grow at rates of from 0.25–0.6 mm/day, which are as high as those reported in any carnivorous species, and that herbivorous lizards such as *Dipsosaurus dorsalis* and *Sauromalus obesus*, with slow rates of growth (see Parker, 1972; Nagy, 1973; Berry, 1974), occupy desert habitats where both herbivores and carnivores typically have reduced rates of growth. Although it is a commonplace that temperature has a marked influence on rate of growth in reptiles, the nature of this interaction has been little analysed and the importance of intrinsic rhythms is again highlighted by studies such as that of Jackson (1970) in which it was found that individuals of the turtle *Pseudemys concinna*, living in a constant-temperature spring, only grew between March and November, despite favourable conditions throughout the whole of the year.

Mention has already been made of attempts to describe growth in terms of various mathematical models (e.g. logistics, Gompertz and von Bertalanffy equations) and Case (1978), in a major review of the adaptive significance of postnatal growth rates in terrestrial vertebrates, found that growth rate is a function of mass in reptiles with an exponent of 0.67. Andrews 1982, however, recalculated this relationship using body mass at the time growth rate was measured (rather than adult body mass as employed by Case [1978]), and derived a relationship as follows:

$$y = 0.016x^{0.75}$$

where: $y$ = maximum rate of mass growth ($d\mathrm{W}/dt$) in g/day, and $x$ = mass in g. The relationship between growth in length and mass is more complicated, however, because of the great variation in *shape* of reptiles, and, arguing from theoretical grounds, Andrews (1982) predicts that rates of growth in length of snakes, crocodiles, lizards and turtles should differ in the ratios of 1.0:0.5:0.4:0.2 respectively. Coefficients for the Mass/Length function $\mathrm{W} = c\mathrm{L}^d$, where W = mass and L = length are shown for a number of representative species of reptiles in Table XLIII from which it is clear that there is little variation in the exponent '*d*', which ranges from 2.16–3.30, and that most of the divergence is produced by group-typical variation in the intercept '*c*' which ranges, for example, from $1.6 \times 10^{-3}$ in *Chelonia mydas* to as low as $6.9 \times 10^{-8}$ in the snake, *Crotalus atrox*. It must be borne in mind however that these equations are purely descriptive devices and, as Parks (1970) has observed, " . . . the explanation of planetary phenomena had to await the universal gravitation concept and the explanation of Mendel's counts had to await the molecular biology of DNA, so the explanation of growth must await the explication of the aetiology of growth."

**Table XLIII:** Patterns of growth in reptiles (from Andrews 1982).

| | c | d | Source ** |
|---|---|---|---|
| Turtles | | | |
| *Chelonia mydas* | $1.6 \cdot 10^{-3}$ | 2.60 | Caldwell, 1962 |
| *Chrysemys picta* (M > 80 mm) | $8.2 \cdot 10^{-3}$ | 2.16 | Wilbur, 1975a |
| *Gopherus agassizi* | $4.9 \cdot 10^{-4}$ | 2.80 | Jackson *et al.*, 1976 |
| Lizards | | | |
| *Anolis limifrons* | $4.8 \cdot 10^{-5}$ | 2.73 | Andrews, 1979b |
| *Basiliscus basiliscus* (F) | $0.9 \cdot 10^{-5}$ | 3.26 | Van Devender, 1978 |
| *Cyclura carinata* | $3.2 \cdot 10^{-5}$ | 3.03 | Iverson, 1979 |
| *Sceloporus occidentalis* (M) | $3.5 \cdot 10^{-5}$ | 3.01 | Davis, 1967 |
| Crocodilians | | | |
| *Crocodylus porosus* ( < 198 mm) | $3.4 \cdot 10^{-4}$ | 2.47 | Webb and Messel, 1978 |
| *Crocodylus porosus* (199–394 mm) | $8.5 \cdot 10^{-6}$ | 3.15 | Webb and Messel, 1978 |
| *Crocodylus porosus* ( > 394 mm) | $3.7 \cdot 10^{-6}$ | 3.29 | Webb and Messel, 1978 |
| Snakes | | | |
| *Carphophis vermis* (F) | $3.8 \cdot 10^{-6}$ | 2.58 | Clark, 1970 |
| *Crotalus atrox** | $6.9 \cdot 10^{-8}$ | 3.30 | Klauber, 1972 |
| *Heterodon platyrhinos* (M)* | $8.5 \cdot 10^{-7}$ | 2.88 | Platt, 1969 |

Coefficients of the mass (W) – length (L) function, $\mathrm{W} = c\mathrm{L}^d$, for representative species of reptiles.

* Total length.

Mass is in grams and lengths are in millimetres. Lengths for turtles are of plastron or carapace and for other groups are snout-vent. M (male) and F (female). ** Cited in Andrews (1982)

Little is known, unfortunately, of *intrinsic* factors which influence rate of growth in reptiles and particularly those which may account for the wide variability of *individual* growth rates which is such a conspicuous feature of the literature on reptilian growth. Some of this variability is undoubtedly due to extrinsic factors such as spatial variation in food availability, but observations showing that such variability in growth may persist when animals are reared under constant conditions in the laboratory suggests a genetic basis. Tinkle (1967), for example, found that *Uta stansburiana* grew at rates varying from 0.02–0.27 mm/day under uniform laboratory conditions with continuous food supply, which were just as variable as those rates measured on individuals in the field. Fig. 4.1 graphically illustrates this variation in the turtle *Chrysemys picta* where hatchling mass is compared with that achieved five months later after the hatchlings were housed individually and fed uniformly. As may be seen, some individuals increased as much as threefold in mass whereas others did not grow at all. Andrews (1976) reported similar variations in rate of growth of anoline lizards and Bradshaw (1971) studied this problem in some detail in the Australian agamid, *Amphibolurus ornatus*. Growth rates of individually-marked animals were recorded over a number of years at Bakers Hill in Western Australia and, as shown in Fig. 4.2, considerable individual variability is manifest. Some individuals of both sexes (about 50% of the cohort) grow rapidly and reach sexual maturity within their first year whereas others (about 40%) do not breed until their second year and an even smaller proportion (< 10%) take three years to reach sexual maturity. Growth rates varied from 0.01–0.23 mm/day and, although such variation appeared to be continuous, Bradshaw (1971) categorised individuals as being either *fast-growers* or *slow-growers*, with the latter taking more than one year to achieve sexual maturity. Survivorship of 'fast' and 'slow-growers' was found to differ markedly in the field in a long-term study of the population dynamics of *Amphibolurus ornatus* at Bakers Hill. Baverstock (1975) attempted to account for this differential mortality through a study of the physiology of individuals with differing rates of growth. No differences were observed in metabolic rate, behavioural temperature regulation or electrolyte concentrations in the erythrocytes, but saline loading affected urinary and evaporative water losses differentially. As seen in Fig. 2.19 'fast-growers' lost body weight when given daily injections of sodium chloride which were either isosmotic or hyperosmotic to the body fluids (150–300 mmol/L NaCl) and voided significantly greater volumes of urine. Urinary sodium concentrations did not differ with the various régimes, but, because of the enhanced urinary losses, total sodium excretory rates were higher in 'fast' than in 'slow-growers'. Evaporative water loss was also enhanced in 'fast-growers' by salt loading and this factor also contributes to the greater weight loss of these individuals when subjected to chronic salt loading.

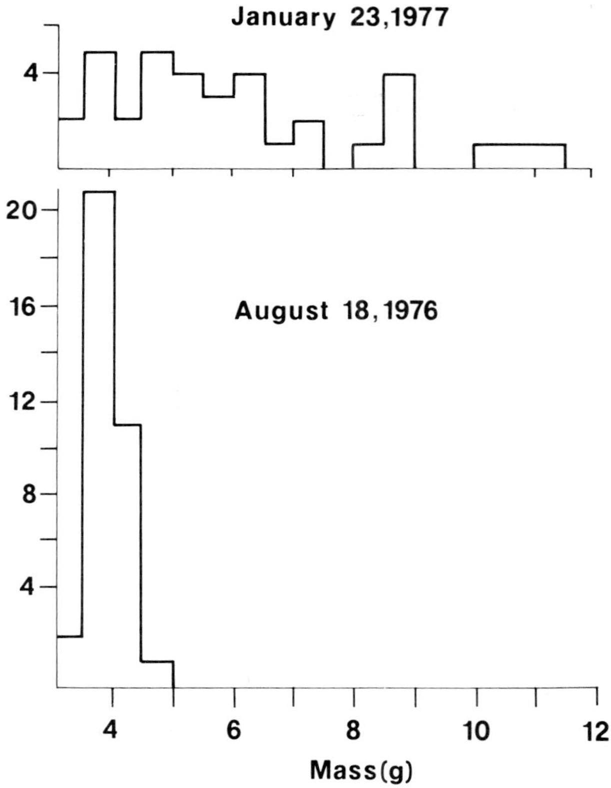


**Fig. 4.1** Mass of *Chrysemys picta* at hatching on August 18, 1976 (lower figure) and on January 23, 1977 (upper figure) (from Andrews, 1982).

Baverstock & Bradshaw (1975) studied adrenal corticosteroidogenesis in 'fast' and 'slow-growing' *Amphibolurus ornatus* under both field and laboratory situations and found that plasma levels of corticosterone increase to a greater extent in 'slow-growers' when subjected to salt loading. As seen in Fig. 4.3, this inverse relationship between corticosterone production and rate of growth is also evident when individuals are stressed by water deprivation and would appear to result from impaired adrenocortical function in the faster-growing animals. These authors investigated the possibility that the higher corticosterone levels in 'slow-growers' were themselves responsible for the slow rate of growth, as steroids in high doses are known to inhibit growth in

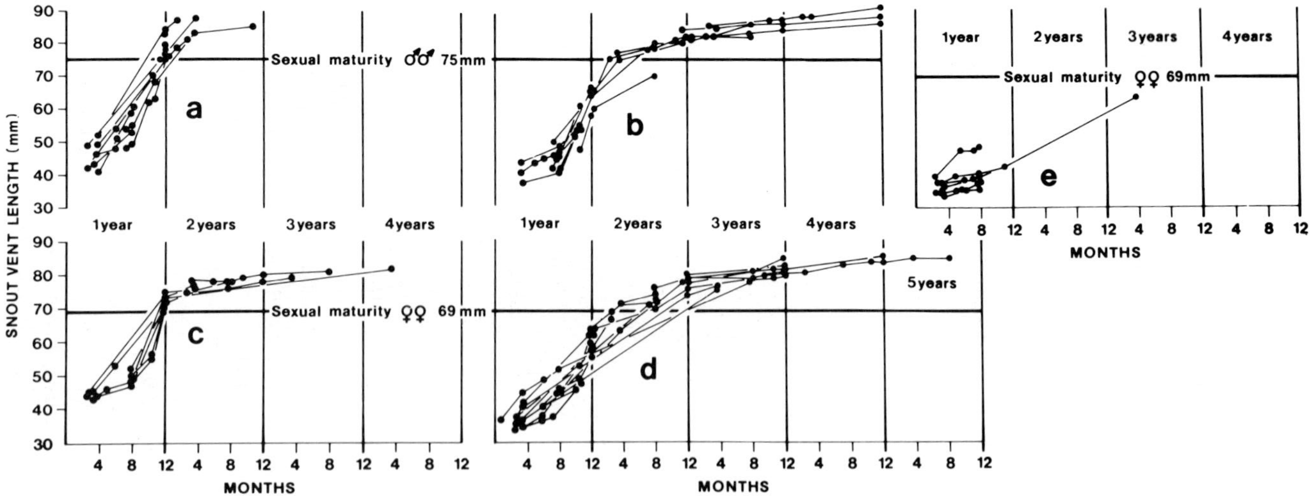


**Fig. 4.2** Growth records from individually-marked *Amphibolurus ornatus* at Bakers Hill from 1961–1964. (a) Type A: fast growers—male; (b) Type B: slow growers—male; (c) Type A: fast growers—female; (d) Type B: slow growers—female; (e) Type C: exceptionally-slow growers—sex unknown (from Bradshaw, 1971).

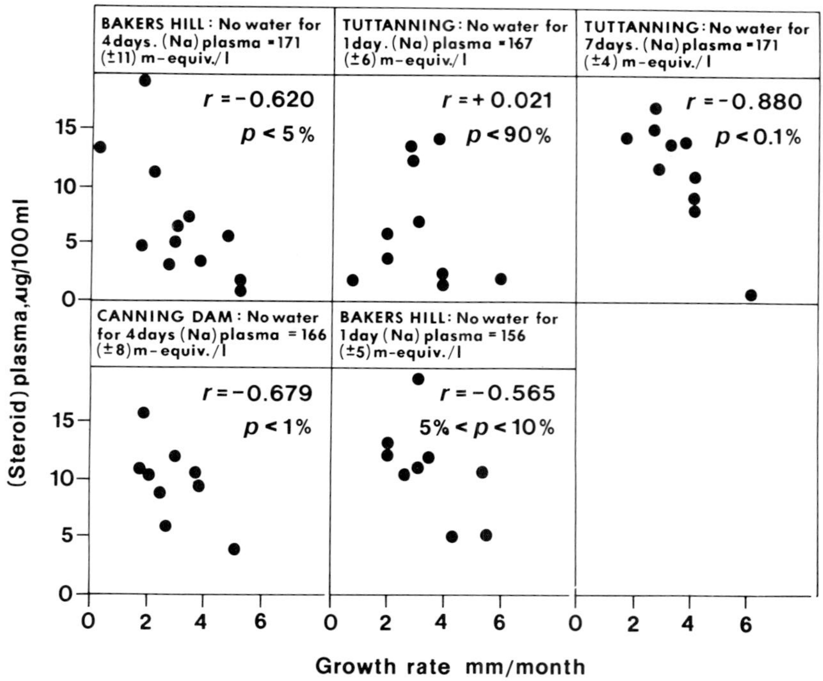


**Fig. 4.3** Relationship between corticosteroid concentration in the plasma and growth rate of juvenile *Amphibolurus ornatus* reared in the laboratory. Values of r indicate the parametric correlation coefficients. P indicates the probability that the observed correlation is due to chance on the null hypothesis that there is no relationship between corticosteroid concentration and growth rate (from Baverstock & Bradshaw, 1975).

mammals, but this proved not to be the case and corticosterone, in physiological doses, stimulated growth in adrenalectomised juveniles. The marked difference between individuals with differing rates of growth thus appeared to result from a more responsive pituitary–adrenal axis in the case of the 'slow-growers' and Baverstock & Bradshaw (1975) speculated that selection for rapid growth in this species, which presumably involves enhanced secretion of growth hormone and prolactin, may have resulted in an impairment in the ability of the pituitary gland to secrete other hormones, such as adrenocorticotrophic hormone (ACTH) and arginine vasotocin (AVT) which are essential for the maintenance of physiological homeostasis. Although there is no evidence to substantiate this hypothesis at the present time, it would account satisfactorily for the inability of these 'fast-growers' to restrict water loss from the kidneys when salt loaded (AVT function) and their attenuated response to stress (ACTH function). Certainly it provides a physiological rationale with which to account for the observed differential mortality of 'fast' and 'slow-growers' of this species when confronted with environmental stresses (see later).

Baverstock (1972) attempted to breed *Amphibolurus ornatus* in the laboratory, in an effort to show that individual differences in rate of growth were genetically based, rather than reflecting environmental influences on a relatively plastic phenotype, but his efforts were unfortunately not successful. Some success, however, was achieved in rearing juveniles in the laboratory from eggs laid by females of known growth rate, and the most important result to emerge from these studies was that growth rates of juveniles reared in the laboratory reflect faithfully the population from which they had been derived. Bradshaw (1971) predicted that enhanced variability would prove a characteristic of *Amphibolurus ornatus* populations inhabiting climatically-unpredictable environments and Baverstock (1972) was able to confirm this. Lizards from Canning Dam, close to Perth, Western Australia, for example, which has a relatively high rainfall but a very predictable dry summer, grew at a slower and less variable rate than individuals from Bakers Hill, which has a lower rainfall but a less predictably dry summer (see Fig. 4.4). Thus the variability in growth rates of juveniles from populations such as Bakers Hill is either genetically determined, or conditioned by the heterogeneity of the environment to which the eggs are briefly exposed before being brought to the laboratory.

Further evidence that variations in rate of growth are genetically conditioned comes from a study of the lizards *Sceloporus graciosus* and *Sceloporus undulatus* by Ferguson & Brockman (1980) in which hatchlings from the two species were reared under identical social, thermal-gradient and feeding régimes, and found to grow at significantly distinct, population-specific rates. Table XLIV shows that the two subspecies of *Sceloporus undulatus* from Texas and Kansas grew at high rates both in the laboratory and in the field whereas growth rates were reduced in the South Carolina subspecies and in the Utah population of *Sceloporus graciosus*. Growth in the field of individuals from these latter two populations was also considerably slower than that measured in the laboratory and indicates perhaps that food was limiting over the period of the field studies. Ferguson & Brockman (1980) suggest therefore that juvenile growth rate differences measured in the field result partly from genetic differences and partly from differences in the 'quality' of the growing environment which may vary both spatially as well as temporally.

There has been considerable speculation in the literature that observed variations in rate and pattern of growth in reptiles are adaptive, fitting animals to particular lifestyles and environments, but quantitative data which might bear on the problem are few. Mautz (1979) has suggested that the extremely low rate of growth of the xantusiid lizards, for example, represents an

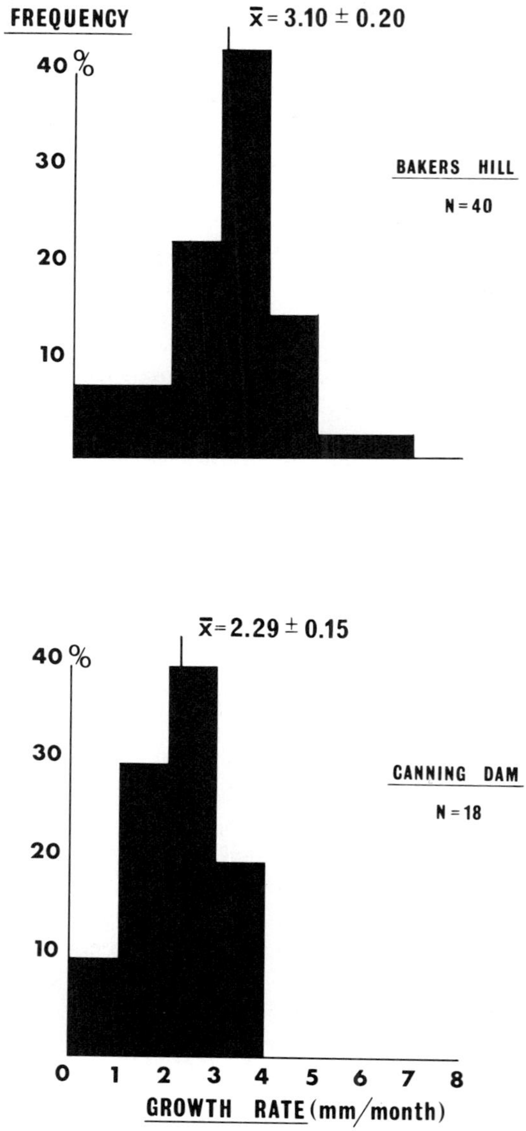


**Fig. 4.4** Frequency distributions for individual growth rates of *Amphibolurus ornatus* in the field at Bakers Hill and Canning Dam in Western Australia (from Bradshaw, 1981).

**Table XLIV:** Growth in laboratory; growth and survival in the field (from Ferguson & Brockman, 1980).

| Origin | Species | Growth in g/day in the laboratory $\bar{x} \pm$ S.E. (n) | A. Growth in mm/day in the laboratory $\bar{x} \pm$ S.E. (n) | B. Growth in mm/day in the field | % B of A | Annual adult survival in field |
|---|---|---|---|---|---|---|
| Texas | *Sceloporus undulatus consobrinus* | 0.038 ± 0.003 (11) | 0.39 ± 0.019 (11) NS | 0.37 | 95 | 0.11 |
| Kansas | *Sceloporus undulatus garmani* | 0.027 ± 0.003 (7) NS | 0.42 ± 0.039 (7) | 0.34 ± 0.01 (111) | 81 | 0.27 |
| South Carolina | *Sceloporus undulatus undulatus* | 0.032 ± 0.001 (16) | 0.32 ± 0.007 (16) | 0.13 | 41 | 0.49 |
| Utah | *Sceloporus graciosus graciosus* | 0.020 ± 0.001 (10) | 0.22 ± 0.014 (10) | 0.14 | 64 | 0.50 |

Values for the first two variables were compared between populations and were different at least at the 0.05 level except where NS appears. (ANOVA and SNK test.)

adaptation to a reclusive mode of existence, but it has not been demonstrated that they lack the potential for more rapid growth in the laboratory. The rhynchocephalian tuatara *Sphenodon punctatus* is another example of a reptile with an extremely slow rate of growth. Crook (1975) has cited rates as low as 10 mm/year, but it is not at all clear that this is not simply a consequence of the cold and wet habitat of the offshore islands of New Zealand where this animal is now found, limiting opportunities for more rapid growth. Although adequate comparative data on growth rates in reptiles are not available, it seems reasonable to predict that desert reptiles, with low rates of metabolism and water turnover (Congdon *et al.*, 1982; Nagy, 1982) will grow, on the average, more slowly than species inhabiting resource-rich environments if they are at all long-lived (examples of such desert lizards are *Uromastix, Varanus, Dipsosaurus*, and *Sauromalus*—all of which are relatively large animals and live for at least a number of years). An equally 'successful' strategy adopted by a number of desert reptiles, on the other hand, is to grow extremely rapidly whilst resources are available, breed, and then die in the ensuing summer with the population being maintained over the summer period by the egg phase. The mean longevity of lizards such as *Uta stansburiana* studied by Tinkle (1967) is no more than four weeks (the time taken for the cohort to be reduced by 50%) and few individuals survive for more than one year. Another example is the Australian agamid lizard *Amphibolurus nuchalis* studied by Bradshaw (1981) at Shark Bay in Western Australia. Many of the *Amphibolurus* appear to be 'annual' species. Thus both '*r*-selected' and '*K*-selected' species, in the terminology of Pianka (1970a), occur in arid environments and both are 'successful' in terms of their persistence. Although high individual variability in growth rate seems, from a perusal of the literature, to be almost a feature of reptiles, *Amphibolurus ornatus* appears to be the only species where the consequences of such variation have been investigated. This study is of interest because, although the *advantages* of rapid growth of individuals are obvious in the context of the population, hitherto unsuspected physiological *disadvantages*, in terms of reduced ability to withstand stresses associated with salt loading and dehydration, have also been revealed. It is impossible to predict at this stage, in the absence of similar studies, whether such 'physiological polymorphism' is common in other species of reptiles. What is clear however is that it is not an adaptation to arid environments *per se*, but appears to be a response of *Amphibolurus ornatus* populations which occupy fluctuating and climatically unpredictable environments. The coefficient of variation (CV) for growth, for example, in the lizard *Sceloporus orcutii*, which occupies an arid environment near Riverside, California, was only 4% (Weintraub, 1969) and Dmi'el (1967) reported a CV of 5% for the snake *Spalerosophis cliffordi* in the Israeli desert. It is evident that more detailed study is required of the growth of survivorship of a much greater variety of reptilian species, and in a specifically ecophysiological context, before it will be possible to generalise.

## II. POPULATION DYNAMICS AND THE INTERPRETATION OF LIFE-HISTORY PHENOMENA

Saint Girons (1965) early identified the major problems confronting herpetologists when attempting to analyse dynamic changes in natural populations of reptiles: sampling difficulties, and the need to know the age of individuals captured. Estimates of population size based upon mark-and-recapture methods require that samples be 'random' (i.e., that each animal in the population has an equal probability of capture, and thus an equal chance of being represented in the sample) and this is almost never achieved in reality. The majority of reptiles studied, and thus represented in the literature, are relatively large, diurnal species which are active foragers. It follows from this that juveniles are usually poorly represented, as are species which hunt their prey by ambush, and samples are frequently biased towards the more active of the two sexes. Estimating the chronological age of reptiles has always been a difficult process and, despite the promise of methods based on annual growth rings which are very successful in fish, early work on this problem with reptiles (Fukada, 1961) has not been substantiated (Griffiths, 1962; Gibbons, 1976) and mark-and-recapture methods over a long period remain the only satisfactory means of determining the age of free-ranging reptiles. Saint Girons (1975), reviewing the same field 10 years later, concluded that little had changed and that the excellent work of many herpetologists was devalued by a lack of appreciation of these fundamental problems. As an example of how critical this problem of ageing animals in the field may be, an examination of individual growth records for the lizard *Amphibolurus ornatus* in Fig. 4.2 shows that some individuals grow so slowly (Type C) that, at the end of one year, they are still no larger than hatchlings, and would be classified as such if they had not been marked individually. In many studies, therefore, where the age of animals collected in the field is estimated from their size, errors of considerable magnitude may be unwittingly promulgated.

Pearl & Miner (1935) first brought into focus the problem of comparing, in an adequate manner, schedules of reproduction and mortality for organisms with vastly different life spans and the subsequent development and application of *life tables* owes much to the imaginative work of Deevey (1947), who first generated these for a variety of natural populations of animals. One of the major limitations in comparing information on the dynamics of different populations of reptiles is that life tables have been erected for so few species and, without this detail, meaningful comparisons are very limited. Turner (1977) has reviewed available data on density and population dynamics of reptiles and, as seen in Fig. 4.5, observed densities vary enormously, from 0.15/ha in the desert agamid *Uromastix acanthinurus* to over 1000/ha in

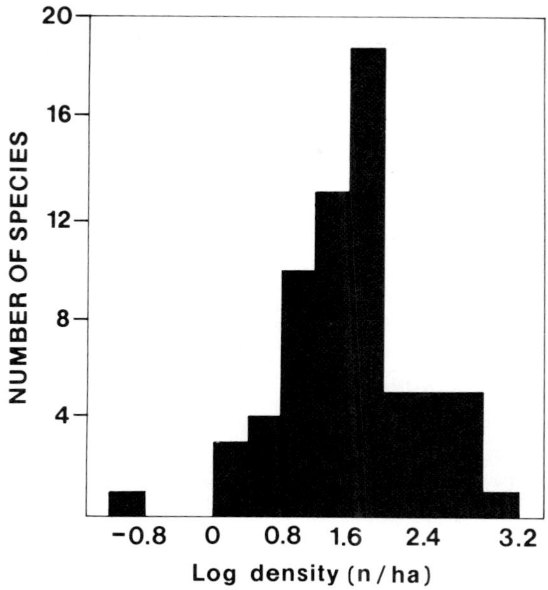


**Fig. 4.5** The range and distribution of estimated densities of lizards. The lowest estimate (0.15/ha) pertains to *Uromastix acanthinurus* in the Sahara Desert; the highest (1000/ha) to tropical and subtroptical anoles. (from Turner, 1977).

tropical and subtropical anoles. The density for a given species will of course vary both spatially and temporally, but Table XLV shows that interspecific differences far exceed intraspecific variation and annual variations for a given species are relatively modest, rarely exceeding 100% increases or 50% reductions in numbers.

Turner (1977) also investigated the proposition that survivorship in reptiles is a function of body size, as it is a truism in the literature that large animals live longer but, after a careful analysis of the available data, he concluded that no significant correlation exists. Taking data from reliable mark-and-return studies on lizards, he regressed the natural logarithm of the number of survivors ($N$) of an initial cohort of 100 individuals at time ($t$) = 0 on $t$ in months, such that:

$$\ln N = \alpha t + \beta \quad \text{or} \quad N = ke^{\alpha t}$$

**Table XLV:** Serial annual estimates of the density of seven species of lizards (from Turner, 1977).

| Species | Reference * | Initial year of series | Annual density estimates (n/ha) | | | | | | | |
|---|---|---|---|---|---|---|---|---|---|---|
| *Cnemidophorus sexlineatus* | Fitch, 1958 | 1953 | 164.7 | 99.7 | 121.3 | 177.7 | | | | |
| *C. tigris* | Turner *et al.*, 1969a, and unpubl. | 1963 | 14.1 | 11.3 | 9.9 | 18.1 | 15.2 | 15.7 | 16.9 | 21.6 |
| *Amphibolurus ornatus* | Bradshaw, 1971 | 1961 | 30.6 | 35.3 | 30.0 | 20.0 | | | | |
| *Crotaphytus wislizenii* | Turner *et al.*, 1969b | 1964 | 2.9 | 2.5 | 3.7 | 2.8 | | | | |
| *Sceloporus olivaceus* | Blair, 1960 | 1952 | 61.8 | 53.4 | 49.6 | 44.0 | 48.7 | | | |
| *Uta stansburiana* (Texas) | Tinkle, 1967a | 1962 | 50.6 | 32.1 | 54.3 | 30.9 | | | | |
| *Uta stansburiana* (Nevada) | Turner *et al.*, 1970, and unpubl. | 1966 | 56.8 | 80.3 | 42.5 | 34.6 | 42.8 | 23.5 | 24.6 | 43.6 |
| *Xantusia vigilis* | Zweifel and Lowe, 1966 | 1950 | 46.9 | 49.4 | 61.8 | 54.3 | 44.5 | 27.2 | | |

Only studies yielding at least four consecutive annual estimates are considered.

* Cited in Turner (1977)

and from this, values of $\alpha$ can be translated to annual percent survival (S) as follows:

$$S = e^{12\alpha}$$

Data from this analysis are shown in Table XLVI and, when annual survival rates were tested for correlation with body size, they were just marginally significant, but almost 90% of the variation remained unexplained. Turner (1977) points out that large lizards almost invariably enjoy high survival rates, but that small lizards may exhibit either low or high rates. This is due primarily to the confounding effect of lizards occurring in the southern latitudes of the United States with exceptionally *poor survival* (e.g. *Uta stansburiana* and *Sceloporus olivaceus* in Texas, *Sceloporus undulatus* in Kansas and Texas and *Mabuya buettneri*), which is balanced by the relatively *high survival* of lizards living in montane situations (e.g. *Sceloporus graciosus* in Utah and California, *Sceloporus virgatatus* and *undulatus* in Utah, and *Eumeces fasciatus* and *Xantusia vigilis*). Turner (1977) attributes such marked differences in survivorship to variations in predation pressure.

The above analysis is of course predicated on the assumption that individual survival is not age-dependent (and hence the significant linear correlation between survivorship and time for each species). Although this seems to be generally the case in reptiles, a number of studies has revealed evidence of age-specific changes in survival rates. Fig. 4.6 shows that survival is poor in juvenile *Sceloporus poinsetti, Uta stansburiana, Amphibolurus fordi* and *Sceloporus jarrovi*, but not in *Lygosoma laterale*. Interestingly, mortality rates ($1000q_x$) also increase in later life in both *Amphibolurus fordi* and *Lygosoma laterale*. The overall mortality curve of *Amphibolurus fordi* is very reminiscent of that for a mammal (see Caughley, 1966). The whole question of physiological senescence, however, is an open one in reptiles (see Gibbons, 1976 for a review) and all that can be gleaned from an analysis of relative life-spans is that the squamates (snakes and lizards) live for substantially shorter periods of time than do turtles and crocodilians, which are amongst the longest-lived of all animals, with a life-span exceeding mammals and even birds of prey (see Fig. 4.7).

Survivorship may also vary between different individuals within the one population and, as mentioned above when discussing growth in reptiles, Bradshaw (1971) found that survivorship of 'fast' and 'slow-growing' *Amphibolurus ornatus* at Bakers Hill in Western Australia was not uniform, and evidence of differential survival was found during periods of seasonal stress. 'Slow growers' were found to be particularly frost-sensitive as juveniles, but drought-hardy as adults. 'Fast growers' on the other hand suffered comparatively little juvenile mortality during winter but succumbed during long

**Table XLV1:** Estimated annual survival rates of lizards (from Turner (1977)).

| Species | Body weight (g) | $\alpha$ | Annual survival ($e^{12\alpha}$) | $r$ | Reference* |
|---|---|---|---|---|---|
| Agamidae | | | | | |
| *Amphibolurus fordi* | 3.5 | −0.1722 | 0.13 | −0.972 | Cogger, 1969 |
| *A. ornatus* | 22.0 | −0.0961 | 0.32 | −0.945 | Bradshaw, 1971 |
| *Phrynocephalus helioscopus* | 3.0 | −0.2966 | 0.03 | −0.907 | Darevsky, 1960 |
| *Draco volans* | | −0.1342 | 0.20 | | Alcala, 1966 |
| Iguanidae | | | | | |
| *Anolis cupreus* | | −0.3990 | 0.01 | −0.928 | Fitch, 1973b |
| *A. humilis* | | −0.5819 | 0.00 | −0.907 | op. cit. |
| *A. intermedius* | | −0.9316 | 0.00 | −0.966 | op. cit. |
| *A. limifrons* | | −0.6721 | 0.00 | −0.947 | op. cit. |
| *A. tropidolepis* | | −0.2602 | 0.04 | −0.957 | op. cit. |
| *Crotaphytus collaris* | 40.0 | −0.0617 | 0.48 | −0.930 | Fitch, 1956 |
| *C. wislizenii* | 27.0 | −0.0510 | 0.54 | | Turner *et al.*, 1969b |
| *Sceloporus graciosus* (Calif.) | 7.5 | −0.0382 | 0.63 | −0.989 | Stebbins, 1944, 1948 |
| *S. graciosus* (Utah) | 7.0 | −0.0634 | 0.47 | −0.995 | Tinkle, 1973 |
| *S. graciosus* (Utah) | 6.0 | −0.0818 | 0.38 | −0.988 | Burkholder & Tanner, 1974 |
| *S. jarrovi* | 25.0 | −0.0914 | 0.36 | −0.985 | Ballinger, 1973 |
| *S. olivaceus* | 30.0 | −0.1507 | 0.16 | −0.996 | Blair, 1960 |
| *S. poinsetti* | 50.0 | −0.0704 | 0.43 | −0.938 | Ballinger, 1973 |
| *S. undulatus* (S. Car.) | 16.0 | −0.0770 | 0.40 | −0.984 | Tinkle & Ballinger, 1972 |
| *S. undulatus* (Texas) | 12.0 | −0.2324 | 0.06 | −0.959 | op. cit. |
| *S. undulatus* (Ohio) | 20.0 | −0.0882 | 0.35 | −0.922 | Tinkle & Ballinger, 1972 |
| *S. undulatus* (Utah) | 13.0 | −0.0703 | 0.43 | −0.981 | Tinkle, 1972a |
| *S. undulatus* (Kansas) | 15.0 | −0.2205 | 0.07 | −0.957 | Ferguson & Bohlen, 1972 |
| *S. variabilis* | | −0.3905 | 0.01 | −0.977 | Fitch, 1973b |
| *S. virgatus* | 4.8 | −0.0763 | 0.40 | | Vinegar, 1975 |

| | | | | | |
|---|---|---|---|---|---|
| *Uta palmeri* | [illegible] | −0.0844 | 0.56 | −0.996 | Tinkle, 1970 |
| *Uta stansburiana* (Texas) | 3.6 | −0.2658 | 0.04 | −0.959 | Tinkle, 1967b |
| *Uta stansburiana* (Nevada) | 3.2 | −0.1274 | 0.22 | −0.997 | Turner *et al.*, 1970 |
| Lacertidae | | | | | |
| *Lacerta taurica* | | −0.0155 | 0.83 | −0.963 | Cruce & Leonte, 1973 |
| *L. vivipara* | 3.3 | −0.1342 | 0.20 | | Avery, 1975 |
| *Takydromus tachydromoides* | 3.8 | −0.1179 | 0.24 | −0.984 | Telford, 1969 |
| Scincidae | | | | | |
| *Lamprolepis smaragdina* | | −0.0498 | 0.55 | | Alcala, 1966 |
| *Emoia atrocostata* | 18.0 | −0.0950 | 0.32 | | Alcala & Brown, 1967 |
| *Eumeces fasciatus* | 6.5 | −0.0578 | 0.50 | | Fitch, 1954 |
| *Eumeces obsoletus* | 30.0 | −0.0477 | 0.56 | −0.998 | Hall, 1971; Hall & Fitch, 1972 |
| *Leiolopisma cherriei* | | −0.6684 | 0.00 | −0.978 | Fitch, 1973b |
| *Mabuya buettneri* | 5.8 | −0.3669 | 0.01 | −0.851 | Barbault, 1973, 1974b |
| *M. maculilabris* | 6.0 | −0.1488 | 0.17 | −0.881 | Barbault, op. cit. |
| *M. multifasciata* | | −0.0529 | 0.53 | | Alcala, 1966 |
| *Panaspis nimbaensis* | 2.0 | −0.1494 | 0.17 | −0.981 | Barbault, 1973, 1974b |
| *Scincella lateralis* | 1.5 | −0.1447 | 0.18 | −0.974 | Brooks, 1967 |
| Teiidae | | | | | |
| *Cnemidophorus deppii* | | −0.3684 | 0.01 | −0.673 | Fitch, 1973b |
| *C. sexlineatus* | 9.5 | −0.0732 | 0.41 | −0.998 | Fitch, 1958 |
| *C. tigris* | 20.0 | −0.0767 | 0.40 | −0.995 | Turner *et al.*, 1969a, and unpubl. |
| Xantusiidae | | | | | |
| *Xantusia vigilis* | 1.0 | −0.0544 | 0.52 | −0.868 | Zweifel & Lowe, 1966 |

Survivorship data have been fitted to an exponential model assuming age-constant mortality.
* cited in Turner (1977).

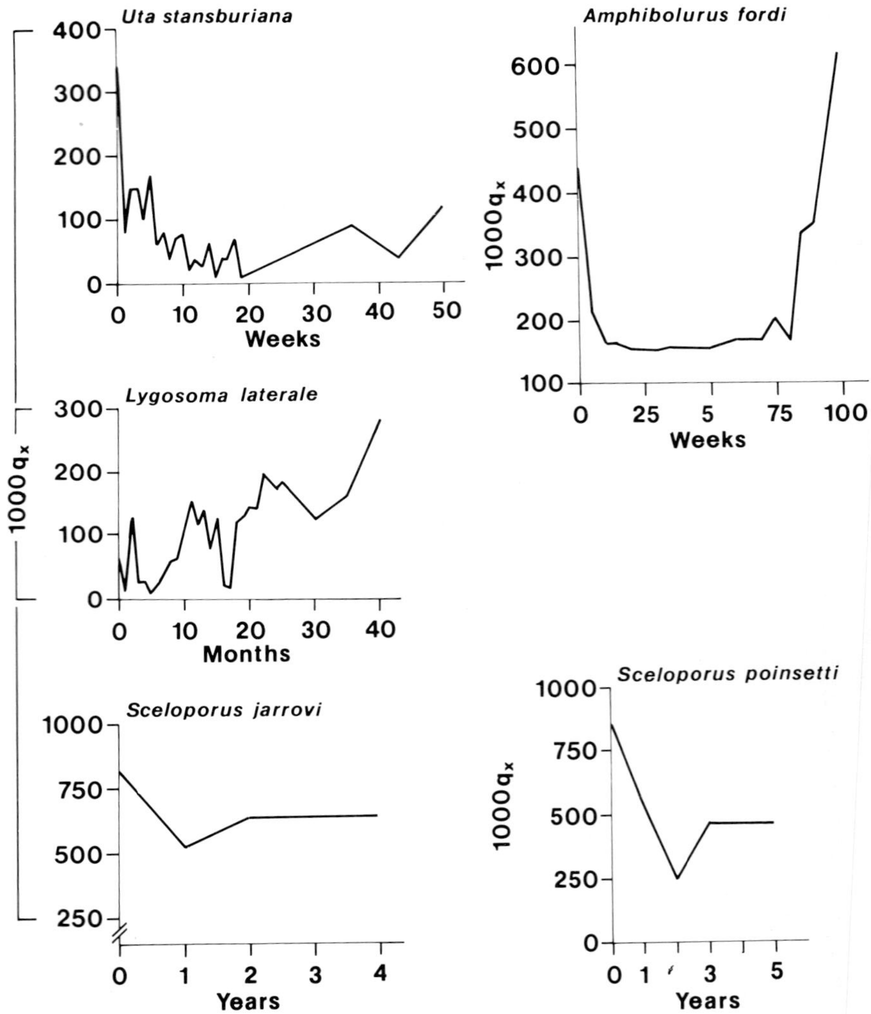


**Fig. 4.6** Age-specific changes in mortality rates ($1000_{q_x}$) of five species of lizards (from Turner, 1977).

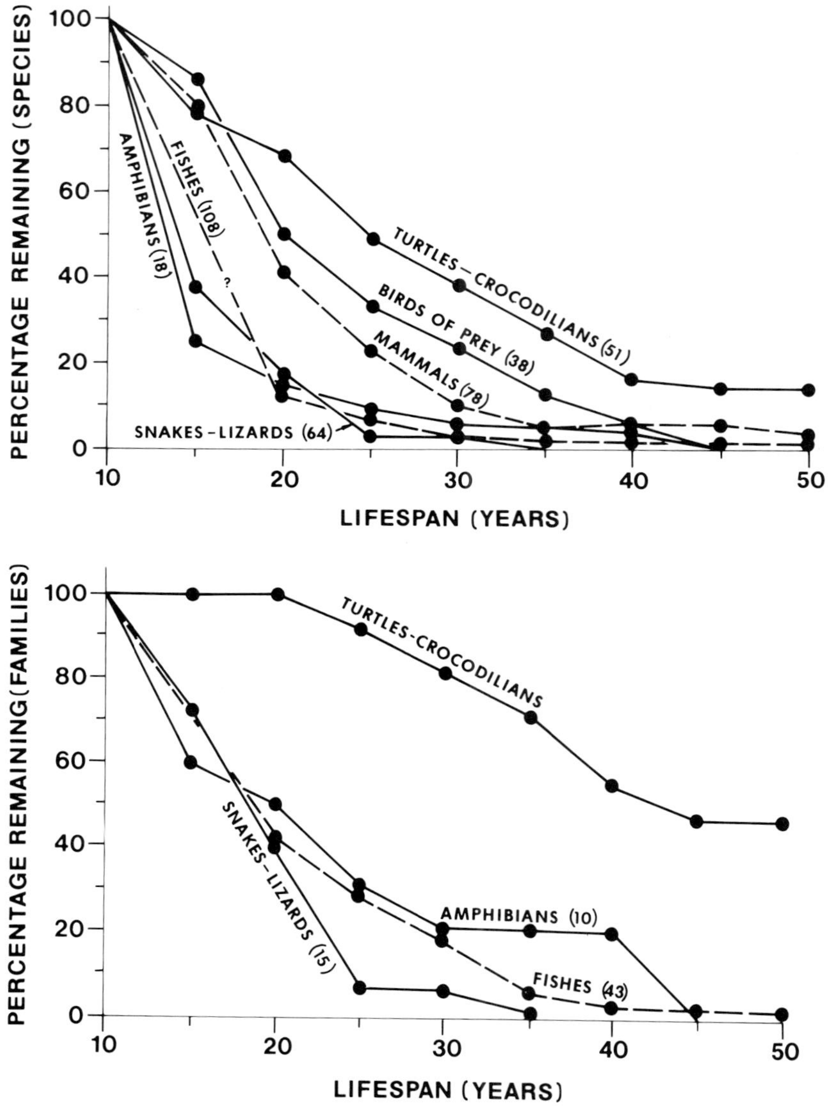


**Fig. 4.7** Comparative life spans, represented by maximum longevity records of individuals of selected vertebrate groups. Numbers in parentheses indicate number of species or families represented by individuals living 10+ years (from Gibbons, 1976).

periods of drought. Baverstock (1978) found that the probability of an individual spending the night in a refuge, safe from the lethal effects of frost, was directly related to the size of the individual and 'fast growers' behaviourally exclude the smaller 'slow growers' from deep rock crevices which protect them at night from the effects of frost. An analysis of the survivorship of successive cohorts of 'fast' and 'slow growers' of *Amphibolurus ornatus* is shown in Fig. 4.8 from which it is evident that 'slow growers' have significantly better survival than 'fast growers' at Bakers Hill and Bradshaw (1971) calculated that mean longevity was 0.98 ± 0.31 *versus* 1.13 ± 0.39 years for 'fast' and 'slow growers' respectively, and the maximum potential longevity of 'slow growers' in this particular habitat was almost three times that of fast growers (7.86 ± 1.57 *versus* 2.91 ± 0.51 years). The alternation of seasons which occurred during this particular study was particularly fortunate because it also revealed that the *potential longevity* of the two forms is equal, as shown by the fact that survival of both 'fast' and 'slow growers' from the 1962 cohort was identical during their first two years of life, because the winters of 1962 and 1963 were both mild and relatively frost-free, and only diverged when this cohort met its first hot, dry summer, at the end of 1963, with the consequent selective elimination of 'fast growers'. Similarly, it was possible to see that a series of early, severe frosts, such as occurred in 1964, could effectively reduce the survivorship of 'slow growers' relative to that of 'fast growers', as shown in Fig. 4.8, although this is not the usual pattern at Bakers Hill.

The interpretation of survivorship in this species was further complicated by the fact that juveniles from any given cohort would be forced off the main rock outcrop by territorial males, and apparently perish, but a substantial number of individuals each year was in fact able to survive in marginal habitats and return to the main rock to breed as two-year-olds. This pattern of emigration and subsequent immigration to the main outcrop is shown in Fig. 4.9 for the 1962 cohort and in greater detail, showing successive waves of immigration in April, August and December, for the 1960 cohort in Fig. 4.10. Whether such a complex pattern of movement to and from what is obviously the optimal habitat for this lizard is a common feature with other species is not known at this stage, but one important observation to come from this study with *Amphibolurus ornatus* was that the breeding population is normally composed of two-year-olds, with a small proportion of older and younger individuals, but if the age-structure deviated from this, overall fecundity was reduced. The breeding potential of the population thus appears to be maximised by a behavioural pattern which ensures, under normal circumstances, that the breeding population is dominated by two-year-olds.

It is evident that the Bakers Hill population of *Amphibolurus ornatus* is extraordinarily well-buffered against the exigencies of the abiotic environment which are manifested in the form of two seasonal stresses; winter frost and

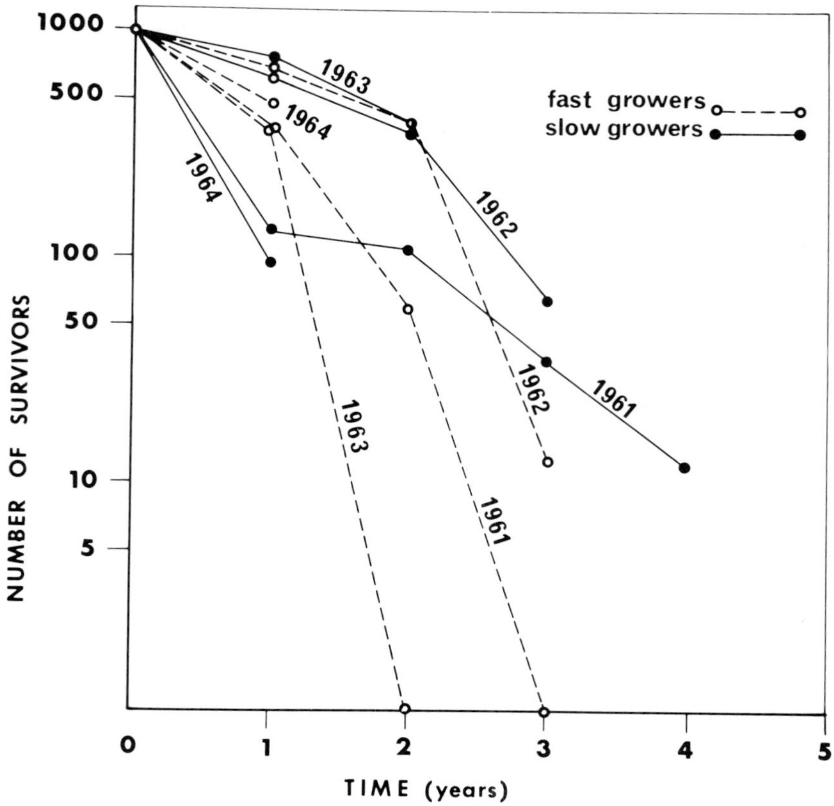


**Fig. 4.8** Survivorship curves for successive cohorts of fast and slow-growing *Amphibolurus ornatus* at Bakers Hill (from Bradshaw, 1981).

summer drought. Given the form of 'physiological polymorphism' which operates in this species through variation in growth rates, the population could only be exterminated by a 'run' of seasons, with severe frost being followed by long droughts. Obviously this does not occur to any significant extent, since the species would be extinct at Bakers Hill if it did, but the dramatic effects of one such alternation of severe seasons, and the consequent changes in size and structure of the breeding season, were in fact documented. In 1961, there was a record frost followed by a long summer with little rain and this combination of climatic events resulted in the virtual elimination of that year's cohort. The 1961 breeding population (breeding is in December in this species) was still composed primarily of two-year-olds (see Fig. 4.11), but the 1962 breeding

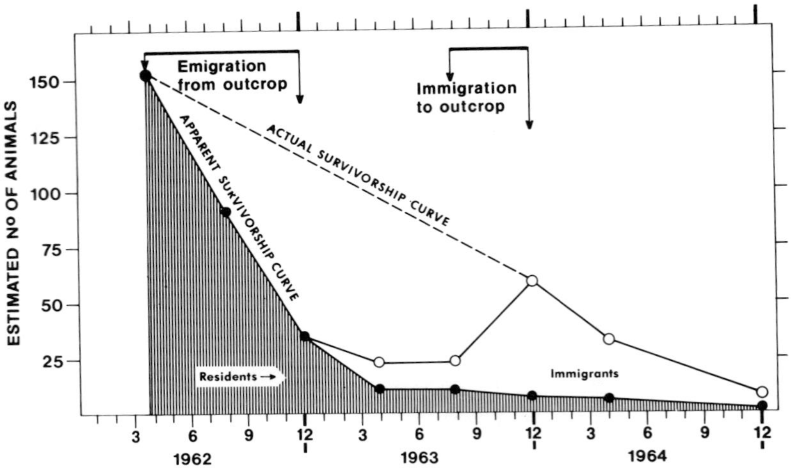


**Fig. 4.9** Survivorship curve for the 1962 cohort of *Amphibolurus ornatus* at Bakers Hill showing the subsequent return to the population of many of the early emigrants after a period of two years (December 1963) (from Bradshaw, 1971).

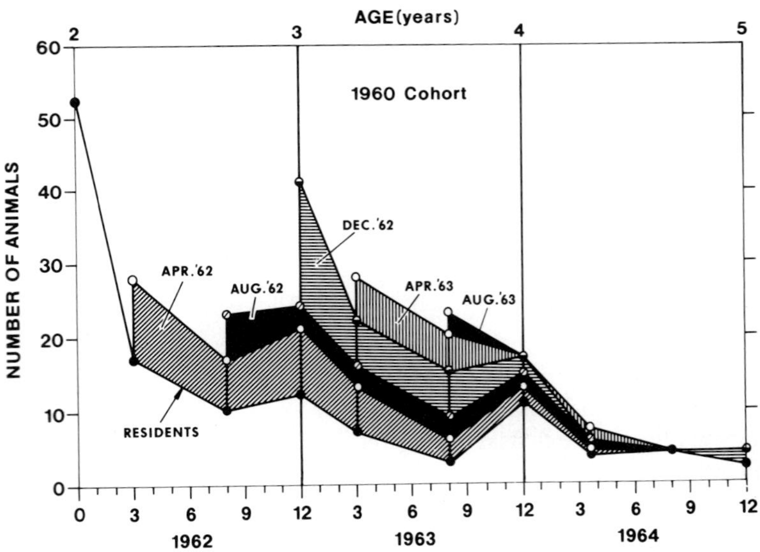


**Fig. 4.10** Characteristic survival pattern for a cohort of *Amphibolurus ornatus* at Bakers Hill showing persistent immigration of animals which emigrated to marginal habitats as juveniles (from Bradshaw, 1971).

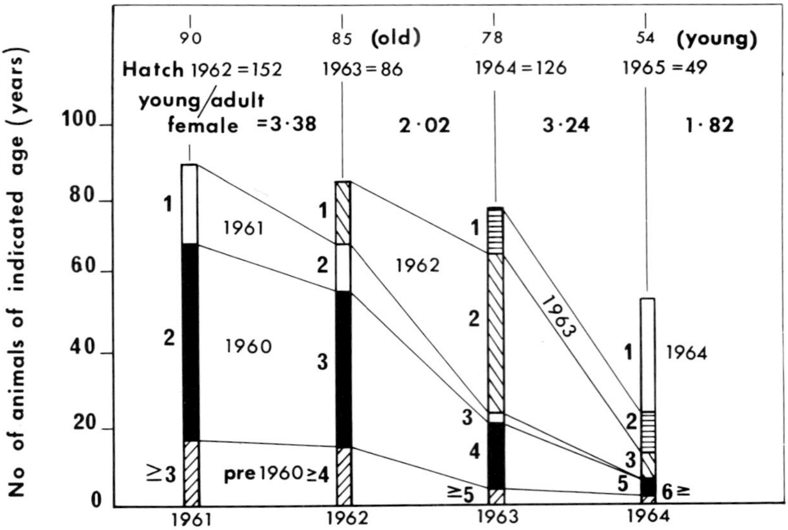


**Fig. 4.11** Changes in the size and the age-structure of the breeding population of *Amphibolurus ornatus* at Bakers Hill from 1961–1964. The number of hatchlings derived from each breeding season is also indicated and the ratio of young produced per breeding female (from Bradshaw, 1971).

population, which should have been sustained by the recruitment of individuals from the 1961 cohorts, was instead an older population, dominated by three-year-olds breeding again. Despite the fact that the population was no smaller (85 *versus* 90 individuals) the number of young hatched per female was substantially reduced from 3.38 in 1961 to 2.02 in 1962, suggesting that fecundity in this species reaches a maximum at two years of age.

The 1963 breeding population was again 'normal', with a preponderance of two-year-olds, drawn from the 1962 cohort, and the number of young hatched per female was again high at 3.24, but by 1964 when animals from the 1963 cohort would have been expected to sustain the breeding population, their numbers had been so reduced by the usual attrition (only 86 individuals were born from the older females breeding in 1962 as compared with 130–150 in 'normal' years) that the overall size of the breeding population was reduced to only 49 individuals, and the majority of these were one-year-olds, breeding for their first time, and fecundity was again reduced to 1.82 young per female (see Fig. 4.11). Very complex interactions may thus be operating in an apparently stable population such as *Amphibolurus ornatus* at Bakers Hill, and their unravelling requires a degree of persistence and a certain amount of luck as concerns the period over which the observations are made.

Bradshaw (1971) interpreted the all-pervading effects of individual variation in growth rates of *Amphibolurus ornatus* as a case of 'physiological polymorphism' which had evolved in populations exposed to highly variable and unpredictable climatic extremes, and predicted that populations of this species occupying more stable habitats would be characterised by reduced variability in this character. This is in fact what Baverstock (1972) found when studying other populations of *Amphibolurus ornatus* in Western Australia and it is therefore tempting to interpret this whole phenomenon of 'growth morphs' as an instance of '*r*' and '*K*' selection, operating intraspecifically and moulding local populations to meet the exigencies of their particular environment. A moment's reflexion however is sufficient to conclude that this would be overly simplistic. One is dealing here with a case of enhanced variability in one phenotypic character, which is also linked with other behavioural and physiological characters, and 'fast' and 'slow growers' are not discrete entities as such but merely conceptual devices enabling one to deal with this unaccustomed variability. Of crucial importance to an understanding of this phenomenon is the fate of 'intermediate growers' during periods of environmental stress, and whether they show reduced or enhanced survivorship when compared with the two extremes of the growth spectrum. Baverstock (1972) accordingly divided his data from Bakers Hill animals into three rather than two groups and, with the inclusion of Bradshaw's data, was able to show that survivorship of individuals with 'intermediate' rates of growth was *superior* to that of 'fast growers' during periods of frost and also to that of 'slow growers' during drought. We are thus not dealing with a case of polymorphism but instead one of stabilising or normal selection. The relevant question is why such variability is maintained in some populations, such as Bakers Hill, and not in others such as Canning Dam (see Fig. 4.4). Either one is dealing with a case of complex heterozygosity at Bakers Hill, with persistent selection for intermediate growth rates, and this genetic diversity has been lost at sites such as Canning Dam, or the situation is more complex and involves the occurrence of heterosis in more-extreme or fluctuating environments such as Bakers Hill (see Lerner, 1958). Clearly the data available at this stage are inadequate to allow a choice between these two alternatives and it is to be hoped that further study will be made of this interesting phenomenon which may be quite widespread in reptiles.

## III. COMPARATIVE DEMOGRAPHY

One of the fundamental problems in attempting to interpret variations in life-history phenomena is whether such variation is generated by geographic, climatic or genetic factors. Data have been accumulated on a number of species of lizards which reveal important spatial and temporal variations in characters such as age at sexual maturity, mean clutch size, number of clutches per year, juvenile survivorship etc., but attempts to show that these are genetically conditioned adaptive responses to selection pressures have been fraught with difficulty. The massive study by Tinkle (1967) on the side-blotched lizard, *Uta stansburiana*, drew attention to the limited nature of the life-history data then available in the herpetological literature and he urged a more systematic approach on the part of field biologists. Turner *et al.* (1970) highlighted the major problem in erecting adequate life tables for free-ranging species of reptiles: that of estimating with some degree of precision egg production by the population and prehatchling mortality. Most studies seem to *underestimate* egg production as judged by the fact that the *reproductive potential* ($R_0$) is often less than 1.0, even though the population may show no evidence of declining. Tinkle (1967), for example, estimated an $R_0$ of only 0.55 for the population of *Uta stansburiana* that he studied in Kermit, Texas, but the population did not decline and Turner *et al.* (1970) concluded that he had grossly under-estimated the number of clutches laid per female each season. Other estimates of $R_0$ in the literature may also be questioned, such as that proposed for *Crotaphytus collaris* by Fitch (1956) for a small, introduced population of these lizards ($R_0 = 1.58$). In fact, despite the high $R_0$, the population did not persist and became extinct within six years, indicating that the $R_0$ must of necessity have been less than 1.0. This problem of how best to estimate life-history parameters that are characteristic of the species as a whole from relatively short-term studies on localised populations is brought sharply into focus by Vinegar's (1975) study of the striped-plateau lizard, *Sceloporus virgatus* in which estimates of $R_0$ varied between 0.495 and 1.769 during two separate time periods, with prehatchling egg mortality apparently being the major factor responsible for the difference. Comparative demographic data for a variety of species of lizards from desert, semi-arid and mesic habitats are grouped together in Table XLVII and analysed according to habitat type. The population parameters include age at first breeding, reproductive potential ($R_0$), cohort generation time ($T_c$), capacity for increase ($r_c$) and the annual rate of increase of the population. No significant differences are apparent when comparing means, and this is perhaps not surprising considering the heterogeneity of the data set, but a trend which is apparent is for the species occupying arid habitats to mature early, have somewhat higher values for $R_0$, lower cohort generation times and slightly higher annual rates of increase.

Tinkle & Ballinger (1972) compared the demographic characteristics of a number of different populations of the wide-ranging fence lizard, *Sceloporus undulatus*, in various parts of the United States and recorded variations in $R_0$ from 0.61 in Ohio, in a relatively mesic situation, to 1.23 in Colorado where the environment was semi-arid. These authors noted that interpopulation variations in life-history characteristics could be as great *within a single species* as those seen between different species, with populations such as that in Texas evidencing '*r*-selection' (high fecundity, early maturity, small size and short life

**Table XLVII:** Comparative demography of lizard populations.

| Species | Habitat | Age at first breeding (yrs) | Reproductive potential ($R_o$) | Cohort generation time ($T_c$)* | Capacity for increase ($r_c$) † | Population increase per year ‡ | Reference |
|---|---|---|---|---|---|---|---|
| *Amphibolurus ornatus* (Bakers Hill, West. Aust.) | Semi-arid (S) | 1.0 | 1.57 | 2.90 yr | 0.16 yr | 1.17 | Bradshaw (1971) |
| *Amphibolurus nuchalis* Shark Bay, West. Aust.) | Arid (A) | 0.75 | 2.02 | 1.22 | 0.58 | 1.79 | Bradshaw (1981) |
| *Amphibolurus fordi* (Nymagee, N.S.W.) | A | 0.50 | 0.83 | 0.82 | −0.23 | 0.79 | Calculated from Cogger (1969) |
| *Uta stansburiana* (Rock Valley, Nevada) | A | 0.80 | 1.65 | 1.28 | 0.39 | 1.48 | Turner *et al.*, (1970) |
| (Kermit, Texas) | A | 0.75 | 0.55 | 0.79 | −0.76 | 0.47 | Tinkle, (1967) |
| *Crotaphytus collaris* (Kansas) | A | 1.0 | 1.58 | 2.51 | 0.18 | 1.20 | Fitch (1956) |
| *Xantusia vigilis* (California) | A | 3.0 | 0.63 | 4.33 | −0.11 | 0.90 | Zwiefel & Lowe (1966) |
| *Lygosoma laterale* (Florida) | Mesic (M) | — | 0.50 | 1.03 | −0.67 | 0.51 | Brooks (1967) |
| *Sceloporus magister* (Utah) | S | 1.88 | 0.99 | 2.78 | ~ 0 | 1.00 | Tinkle (1976) |

| | | | | | | | |
|---|---|---|---|---|---|---|---|
| *Sceloporus virgatus* | S | | | | | | |
| (Arizona) 1970–71 | | 1.5 | 0.50 | 2.61 | −0.27 | 0.77 | Vinegar (1975) |
| 1971–72 | | | 1.77 | 2.29 | 0.25 | 1.28 | Vinegar (1975) |
| *Sceloporus undulatus* | | | | | | | |
| (S. Carolina) | M | 1.0 | 1.14 | 1.32 | 0.10 | 1.10 | Tinkle & Balinger (1972) |
| (Texas) | A | 1.0 | 1.05 | 0.93 | 0.05 | 1.05 | Tinkle & Balinger (1972) |
| (Ohio) | M | 2.0 | 0.61 | 2.39 | −0.21 | 0.81 | Tinkle & Balinger (1972) |
| (Colorado) | S | 2.0 | 1.23 | 2.17 | 0.10 | 1.10 | Tinkle & Balinger (1972) |
| (Utah) | S | 1.8 | 0.83 | 2.70 | −0.07 | 0.93 | Tinkle (1972) |
| (Kansas) | S | 1.0 | 0.87 | 1.15 | −0.12 | 0.89 | Ferguson, Bohlen & Woolley (1980) |
| *Sceloporus merriami* (Texas) | A | 0.83 | 1.17 | 1.56 | 0.10 | 1.11 | Dunham (1981) |
| *Urosaurus ornatus* | | | | | | | |
| (Texas) | A | 0.83 | 1.06 | 1.54 | 0.04 | 1.04 | Dunham (1981) |
| (Sunflower, Arizona) | S | 0.71 | 0.50 | 1.20 | −0.58 | 0.56 | Tinkle & Dunham (1983) |
| (Chiricahua Mts., Arizona) | S | 0.79 | — | 2.24 | — | — | Dunham (1982) |
| (Animas, New Mexico) | A | 0.92 | — | — | — | — | Dunham (1982) |
| MEANS | | | | | | | |
| Arid (A) | | 1.04 ± 0.22 | 1.17 ± 0.16 | 1.66 ± 0.38 | 0.03 ± 0.13 | 1.09 ± 0.13 | |
| Semi-arid (S) | | 1.34 ± 0.18 | 1.03 ± 0.16 | 2.23 ± 0.22 | 0.06 ± 0.16 | 0.96 ± 0.08 | |
| Mesic (M) | | 1.50 ± 0.50 | 0.75 ± 0.20 | 1.58 ± 0.41 | −0.26 ± 0.22 | 0.81 ± 0.17 | |

* $\mathbf{T_c} = \frac{\Sigma x l_x m_x}{\Sigma l_x m_x}$ † $r_c = \frac{\ln R_o}{T_c}$ ‡ Population increase/year $= e^{r_c}$

expectancy) in contrast to more northern populations which displayed the characteristics of '*K*-selected' species. Tinkle & Ballinger (1972) speculated that such differences resulted from selection that favours one set of life-history 'genotypes' in one population and others in another. Little evidence has been presented however to suggest that such differences are the result of changes in gene frequency, rather than being instead relatively plastic phenotypic responses to temporal and spatial variations of the immediate environment. Tinkle (1972a) was forced to admit this possibility when studying a western population of *Sceloporus undulatus* in Utah which showed unexpected delayed maturity. Tinkle & Ballinger (1972) had previously argued that delayed maturity in northern populations of this species in Ohio and Colorado was adaptive, because they were able to calculate that replacement rates in these populations were higher than would have been the case had maturity not been delayed. This was not the case in Utah, however, and Tinkle (1972a) was forced to conclude that proximate ecological factors, rather than ultimate evolutionary ones, were responsible for the observed variation.

Ferguson *et al.* (1980) studied another population of *Sceloporus undulatus* in eastern Kansas and compared their results with measurements of life-history traits derived from 10 other populations of this species in the United States (see Table XLVIII), and then ranked the various criteria in an effort to find latitudinal trends. These rankings are shown in Table XLIX and it is of interest that certain traits covary in some populations (small body size, early maturity, short life span, small eggs and high clutch mass–body mass ratios) whilst the opposite suite of characters are found grouped in other populations. Three of the four low-ranking populations (large, late-maturing, long-lived etc.) were found to be northern in distribution, whereas the desert grassland populations ranked highly and have a southern distribution (Texas and Lordsburg, New Mexico). The Kansas population was exposed to high predation rates, relatively unlimited food supply, as well as considerable variation in breeding success as a result of flooding, and ranked highly as would be predicted from '*r* and *K*' selection theory. Many workers thus have established that significant latitudinal trends are found in life-history characteristics of various iguanid lizards in North America, and these have usually been interpreted as being adaptive responses to variations in predation pressure, competition and availability of resources (see Nussbaum & Diller, 1976; Pianka, 1970b). Ferguson *et al.* (1980) were more conservative in the interpretation of their data from *Sceloporus undulatus*, however, and suggested that more quantitative data are needed on predation rates and competition before these can be invoked as universal explanations for the phenomena. They further pointed out that . . . "We know nothing of the genetic basis of any of these population differences . . . and we need long-term, in-depth studies on single populations to weed out the adaptive significances and genetic bases of these life history strategies . . .".

**Table XLVIII:** Comparison of life history traits of ten different populations of *Sceloporus undulatus* (from Ferguson *et al.* (1980)).

| Population | Texas | Lordsburg New Mexico | Kansas | South Carolina | Georgia | Pinos Altos, New Mexico | Ohio | Missouri | Colorado | Utah |
|---|---|---|---|---|---|---|---|---|---|---|
| Minimum SVL* (mm) adult females | 47.0 | 54.0 | 47.0 | 55.0 | 52.0 | 53.0 | 66.0 | . . . | 58.0 | 58.0 |
| Mean SVL* (mm) adult females | 57.0 | 68.0 | 57.0 | 63.0 | 62.0 | 63.0 | 75.0 | . . . | 70.0 | 69.0 |
| Mean clutch size | 9.5 | 9.9 | 7.0 | 7.4 | 7.6 | 7.2 | 11.8 | 11.3 | 7.9 | 6.3 |
| Mean egg mass (g) | 0.22 | 0.24 | 0.26 | 0.33 | . . . | 0.29 | 0.35 | 0.38 | 0.42 | 0.36 |
| Egg mass/body mass | 0.029 | 0.022 | 0.044 | 0.033 | . . . | 0.032 | 0.021 | . . . | 0.029 | . . . |
| Number of clutches per year | 3.0 | 4.0 | 2.0 | 3.0 | 3.0 | 2–3 | 2.0 | 2.0 | 2.0 | 3.0 |
| Clutch mass/female mass | 0.27 | 0.21 | 0.28 | 0.23 | . . . | 0.22 | 0.25 | 0.28 | 0.23 | 0.21 |
| Age at first breeding | 1.0 | 1.0 | 1.0 | 1.0 | 1.0 | 1 yr– 50%<br>2 yr–100% | 2.0 | . . . | 2.0 | 1 yr–10%<br>2 yr–90% |
| Survivorship to age of first breeding | 0.06 | 0.03 | 0.10 | 0.11 | . . . | 0.02 | 0.03 | . . . | 0.11 | 0.05 |
| Average adult annual survivorship | 0.11 | 0.20 | 0.27 | 0.49 | 0.07 | 0.34 | 0.44 | . . . | 0.37 | 0.48 |
| Mean number per hectare | 2.0 | 10.0 | 48.0 | 8.0 | . . . | 35.0 | 22.0 | . . . | 14.0 | 33.0 |
| Authority | Tinkle & Ballinger (1972) | Vinegar (1975) | Ferguson *et al,* (1980) | Tinkle & Ballinger (1972) | Crenshaw (1955) | Vinegar (1975) | Tinkle & Ballinger (1972) | Tinkle (1972a) | Tinkle & Ballinger (1972) | Tinkle (1972a) |

. . . means not available.

* SVL = snout-vent length.

**Table XLIX:** Rank and rank totals for five correlated parameters for several populations of *Sceloporus undulatus* (from Ferguson *et al.* 1980).

| Population | Rank body size | Rank age of maturing | Rank annual adult survival | Rank egg size | Rank female mass/ clutch mass | Total of ranks | Rank of rank totals |
|---|---|---|---|---|---|---|---|
| Texas | 7.5 | 6.5 | 8 | 8 | 7 | 37 | 1 |
| Kansas | 7.5 | 6.5 | 6 | 6 | 8 | 34 | 2 |
| Lordsburg, New Mexico | 4 | 6.5 | 7 | 7 | 1.5 | 26 | 3 |
| South Carolina—Georgia | 5.5 | 6.5 | 5* | 4 | 4.5 | 25.5 | 4 |
| Pinos Altos, New Mexico | 5.5 | 4 | 4 | 5 | 3 | 21.5 | 5 |
| Ohio | 1 | 1.5 | 2 | 3 | 6 | 13.5 | 6 |
| Colorado | 2 | 1.5 | 3 | 1 | 4.5 | 12 | 7 |
| Utah | 3 | 3 | 1 | 2 | 1.5 | 10.5 | 8 |

* The rather disparate values of this parameter from South Carolina and Georgia were averaged for this analysis. Tied values were given the average value of the two or more tied ranks.

*Urosaurus ornatus* is another lizard species which has been the subject of intensive study by herpetologists in the United States and, like *Sceloporus undulatus*, interpopulation variation is a feature of these studies. Some populations have also been manipulated experimentally, in an effort to determine whether changes in density affect survivorship, but Ballinger (1976) found that, in many cases, the results were the opposite of those anticipated. Tinkle & Dunham (1983) stressed that ". . . variation in observed life history traits within each population studied is as great as that observed among populations", unlike the situation with *Sceloporus undulatus*, and they were at a loss to identify ". . . What, if any, are the common environmental pressures responsible for this life history pattern" Dunham (1981) undertook a detailed study of the demography of two arid-living species of lizards, *Sceloporus merriami* and *Urosaurus ornatus*, in the Big Bend National Park, Texas, in an effort to determine whether " . . variation in life history traits such as age-specific survival, individual growth rates, and clutch frequency (is) in response to temporal fluctuation in food resource abundance or other environmental variables that are affected by temporal variation in the amount of precipitation". He found that, in years of low rainfall and reduced prey availability, individuals *of both species* exhibit lowered size-specific fecundity, growth rates, and prehibernation lipid levels, and displayed a remarkable similarity in life-history characteristics and population dynamics which was unexpected. Table L summarises the effects of lower food availability in dry seasons of life-history parameters for the two species, and the overall similarity of their responses is quite evident. Considerable convergence in terms of population responses is thus apparent with these two species of sceloporine lizards. Dunham (1981) commented that ". . . The life history characteristics and population dynamics of the population of *Urosaurus ornatus* from the Grapevine Hills (Texas) are more similar to those of synoptic populations of canyon lizards (*Sceloporus merriami*) than to populations of *Urosaurus ornatus* from different parts of the geographic range". We are still left, however, with the problem of whether these differences are in any way genetically conditioned, as they must be if they are to be considered as 'adaptations', or merely environmentally-induced changes in parameters such as rate of growth which are a direct consequence of a decrease in the availability of resources. Dunham (1981) argues, for example, that in dry years, when food is scarce, one might expect individuals to increase their foraging effort in order to compensate for the lack of food, and this would be considered 'adaptive'. In fact the lizards do the exact opposite, maintaining or even reducing, their foraging intensity and catabolise their lipid reserves, but this too is interpreted as being 'adaptive' because it is a strategy which presumably reduces the risk of predation on animals that are in a weakened condition. Clearly, no matter what response the animals show to changes in resource availability, a sufficiently fertile mind, assisted by a modicum of sophistry, can interpret almost any change as being 'adaptive' and leading to an enhancement of Darwinian fitness.

**Table L:** Summary of the effects of lower food availability in dry years on selected life history characteristics of *Urosaurus ornatus* and *Sceloporus merriami* [*from Dunham, 1981*].

| Life history characteristics affected (1975, 1977) | *Urosaurus ornatus* | *Sceloporus merriami* |
|---|---|---|
| Individual foraging success | lower | lower |
| Individual growth rates | lower | lower |
| Size-specific fecundity of females | lower | lower |
| Survival of | | |
| Eggs | higher | higher |
| Hatchlings | no effect | no effect |
| Adults | no effect | no effect |
| Size-specific body weight (Prehibernation) | lower | lower |
| Size-specific Prehibernation Lipid Storage | lower | lower |
| Population density (early June) | lower | lower |
| Age structure | excess of animals over 2 yrs old | excess of animals over 2 yrs old |

This problem of concordance with current theoretical expectations has been analysed in detail by Dunham (1982). He looks at data on *Urosaurus ornatus* populations and tests the predictions of '*r* and *K*' selection theory. This theory argues (see Stearns, 1976) that life-history traits such as early maturity, high age-specific fecundity, small offspring, high age-specific reproductive effort, and short life expectancy are expected to evolve in organisms inhabiting environments with high resource supply/demand ratios—leading to high intrinsic rates of increase or '*r*-selection'. Conversely, life-history traits such as delayed maturity, slow growth, low age-specific fecundity and reproductive effort, large offspring and longer life expectancy were expected to evolve in organisms inhabiting highly-competitive environments, with low resource supply/demand ratios where the animals would be found at, or near, saturation densities, close to the carrying capacity of the environment. Hence the use of the term '*K*-selection' with reference to the well-known logistic equation describing population growth (see Table LI). Several authors have also argued that the *demographic environment* is also an important selective factor in life-history evolution (e.g. Murphy, 1968; Hirshfield & Tinkle, 1975; Stearns, 1977) and that in environments in which juvenile survivorship is variable or unpredictable, selection can favour traits such as delayed maturity, low levels of age-specific reproductive effort and increased adult survivorship—or in effect, '*K*-selection' can be operative. Similarly, environments characterised by high extrinsic adult mortality and poor juvenile survival should select for rapid

**Table LI:** Some of the correlates of $r$- and $K$-selection (from Pianka, 1970a).

| | $r$-selection | $K$-selection |
|---|---|---|
| Climate | Variable and/or unpredictable: uncertain | Fairly constant and/or predictable: more certain |
| Mortality | Often catastrophic, non-directed, density-independent | More directed, density-dependent |
| Survivorship | Often Type III (Deevey 1947) | Usually Type I and II (Deevey 1947) |
| Population size | Variable in time, non-equilibrium; usually well below carrying capacity of environment; unsaturated communities or portions thereof; ecological vacuums; recolonisation each year | Fairly constant in time, equilibrium; at or near carrying capacity of the environment; saturated communities; no recolonisation necessary |
| Intra- and interspecific competition | Variable, often lax | Usually keen |
| Relative abundance | Often does not fit MacArthur's broken stick model | Frequently fits the MacArthur model |
| Selection favors | 1. Rapid development<br>2. High $r_{max}$<br>3. Early reproduction<br>4. Small body size<br>5. Semelparity; single reproduction | 1. Slower development, greater competitive ability<br>2. Lower resource thresholds<br>3. Delayed reproduction<br>4. Larger body size<br>5. Iteroparity: repeated reproductions |
| Length of life | Short, usually less than 1 year | Longer, usually more than 1 year |
| Leads to | Productivity | Efficiency |

maturity, large numbers of small offspring and a high age-specific reproductive effort (see also Table XL).

Dunham (1982) estimated resource supply/demand ratios for four populations of *Urosaurus ornatus* from Texas (Grapevine Hills), New Mexico (Animas), and Arizona (Chiricahua and Sunflower), with resource supply being estimated simply as the mean annual precipitation which Dunham (1981) had previously found to be strongly correlated with the abundance of arthropod prey. These ranged from 0.21 for the Animas population, which would be expected therefore to evidence strong '*K*-selection', to 3.19 for the Grapevine Hills population which should show characteristics of '*r*-selection'. The observed and predicted rankings of these four populations with regard to life-history characteristics are shown in Table LII and it is evident that the concordance is very poor with Kendall's Coefficient of Concordance (KCC) = 0.04, $P > 0.85$. Similarly, the ranking of these populations by life-history characteristics is also inconsistent, with KCC = 0.02 $P > 0.93$, and using variability in rainfall as a predictor of juvenile survivorship shows, as Dunham remarks, "terrible" concordance (KCC = 0.02 $P > 0.93$). Dunham (1982) further went on to test the fit with theoretical expectations for those populations for which there were adequate data to ascertain variation in juvenile survivorship; high and unpredictable juvenile mortality being expected to lead to '*K*-selection'. The rankings are shown in Table LIII and the fit is again poor with KCC = 0.02, $P > 0.87$. Dunham (1982) concluded that ". . . These tests employed here are admittedly not powerful, but they do suggest that neither demographic theory nor the theory of "*r*-" and "*K*-selection" is adequate to predict the differences in life histories observed among these populations". Dunham (1982) went on to stress that tests of theories of life-history evolution based on selection arising from resource availability require studies in which *variations in resource abundance are measured simultaneously with life history and demographic parameters*, just as tests of demographic theories require comparable data on critical factors such as age-specific survivorship of juveniles and adults. Dramatically different selective forces can result in the evolution of very similar life histories, as emphasised by Wilbur *et al.* (1974), and seen in Dunham's (1981) study of *Sceloporus merriami* and *Urosaurus ornatus*. Experiments that will allow partitioning of life-history variation within and among populations into genetic and environmental components are essential if these problems are ever to be resolved.

In a series of important papers Stearns (1976, 1977, 1980, 1982, 1984) has developed the thesis that concepts such as '*r*-' and '*K*-selection' are overly simplistic, difficult to test, and in many cases were formulated with little regard for the data base available on natural populations of organisms and have served since as 'paradigms' directing or misdirecting the thoughts of countless ecologists. He argues eloquently that physiological problems can overwhelm the

**Table LII:** Rankings of four populations of *Urosaurus ornatus* based on the predictions of the theory of *r*-and *K*-selection, demographic theory, and observed life-history characteristics (from Dunham (1982)).

| Basis for ranking | Grapevine Hills | Sunflower | Chiricahua Mts. | Animas |
|---|---|---|---|---|
| Predicted | | | | |
| *r*- and *K*-selection | | | | |
| (1 = most *r*-selected) | 1 | 2 | 3 | 4 |
| Variation in annual rainfall | | | | |
| (1 = least variable) | 4 | 1 | 3 | 2 |
| Observed | | | | |
| Size at maturity | 4 | 3 | 2 | 1 |
| Age at maturity | 3 | 1 | 2 | 4 |
| Egg dry mass | 2 | 3 | 4 | 1 |
| Juvenile growth rate | 3 | 4 | 2 | 1 |
| Size-specific clutch size | 1 | 3 | 2 | 4 |
| Relative clutch mass | 1 | 2 | 4 | 3 |

**Table LIII:** Ranking of three populations of *Urosaurus ornatus* based on range of variation in juvenile survivorship and observed life history characteristics (from Dunham, 1982).

| Basis for ranking | Grapevine Hills | Sunflower | Chiricahua Mts |
|---|---|---|---|
| Predicted | 2 | 1 | 3 |
| Observed | | | |
| Size at maturity | 3 | 2 | 1 |
| Age at maturity | 3 | 1 | 2 |
| Egg dry mass | 1 | 2 | 3 |
| Juvenile growth rate | 2 | 3 | 1 |
| Size-specific clutch size | 1 | 3 | 2 |
| Relative clutch mass | 1 | 2 | 3 |
| Cohort generation time | 2 | 3 | 1 |

expected co-adaptations of life history traits, citing his own work on *Gambusia affinis*, a poeciliid fish, and that in many classic instances of 'co-adaptation', such as for example the Euler-Lotka equation:

$$1 = \int_{\alpha}^{\infty} e^{-rxl_x b_x dx}$$

which link age ($x$), age-specific survival ($l_x$), fecundity ($b_x$) and population growth rate ($r$), the co-adaptation of these traits is an *assumption of the model* rather than a conclusion to be drawn from its application. Stearns (1980) points out that allometric constraints, which are a feature of the development of most organisms, violate one of the implicit assumptions of much life-history theory: that life-history traits are 'free' to evolve under purely demographic forces. He further suggests that a training in quantitative genetics, development and physiology is just as necessary for those who would study life-history evolution as is a knowledge of demography and population genetics, and that many of the problems in the literature reflect the overly theoretical approach of recent practitioners.

Stearns (1984) looks specifically at the effects of size and phylogeny on patterns of covariation in the life-history traits of two orders and nine families of reptiles and his conclusions are of some interest. Five 'traits' were analysed: overall length, clutch size, age at maturity, mode of reproduction (oviparous *versus* viviparous) and the number of broods per year. He found that the distinction between the life histories of lizards and snakes results mostly from their different average sizes. Once the effects of size are removed the differences are no longer significant. Significant Family effects on clutch size and mode of reproduction were found, however, which were not altered by the removal of length effects. These Family effects accounted for about 20% of the variation observed in clutch size and 40% of the variation in mode of reproduction. Classwide correlations with size thus explain about half the variation in age at maturity and a quarter of the variation in brood frequency, but have little impact on either clutch size or mode of reproduction. The single pattern of covariation that explained most of the variability in the data remained throughout an axis that ranked species from small, oviparous, early-maturing organisms with many broods per year and small clutches, to large, viviparous, late-maturing organisms with few broods per year and large clutches. Thus, two components of reproductive effort—clutch size and brood frequency—appear to compensate in a manner which reduces the importance of any covariation of reproductive effort itself with other life-history traits. Stearn's work with reptiles thus reinforces the philosophical views of Lauder (1982) who contrasts the difference between 'historical' and 'equilibrium' analyses in functional morphology and shows that history plays an important rôle in shaping the properties that organisms exhibit today. In the case of reptiles, Stearns (1984) concludes that ". . . history, as seen in lineage-specific patterns of covariation, has played an important rôle in shaping suprageneric differences in . . . life histories . . ." This paper of Stearns' has been criticised recently by Vitt and Seigel (1985), not because of the theoretical approach employed, but because of inaccuracies in the data set analysed. Stearns (1985) acknowledges these errors but has yet to admit that they will seriously affect his conclusions.

Given the variety of problems which beset the study of population dynamics of reptiles—problems of ageing individuals in the field, random sampling of populations, estimation of egg production and prehatchling mortality—plus the various inherent biases that different workers have brought to the discipline, it is not surprising that it is difficult to discern any coherent adaptive patterns in the life-history 'tactics' of desert lizards. The data collated in Table XLVII suggest that desert-adapted species of lizards all mature at approximately one year of age and show a relatively rapid population turnover. This merely reflects the bias inherent in the literature and the tendency of most herpetologists to concentrate their studies on small, short-lived, multiple-brooding species of lizards which are found in great abundance in the deserts of North America. There are no adequate population studies of large, long-lived desert species such as the desert iguana, *Dipsosaurus dorsalis* which takes 31–33 months to reach sexual maturity according to Krekorian (1984), or the chuckwalla, *Sauromalus obesus* which grows even more slowly and probably lives far longer than the desert iguana. Populations studies on the 'Fouette-Queue' of North Africa, *Uromastix acanthinurus*, which is one of the most 'desert-adapted' of all reptiles, or the varanid *Varanus griseus* which lives in the mobile dunes of the Sahara would almost certainly change this picture radically and dispel any impression that all desert reptiles have opted for rapid turnover and reduced longevity as a means of survival in such regions. Desert environments are sufficiently heterogeneous to permit co-existence of both extremes and thus neither one qualifies as a specific 'adaptation' in the sense that the word has been employed throughout.

## IV. CONCLUSION

In a paper concerned with the rôle of environment in the evolution of life-history differences within and between lizard species, the late Don Tinkle clearly stated the major problem then confronting population biologists: ". . . whether these life history differences are attributable to the evolution of different gene complexes in the various populations because of the wide differences in selection régimes, or whether the differences simply reflect ecological constraints on essentially the same genetic stock" (Tinkle, 1972b). Unfortunately, despite a burgeoning literature on 'life-history tactics', especially of desert reptiles, we are no closer to resolving this problem. To a large extent this is because much of the data is observational rather than experimental in nature, and genetic bases for observed patterns are inferred from correlations with perceived selection régimes, rather than being deduced from controlled experiments on animals maintained under reproducible conditions, and the overall 'adaptationist' bias of most workers in the field is manifest. Stearns (1984) comments that ". . . Explanations of patterns of covariation in life-history traits have been couched in terms of processes acting within species to produce adaptations to local habitats and mortality schedules. Analysis of the impact of *size* and *lineage* on patterns of covariation in the life-history traits of reptiles suggest that microevolutionary explanations, while perhaps *necessary*, are not *sufficient* to account for the patterns in the data" (my emphases).

Despite this, some progress has been made, and there is evidence to suggest that individual rates of growth in lizards are highly variable and that, in some instances, this variation may be genetically conditioned as it persists when juveniles are reared under constant laboratory conditions with equal access to nourishment (e.g. *Amphibolurus ornatus, Uta stansburiana* and *Chrysemys picta*). The demographic significance of such variation has only been analysed in the case of the Australian lizard *Amphibolurus ornatus*, but the linkage revealed between rate of growth and other physiological parameters, such as renal and adrenocortical function, makes it likely that this is not just an isolated instance. Further study of this general problem is desirable as most workers see accelerated growth as always being of advantage to the population, but the work with *Amphibolurus ornatus* shows that rapidly growing individuals may be substantially disadvantaged in a physiological sense during periods of environmental stress.

Long-term studies of the dynamics of reptilian populations are still beset by the difficulties which attend attempts to age unmarked animals in the field and the problem of accurately estimating critical parameters, such as overall egg production and prehatchling mortality, has proven difficult of resolution, if not virtually intractable in some circumstances. Such data as may be amassed on survivorship and mortality schedules of natural populations of reptiles reveal an inherent bias in the literature towards the study of small, relatively short-lived, multiple-brooded species of lizards, which are of course the most numerous and easy of capture, and studies are urgently needed of larger longer-lived desert species, with different population strategies, so as to redress this obvious imbalance. Although it is perhaps premature to anticipate the results of such studies, it seems unlikely that desert reptiles are typified by any unique set of demographic characters which account for their apparent success in such environments.

# 5

# Causes and Incidence of Mortality in Desert Lizards: Three Case Studies

The generation of life tables for the relatively few species of desert reptiles that have been studied in depth provides precise information on the schedule of births and deaths, *as a function of age*, for these species but gives little idea of the various causes of death, identifying only the time at which it strikes. As we have seen in Chapter 4, the data available are far too scanty to support any major generalisations but a common expectation, if not a working hypothesis, is that survival is not age-dependent, at least in lizards, and that survivorship curves will therefore show a linear decline with ageing of the cohort. We have already seen that this is not the case in at least two lizards (see Fig. 4.6), but most other studies show no significant deviation from linearity. This means that whatever factor or factors are responsible for killing animals in the field, they operate more-or-less uniformly throughout life and 'mortality pressure' is thus constant on both juveniles and adults. At first sight this seems an unlikely conclusion—juveniles would be expected for example to be more subject to predation than adults. There is evidence that this is the case, and the effectiveness of competition for scant resources might be expected to be influenced by metabolic body size, and hence age. But if death results primarily from other causes, such as drought and cold, to which reptiles are particularly susceptible, it becomes more reasonable to imagine that death rates would vary substantially from year to year, depending upon the climate, but that the *vulnerability* of a given individual to the excesses of that climate would not change substantially with age. Death in this instance would occur whenever the exigencies of the local environment exceed the capacities of the individual reptile to maintain physiological homeostasis, and there are as yet no data to suggest that these capacities decline with age, as in mammals where 'senescence' is commonly observed. This hypothesis can be no more than a working one at present, but it may be useful to examine, from this point of view, a number of species of desert reptiles which have been studied in sufficient detail to support intelligent speculation on the causes of death in the field. The cases chosen include species from deserts in Australia, North America and North Africa. This should permit any pattern or commonality of response on the part of these desert reptiles to be discerned.

## I. *AMPHIBOLURUS NUCHALIS* AT SHARK BAY IN WESTERN AUSTRALIA

This agamid lizard (see Frontispiece) has been studied from 1969 to 1983 at Shark Bay, approximately 800 km north of Perth, Western Australia, and some preliminary data from this long-term study have been published (see Bradshaw, 1975, 1978, 1981). *Amphibolurus nuchalis* (De Vis) (formerly *Amphibolurus inermis*; see Cogger, 1975*) is a small (20–50 g) lizard which is extremely common in the arid and semi-arid regions of Australia, extending completely across the continent from approximately 29° northwards. It is a diurnal, heliothermic insectivore, living characteristically in sandy, disturbed areas, such as road verges (see Plate XI), where it digs a number of simple burrows (Plate XII). Aspects of its thermoregulation and water and electrolyte physiology have already been discussed in Chapter 2. An aerial view of the study area is shown in Plate XIII and two distinct habitats occur: Pleistocene mudflats which are now vegetated only by halophytes such as *Arthrocnemum arbuscula* (so-called 'samphire flats' or 'birridas'; see Plate XIV) and low sand dunes overlying these soils which support a more diverse flora dominated by *Acacia* species (see Plate XV). *Amphibolurus nuchalis* occurs in both habitats but, over the period of the study, the density, which was originally quite high in the *Acacia* habitat, has declined considerably but increased relatively in the samphire flats. At Shark Bay, *Amphibolurus nuchalis* is essentially an annual species and, from over 1000 individuals marked, less than 2% have lived for more than one year.

Circulating levels of adrenal corticosteroids have been monitored seasonally in this animal in an attempt to delineate periods when individuals might demonstrate stress, as a result of chronic exposure to unfavourable environmental conditions. Fig. 5.1 shows that *Amphibolurus nuchalis* differs markedly from the related species, *Amphibolurus ornatus*, where steroid levels

---

* Now referred to the revived genus *Ctenophora* (Fitzinger, 1843) by Storr (1982).

**Plate XI** Typical habitat of *Amphibolurus nuchalis* at Shark Bay along a disturbed road verge where burrows are dug in the loose sand.

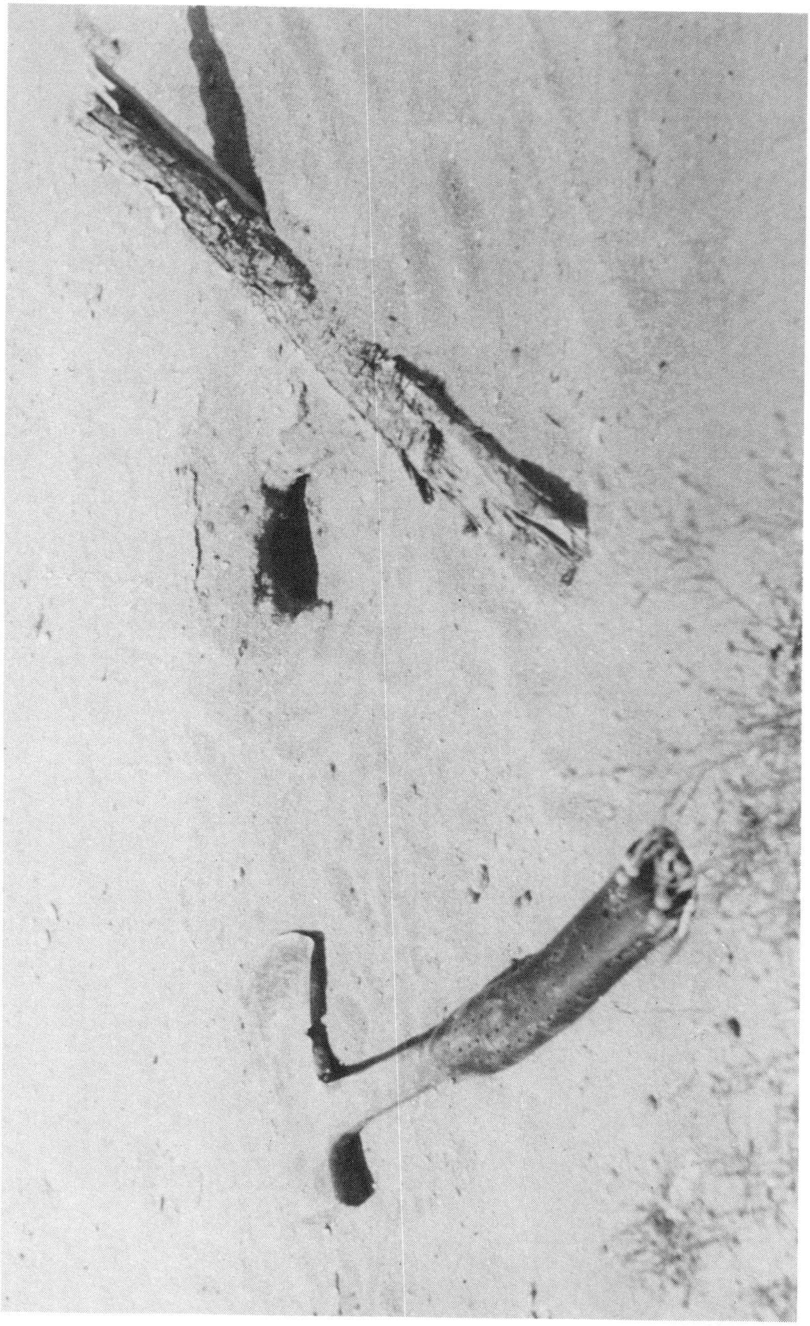

**Plate XII** Entrance to the shallow and simple burrow system characteristically dug by *Amphibolurus nuchalis* in sandy soil at Shark Bay.

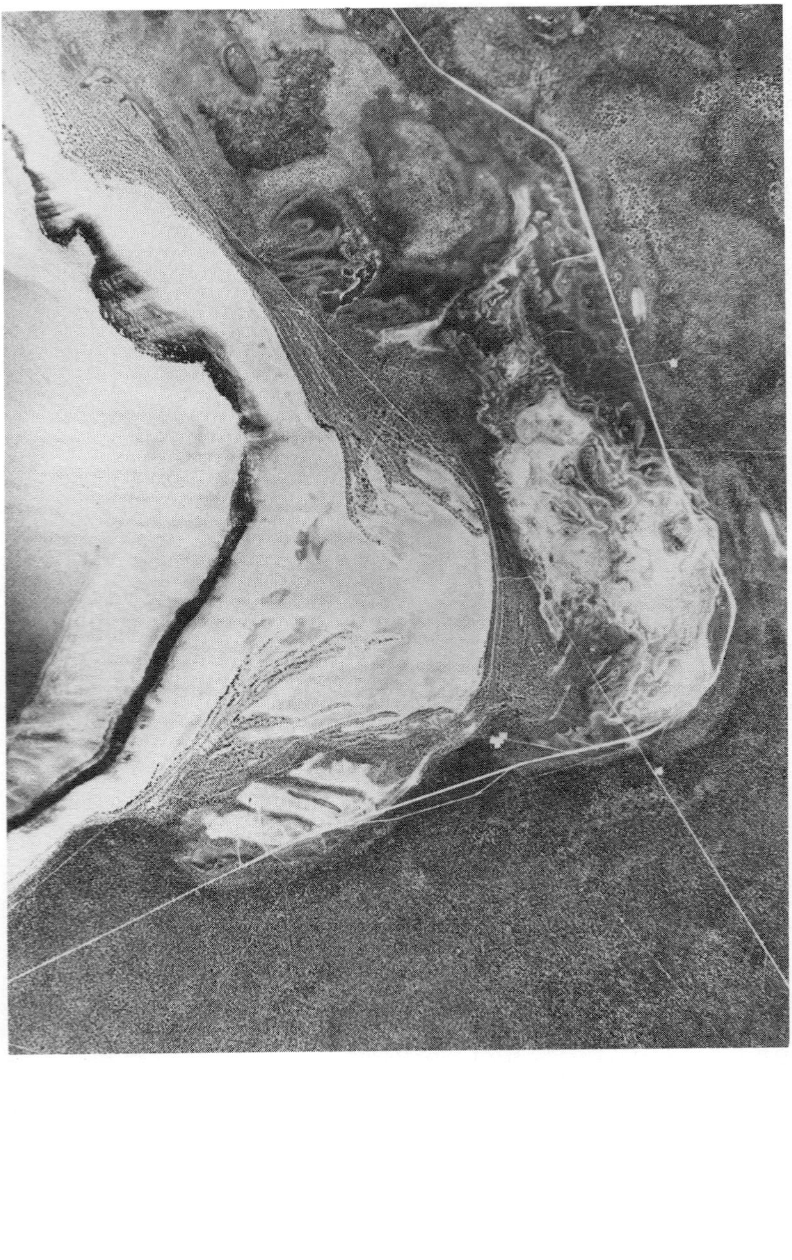

**Plate XIII** Aerial view of study site (alongside road seen in lower right of photograph) at Hamelin Pool, Shark Bay, in Western Australia. Note tidal mudflats and encroaching sand dunes which support 'mulga' vegetation (*Acacia aneura*).

**Plate XIV** Sand track crossing a 'birrida' or 'samphire flat'. These are Pleistocene mudflats which have been colonized by halophytes such as *Arthrocnemum arbuscula*, *Salicornia* and *Atriplex*, and *Amphibolurus nuchalis* is invariably found in low densities in these habitats.

**Plate XV** Commencement of sand dune habitat at the edge of a 'birrida' at Shark Bay. The deeper sandy soil supports a more dense vegetation including the Desert Kurrajong (*Brachychiton gregorii*), snake wood (*Acacia xiphophylla*) and *Acacia sclerosperma*, all seen here. Dense populations of *Amphibolurus nuchalis* typically occur in such habitats whilst they remain disturbed and in a low seral state.

in the field are positively correlated with plasma sodium concentrations. Total corticosteroid levels vary considerably but show no trend with increasing blood sodium concentrations and Bradshaw (1975) compared the response of individuals in spring and late summer to exogenous injections of mammalian adrenocorticotrophic hormone (ACTH), following dexamethasone blockade. As seen in Fig. 5.2, the response was essentially 'normal' in spring, with 60% of the animals showing an increase of plasma corticosteroid concentrations from 2–4 μg/dL, but in late summer the response was noticeably more variable, with a more than three-fold increase in the number of individuals showing little or no response to either stress or ACTH injection.

With the progression of summer at Shark Bay, *Amphibolurus nuchalis* lose body condition, plasma glucose levels fall and plasma corticosteroid concentrations increase significantly (see Table LIV). Measurements of rates of water turnover, using tritiated water, show that water influx and efflux rates are balanced in spring, but efflux rates are almost double influx rates in late summer, leading to chronic dehydration, as seen in Table LV. Incubations of adrenocortical tissue *in vitro* and *in situ* by Bradshaw (1978b) revealed that, although the response of summer animals is more variable than that of individuals in spring, adrenal exhaustion is not a factor and rates of corticosterone production are twice as high in summer, as shown in Table LVI. An interesting positive correlation was also found between circulating corticosterone levels in summer animals and the subsequent production rate of the combined adrenals when incubated *in vitro*, which was not evident in spring, suggesting that chronic exposure to stressful conditions has resulted in an enhanced capacity of the adrenals to secrete corticosterone (see Fig. 5.3).

Bradshaw (1970) developed a 'Condition Index' relating body weight to snout–vent length of the lizard *Amphibolurus ornatus* in order to quantify weight loss in late summer of this species, whether it be due to dehydration, starvation, or a combination of both, and this approach has been extended with *Amphibolurus nuchalis* where three separate 'Condition Indices' are used: one for body weight ($CI_{BW}$), one for total body water content ($CI_{TBW}$) and the third for body solids ($CI_S$). By means of these Condition Indices one may document seasonal changes in body weight and determine whether losses of condition, which are characteristic in late summer, result from dehydration and/or starvation and thus estimate the resulting deviation of the body composition from an optimal state. Fig. 5.4 to 5.6 show the three regressions from which the Condition Indices have been derived, together with the fitted equations for males, females and juveniles. The percentage of the variance in body weight and body water accounted for by each model is respectively 95.5% and 92.7%, but only 85.7% in the case of body solids. The $CI_S$ has less predictive value than the other two CIs because it is derived as the *difference* between weight and water. One would expect errors in the measurement of

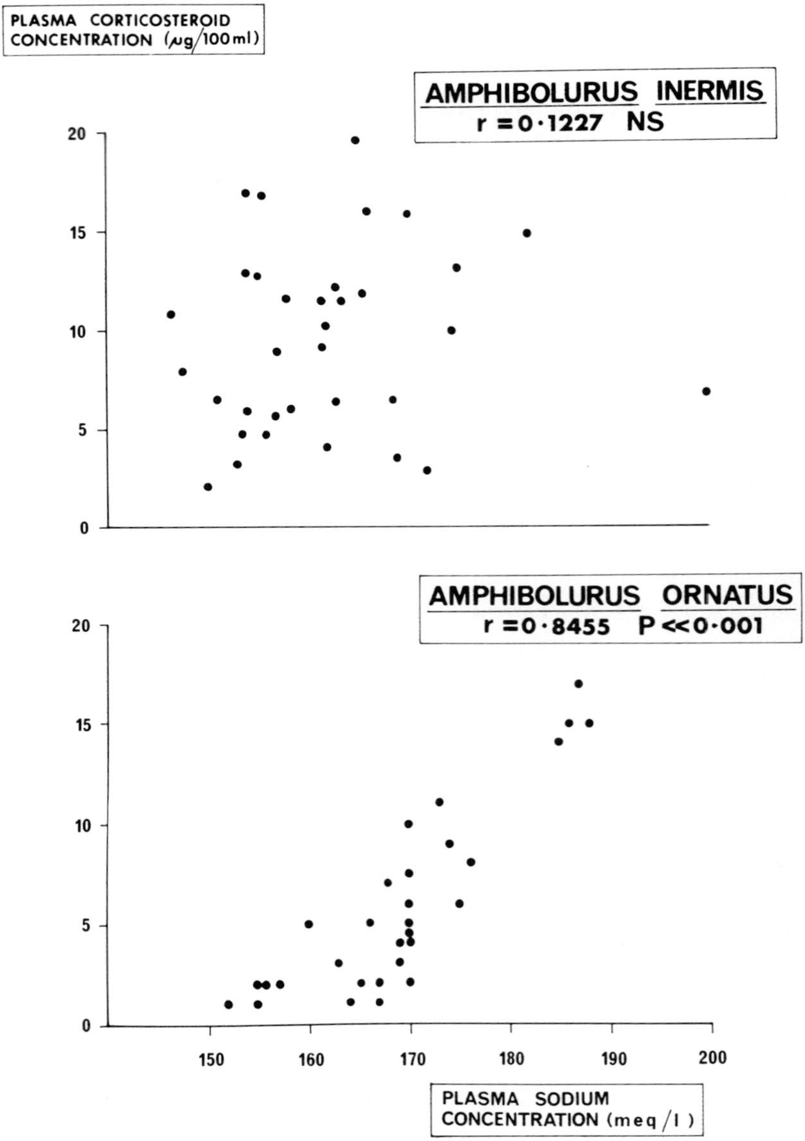


**Fig. 5.1** Relationship between the concentrations of adrenal corticosteroids and sodium in the plasma from *Amphibolurus nuchalis* and *Amphibolurus ornatus* lizards collected in the field during periods of water deprivation (from Bradshaw, 1975).

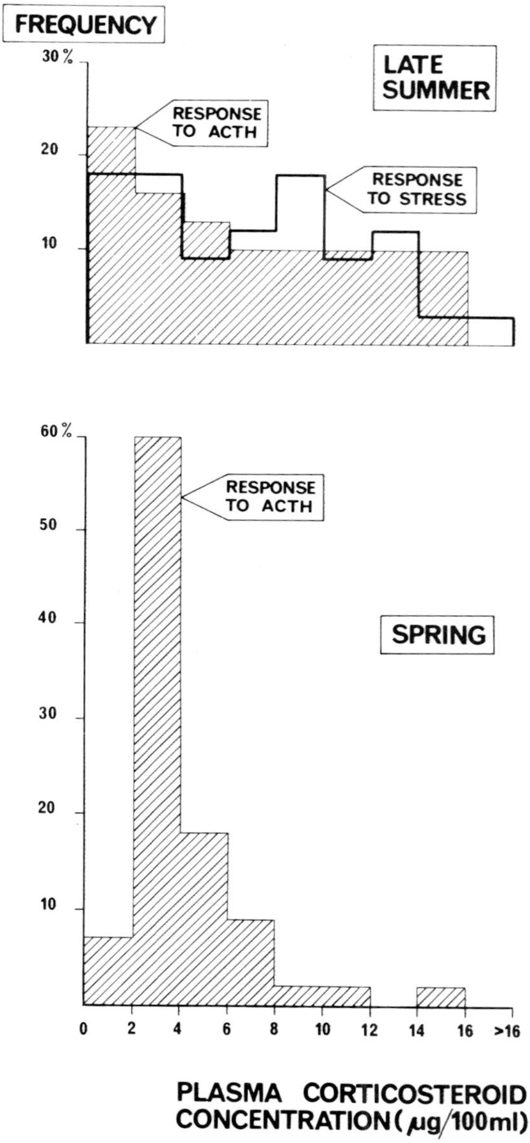


**Fig. 5.2** Frequency distribution of plasma corticosteroid concentrations in *Amphibolurus nuchalis* lizards injected with ACTH after dexamethasone blockade in spring and late summer at Shark Bay in Western Australia. The adrenal response of animals to the stress of confinement in late summer is also shown for comparison (from Bradshaw, 1975).

**Table LIV:** Seasonal changes in physiological parameters in *Amphibolurus nuchalis* at Shark Bay in Western Australia (from Bradshaw, 1978b).

| Season | Condition index* | Total body water content (mL/100g) | Plasma concentration $Na^+$ | Plasma concentration $K^+$ (mmol/L) | Plasma concentration $Cl'$ | Plasma concentration Glucose (mg/dL) | $\pi$ (mOsm/kg) | Plasma corticosteroids (μg/dL) Pre-stress | Plasma corticosteroids (μg/dL) Stressed |
|---|---|---|---|---|---|---|---|---|---|
| Spring | 100.5% | 82.1 | 170.4 | 5.1 | 128.6 | 320.1 | 328.0 | 0.4 | 8.74 |
| (Oct. 1976) | ±1.6 | ±0.9 | ±1.4 | ±0.1 | ±2.6 | ±9.2 | ±8.2 | ±0.1 | ±0.84 |
| Summer | 85.6% | 74.7 | 171.7 | 5.0 | 128.2 | 241.0 | 294.2 | 4.52 | 9.72 |
| (Mar. 1977) | ±2.14 | ±1.32 | ±1.5 | ±0.1 | ±2.5 | ±14.0 | ±2.5 | ±0.48 | ±0.6 |
| Significance | $P < 0.001$ | $P < 0.001$ | NS | NS | NS | $P < 0.001$ | NS | $P > 0.001$ | NS |

* Body weight expressed as a percentage of the fully-hydrated weight

**Table LV:** Seasonal changes in total body water content (TBW) and turnover of tritiated water (HTO) in *Amphibolurus nuchalis* at Shark Bay in Western Australia (from Bradshaw, 1978b).

| Season | Total body water content (mL/100g) | Turnover of HTO | | | |
|---|---|---|---|---|---|
| | | $mL\ (kg.day)^{-1}$ | | % TBW/day | |
| | | Influx | Efflux | Influx | Efflux |
| Spring | 72.4 | 37.8 | 38.3 | 5.3 | 5.4 |
| (Oct. 1969) | ±1.0 | ±5.0 | ±4.3 | ±0.7 | ±0.6 |
| Summer | 62.5 | 44.9 | 68.4 | 6.9 | 10.6 |
| (Feb. 1970) | ±1.1 | ±9.3 | ±9.4 | ±1.2 | ±1.2 |
| Significance | $P < 0.001$ | NS | $P < 0.01$ | NS | $P < 0.001$ |

**Table LVI:** Results of *in vitro* incubation of adrenal glands from *Amphibolurus nuchalis* in spring and summer at Shark Bay in Western Australia (from Bradshaw, 1978b).

| Season | N | Adrenal weight (mg/100g) | Corticosteroid production rate *in vitro* $ng(mg.hr)^{-1}$ | |
|---|---|---|---|---|
| | | | Aldosterone | Corticosterone |
| Spring | 11 | 5.4 | 4.1 | 58.3 |
| (October 1976) | | ±0.4 | ±0.58 | ±6.4 |
| Summer | 12 | 5.4 | 3.2 | 120.8 |
| (March 1977) | | ±0.5 | ±0.65 | ±23.2 |
| Significance | | NS | NS | $P < 0.05$ |

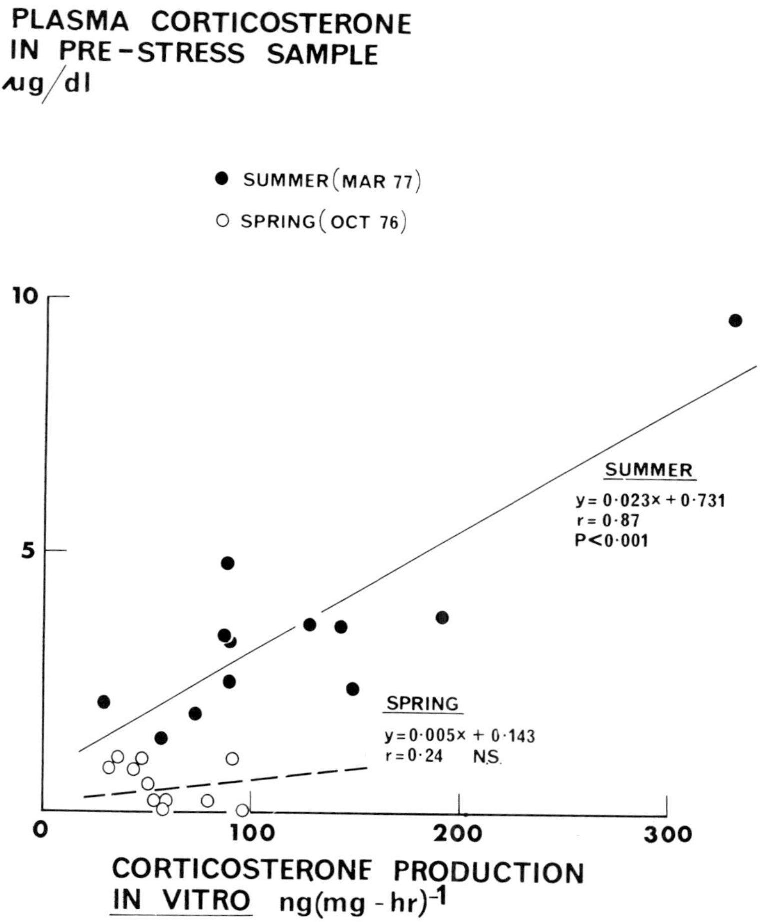


**Fig. 5.3** Correlation between pre-stress plasma corticosterone levels and subsequent *in vitro* production rates of corticosterone by the adrenal glands of the agamid lizard *Amphibolurus nuchalis* incubated *in situ* in spring and in summer at Shark Bay in Western Australia (from Bradshaw, 1978b).

both body weight and body water to be compounded in the data for body solids. Furthermore, since the total body water content is, on average, 80% of the body weight, *proportionally* the errors of measurement of body solids would be expected to be eight to 10 times as large as those of body weight and total body water. This does not mean that the $CI_S$ is useless, only that its predictive value is less than that of the other two Condition Indices.

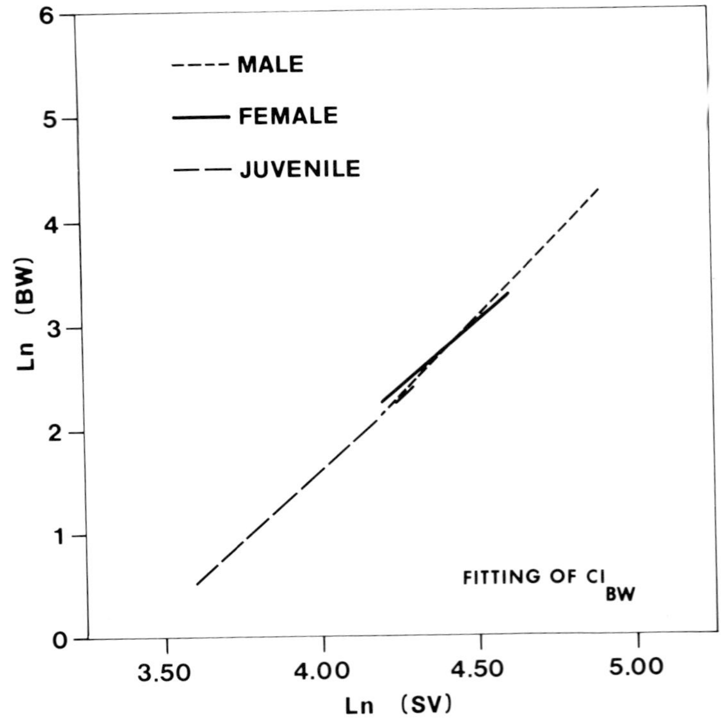


**Fig. 5.4** Derivation of the Condition Index for body weight ($CI_{BW}$) for the agamid lizard *Amphibolurus nuchalis* at Shark Bay. Equations for the three regressions are as follows:

| | | |
|---|---|---|
| Males: | ln(BW) | = 3.05 (lnSV) − 10.68 |
| Females: | | = 2.53 (lnSV) − 8.38 |
| Juveniles: | | = 2.67 (lnSV) − 9.09 |

where BW = body weight in g, SV = snout-vent length in mm, (from Bradshaw & De'ath, unpublished).

If one analyses the data from Shark Bay by season, defining '*spring*' as the period before November 31 (no data were collected prior to the beginning of October in any year) and '*summer*' as the period after December 1, then Table LVII shows that $CI_{BW}$ and $CI_{TBW}$ are both significantly lower in summer than in spring for males, females and juveniles, but $CI_S$ only declines in summer in females, indicating that loss of condition in summer results primarily from dehydration. A temporal analysis of the complete data set shows that $CI_{BW}$ and $CI_{TBW}$ decline progressively with time, whereas $CI_S$ shows no obvious trend. The extent to which Condition Indices are influenced by climatic variables (temperature and rainfall) and activity of the lizards can also be analysed through the development of an appropriate model, such as the following:

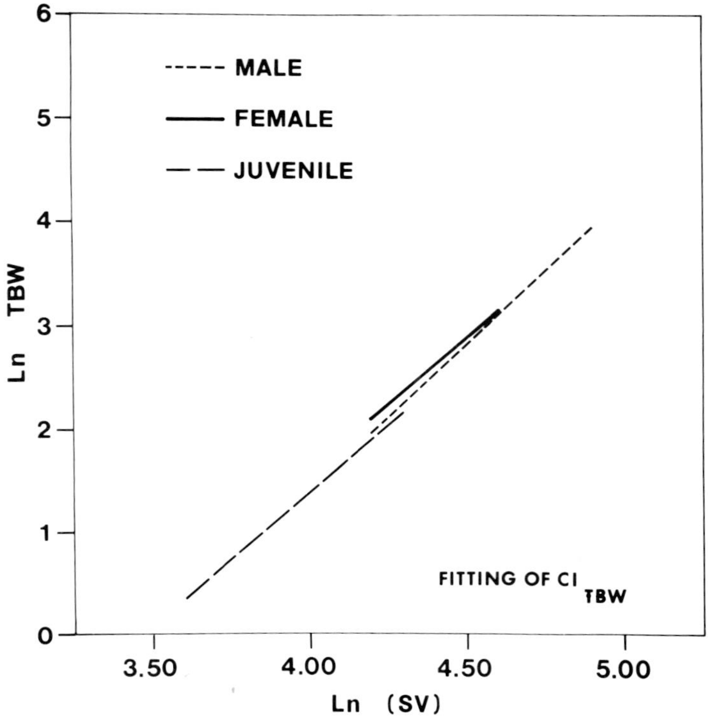


**Fig. 5.5** Derivation of the Condition Index for total body water ($CI_{TBW}$) for the lizard *Amphibolurus nuchalis* at Shark Bay. Equations for the three regressions are as follows:

| | | |
|---|---|---|
| Males: | ln(TBW) | = 2.48 (lnSV) − 9.96 |
| Females: | | = 2.60 (lnSV) − 8.92 |
| Juveniles: | | = 2.61 (lnSV) − 9.05 |

where TBW = total body water content (tritium space) in mL, SV = snout-vent length in mm. (from Bradshaw & De'ath, unpublished).

$$CI_{ij} = a_i + b_{1i}(mxt5)_{ij} + b_{2i}(mnt5)_{ij} + b_{3i}(ac5)_{ij} + b_{4i}(r5)_{ij} + e_{ij}$$

where $CI_{ij}$ is the $j^{th}$ observation of the Condition Index CI for season '$i$'; $a_i$ is the fitted intercept for season '$i$'; $(mxt5)_{ij}$, $(mnt5)_{ij}$, $(ac5)_{ij}$ and $(r5)_{ij}$ are the $j^{th}$ observations of the five-day means of maximum temperature, minimum temperature, activity and rainfall respectively for season '$i$'; $b_{1i}$, $b_{2i}$, $b_{3i}$ and $b_{4i}$ are respective regression coefficients (slopes) for season '$i$'; and $e_{ij} \sim N(0,\sigma^2)$ is the error term.

The means for temperature and rainfall were those calculated for the five days immediately preceding the capture of any given individual. An approximate

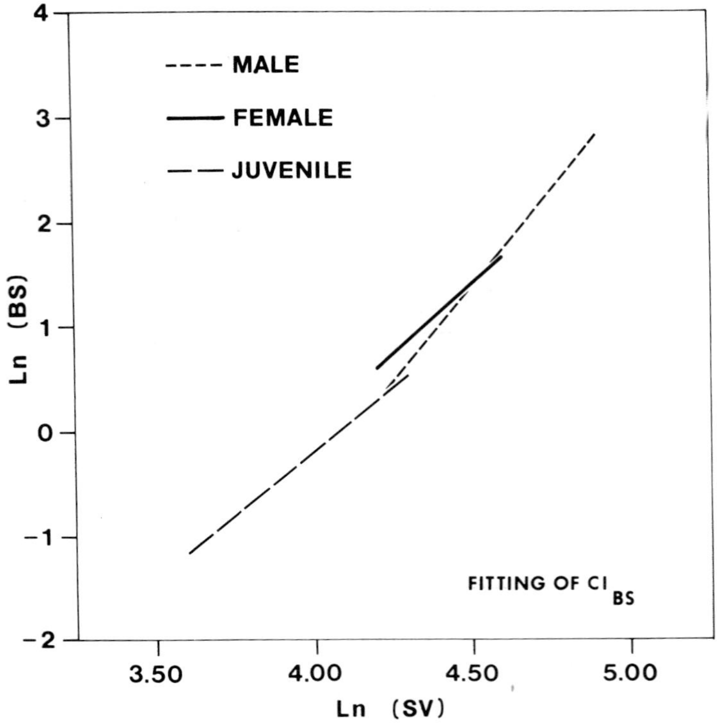


**Fig. 5.6** Derivation of the Condition Index for body solids ($CI_S$) for the agamid lizard *Amphibolurus nuchalis* at Shark Bay. Equations for the three regressions are as follows:

| | | |
|---|---|---|
| Males: | ln(BS) | = 3.63 (lnSV) − 14.96 |
| Females: | | = 2.68 (lnSV) − 10.66 |
| Juveniles: | | = 2.42 (lnSV) − 9.87 |

where BS = body solids in g, SV = snout-vent length in mm. (from Bradshaw & De'ath, unpublished).

activity index for *Amphibolurus nuchalis* was constructed by fitting a sinusoidal curve, with a period of 12 hours, to maximum and minimum temperatures recorded on each day. Potential activity was then estimated, knowing that this species' thermoregulatory sequence is bounded by a lower temperature of 28°C and an upper of 42°C (see Bradshaw & Main, 1968), by measuring the portion of the abscissa whose ordinate lay between 28°C and 42°C on that particular day (Bradshaw & De'ath, in prep.). The variation of this activity potential (*ac*5) with time is shown in Fig. 5.7 and appears to peak in about January (Day 270 with May 1 set as Day 1) and then decline during February and March.

**Table LVII:** Seasonal variation in condition indices for body weight, total body water and body solids for *Amphibolurus nuchalis* at Shark Bay (from Bradshaw and De'ath, unpublished).

| Season | $CI_{BW}$ | | | $CI_{TBW}$ | | | $CI_S$ | | |
|---|---|---|---|---|---|---|---|---|---|
| | ♂♂ | ♀♀ | Juveniles | ♂♂ | ♀♀ | Juveniles | ♂♂ | ♀♀ | Juveniles |
| Spring | 101.68<br>±0.243<br>(154) | 102.23<br>±0.410<br>(120) | 102.54<br>±1.24<br>(8) | 102.16<br>±0.354<br>(152) | 101.68<br>±0.558<br>(117) | 102.00<br>±2.67<br>(8) | 99.91<br>±1.47<br>(127) | 103.23<br>±2.78<br>(95) | 101.38<br>±19.68<br>(7) |
| Summer | 96.61<br>±0.220<br>(188) | 98.18<br>±0.359<br>(147) | 99.12<br>±1.13<br>(40) | 97.46<br>±0.342<br>(130) | 97.49<br>±0.658<br>(79) | 99.13<br>±2.86<br>(32) | 99.93<br>±1.16<br>(105) | 95.43<br>±2.06<br>(65) | 102.37<br>±9.02<br>(28) |
| Statistical significance of difference between means | $P < 0.001$ | $P < 0.001$ | NS | $P < 0.001$ | $P < 0.001$ | NS | NS | $P < 0.05$ | NS |

Means ± S.E., (N)

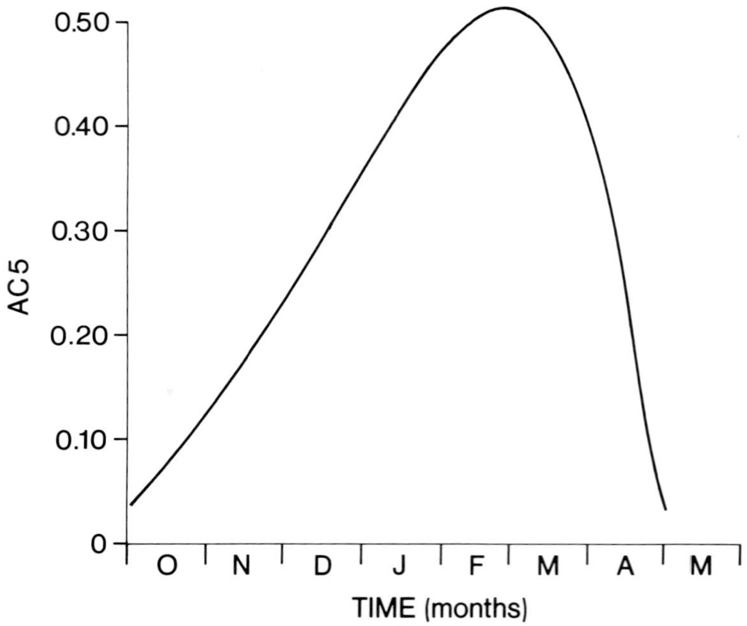


**Fig. 5.7** Potential Activity Index (ac5) for the lizard *Amphibolurus nuchalis* at Shark Bay estimated from a knowledge of this species' thermoregulatory behaviour and using maximum and minimum temperature records for the 5 days immediately prior to capture. The ordinate is expressed in fractions of a day and peak potential activity occurs in mid-summer (February) and thereafter declines rapidly (from Bradshaw & De'ath, unpublished).

The models fitted to the data for spring and summer are as follows:

$$\begin{aligned} CI_{BW} &= 134.34 - 1.250(mxt5) + 30.83(ac5) && \text{SPRING} \\ &= 86.89 + 0.495(mxt5) - 13.35(ac5) && \text{SUMMER} \\ CI_{TBW} &= 143.23 - 1.506(mxt5) + 21.40(ac5) - 0.468(r5) && \text{SPRING} \\ &= 87.83 + 0.478(mxt5) - 15.40(ac5) - 0.468(r5) && \text{SUMMER} \end{aligned}$$

Graphically, the essential features of the model for $CI_{BW}$ are shown in Fig. 5.8 where the spring temperature range has been set at 24–32°C and summer at 32–40°C and the effect of the activity potential (*ac*5) has been investigated by setting it at 0 and 0.3 for spring, and at 0.3 and 0.6 for summer, which are typical high and low values for each season. It is of interest to note that the patterns for spring and summer are *opposed*, with high $CI_{BW}$ being associated

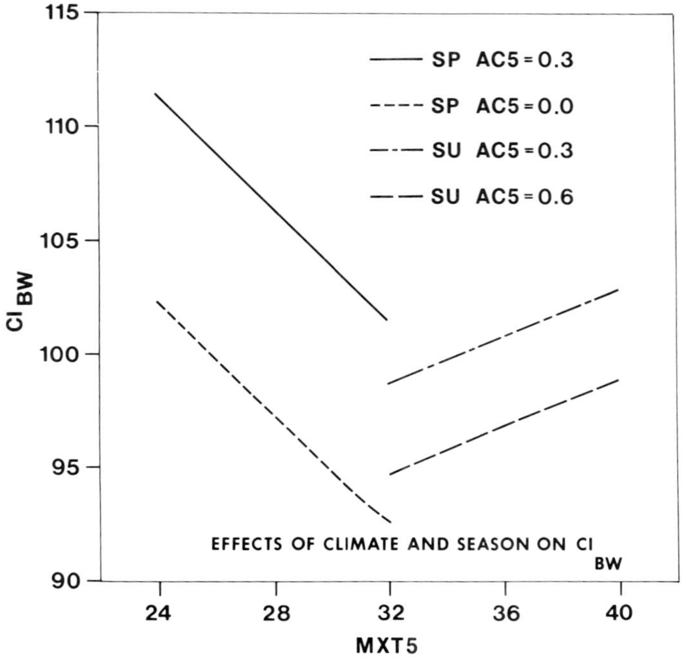


**Fig. 5.8** Results of a model generated to account for seasonal and climatic variations in $CI_{BW}$ of the agamid lizard *Amphibolurus nuchalis* at Shark Bay in Weatern Australia. Variations of $CI_{BW}$ (expressed as a percentage) with changes in mean maximum temperature over the 5 days immediately preceding capture are shown in both spring and summer for differing levels of activity (AC5) varying from 0.0–0.6 (from Bradshaw & De'ath, published).

with low maximum temperatures and high activity. This suggests that when *mxt*5 is less than 28°C and *ac*5 is forcibly equal to 0, then the higher the *mxt*5, the more the inactive animal in its burrow will decline in condition through being unable to feed. The actual mean value for *ac*5 in spring was 0.11, considerably lower than the summer mean of 0.48, and it appears that the relatively short period of time available each day in spring is used by the animals for intensive foraging. This is reflected in the very large effect that activity has on the $CI_{BW}$ in this season. The situation in summer is reversed, with increased activity leading to a *decline* in condition and high maximum temperatures up to 40°C enhancing condition.

The model for $CI_{TBW}$ is similar to that for $CI_{BW}$ so far as *mxt*5 and *ac*5 are concerned (negative effects on the condition index in spring and summer respectively), but a surprising *negative* effect of rainfall is evident in both

**Table LVIII:** Precipitation records from 1969–1982 at Shark Bay (Hamelin Pool Station), showing total rainfall plus summer and winter components and percentage deviation from the respective long-term means.

| Year | Total (mm) | % Deviation from mean* | Winter (May-Aug) (mm) | % Deviation from mean | Summer (Nov-Apr) (mm) | % Deviation from mean |
|---|---|---|---|---|---|---|
| 1969 | 135 | −36.3 | 120 | −14.7 | 189** | +245.0 |
| 1970 | 310 | +46.2 | 103 | −29.3 | 175** | +219.5 |
| 1971 | 409† | +92.9 | 231 | +58.5 | 16 | −70.8 |
| 1972 | 243 | +14.6 | 210 | +44.1 | 9 | −83.6 |
| 1973 | 185 | −12.7 | 141 | −3.3 | 35 | −36.1 |
| 1974 | 343 | +61.8 | 278 | +90.7 | 131** | +139.1 |
| 1975 | 353 | +66.5 | 142 | −2.6 | 34 | −37.9 |
| 1976 | 129 | −39.2 | 112 | −23.2 | 30 | −45.2 |
| 1977 | 138 | −34 9 | 100 | −31.4 | 107** | +95.3 |
| 1978 | 190 | −10.4 | 70 | −52.0 | 46 | −16.1 |
| 1979 | 100 | −52.7 | 54 | −63.0 | 37 | −32.5 |
| 1980 | 199 | −6.1 | 138 | −5.4 | 0 | −∞ |
| 1981 | 138 | −35.1 | 128 | −12.2 | 62 | +13.9 |
| 1982 | 141 | −33.6 | 64.8 | −55.6 | 8 | −85.4 |

* Mean rainfall (1885–1973) = 212 mm

† Highest rainfall ever recorded; lowest = 44 mm in 1891

Mean Winter Rainfall (1885–1977) = 145.8 ± 6.25 mm (CV = 41.2%)

Mean Summer Rainfall (1885–1977) = 54.8 ± 5.30 mm (CV = 93.3%)

** Cyclonic rain recorded.

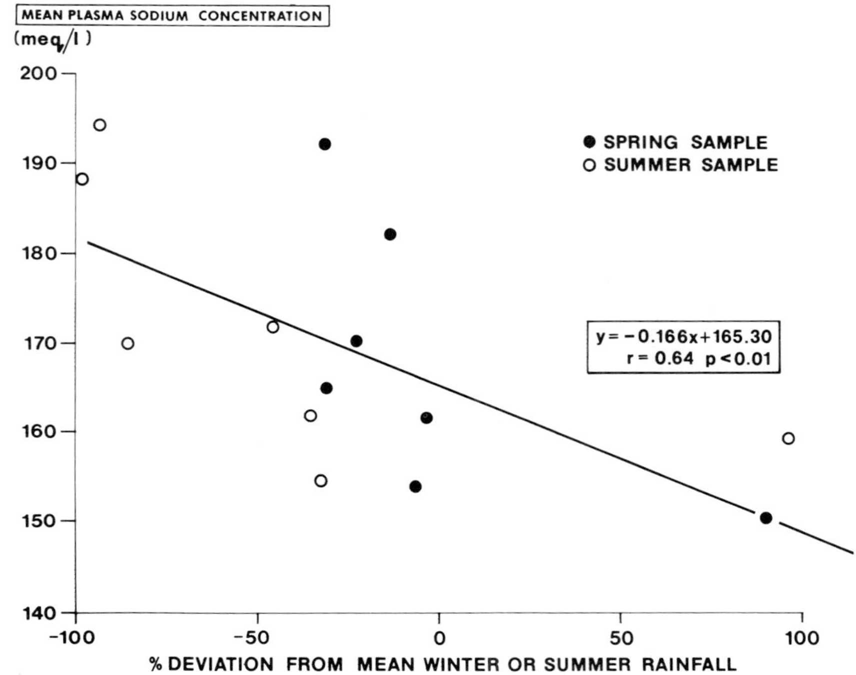


**Fig. 5.9** Variation in mean plasma sodium concentration of *Amphibolurus nuchalis* at Shark Bay as a function of rainfall. Samples from both spring and summer are graphed and rainfall is expressed as the % deviation of the rain falling in the season preceding the sample, from the long-term mean for that season (from Bradshaw, unpublished).

seasons. One of the inherent problems in attempting to interpret the effect of rainfall on physiological parameters and body condition is the extreme variability of the precipitation recorded at Shark Bay over the period of the study. Rainfall records for Hamelin Pool Station at Shark Bay from 1969 to 1983 are tabulated in Table LVIII and the percentage deviation from the overall mean (from 1885 until 1973) and the calculated means for the winter and the summer periods are also listed. Over the 15 years of the study, total rainfall has varied from −52.7% to +92.9% of the average and winter and summer deviations were −63.0% to +90.7% and − ∞% to +245.0% respectively, indicating how unpredictable the amount of rain that will fall in either season is. When this variation is taken into account, however, and plasma sodium concentrations for the individual samples are plotted as a function of the percentage deviation of the rainfall, whether summer or winter for that particular year, a significant negative correlation emerges (see Fig. 5.9) and it is clear that hypernatraemia may be present in either summer or winter, depending upon the extent of the rainfall in the preceding season. Similarly, plasma sodium concentrations are not necessarily elevated every summer (although this is the case *on average*—see Table LIII), and they may be quite low if the animals have been exposed to cyclonic rains, as in 1970 and in 1971 when heavy falls were recorded in January and February.

The picture which thus emerges from the analysis of changes in Condition Indices is one of progressive physiological deterioration throughout the summer period following the major reproductive effort of this species in spring. Body weight declines slowly, due primarily to dehydration. In later summer, the animals are characteristically in negative water balance and females show evidence of an additional loss of body solids. Plasma glucose concentrations fall, and the circulating levels of adrenal corticosteroids rise, but there is no evidence of 'adrenal exhaustion'. Individuals in poor condition typically show enhanced levels of steroidogenesis when their adrenal tissue is incubated *in vitro* with radioactive precursors (see Fig. 5.10). The potential activity index (*ac*5), shown in Fig. 5.7, suggests that activity declines in late summer at Shark Bay, due to the predictable interaction between the animal's stereotyped thermoregulatory behaviour and seasonal changes in environmental temperature. Reduced opportunities for foraging may account for the negative water balance at a time when are temperatures are still high and rates of evaporative water loss would be elevated. In an attempt to clarify this situation, measurements were made in spring and summer of field metabolic rates (FMR), using doubly-labelled water ($^{3}HH^{18}O$) in conjunction with $^{22}Na$, the turnover of which may be used, under appropriate circumstances, to estimate rates of food intake in the field (Bradshaw & Nagy, in prep.).

Field metabolic rates and rates of water turnover for the Shark Bay population of *Amphibolurus nuchalis* during spring and summer periods are shown in Fig. 5.11. It is clear that rates of $CO_2$ production fall as summer progresses, (from 0.51 ± 0.07 mL$(g.h)^{-1}$ in November, to 0.33 ± 0.04 mL$(g.h)^{-1}$ in February ($P < 0.01$) and then to 0.12 ± 0.02 mL$(g.h)^{-1}$ in March ($P < 0.001$). Rates of water turnover similarly fall (influx from 42.8 ± 5.0 mL$(kg.d)^{-1}$ in spring to 18.3 ± 3.0 mL$(kg.d)^{-1}$ in late summer, $P < 0.001$)) and it is important to note that summer animals are in *negative water balance* with rates of water efflux significantly exceeding rates of influx. Rates of sodium turnover in general mirror those for water (see Fig. 5.12) with summer animals exhibiting negative sodium balance, but the decline in sodium influx in late summer, from 1.96 ± 0.24 to 1.35 ± 0.34 mmol$(kg.d)^{-1}$, is not statistically significant.

These rates of turnover of water and sodium in the field can be used to estimate dry-matter intake (DMI), if a correlation can be established between DMI and turnover under controlled laboratory conditions where the lizards are fed a measurable amount of a diet of known composition. Dry-matter intake

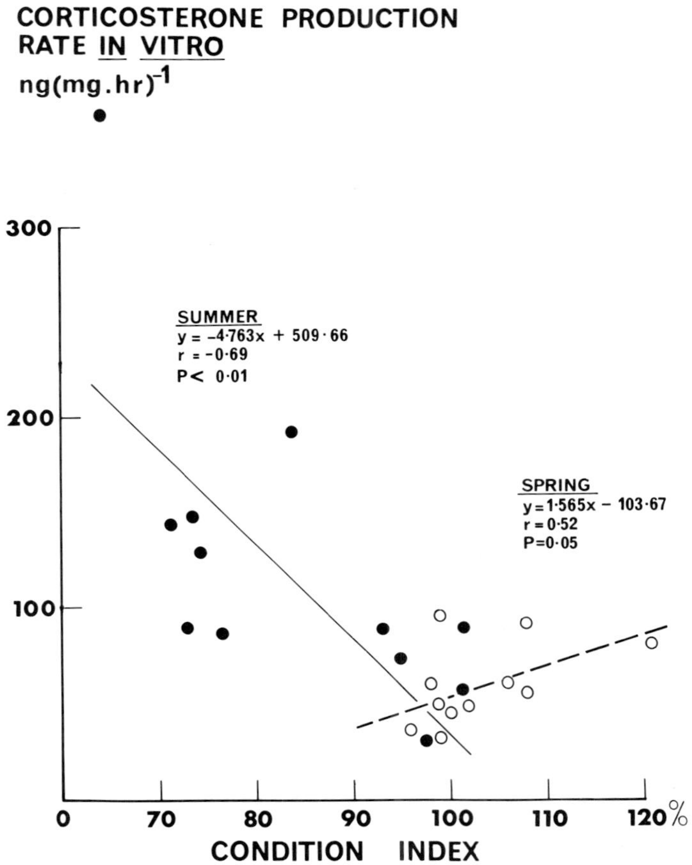


**Fig. 5.10** Seasonal variations in rate of adrenal *in vitro* corticosteroidogenesis in the lizard *Amphibolurus nuchalis* at Shark Bay as a function of the body weight Condition Index. Enhanced rates of corticosterone production in individuals in poor condition in summer are evident (from Bradshaw, unpublished).

and rates of both sodium and water turnover were found to be significantly correlated in a series of feeding trials with *Amphibolurus nuchalis* fed a mixed diet of ants, beetles, grasshoppers and with 13% plant material (to approximate stomach contents of animals in the field). The respective equations were as follows:

$$\text{Water influx} = 2.955(\text{DMI}) + 3.114 \quad (r = 0.91, P < 0.001)$$
$$\text{Sodium influx} = 0.140(\text{DMI}) + 0.233 \quad (r = 0.66, P < 0.01)$$

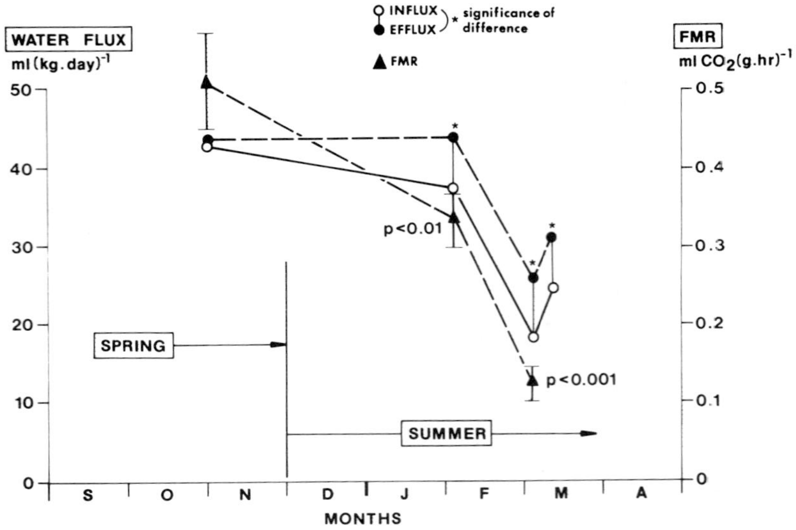


**Fig. 5.11** Seasonal changes in rate of turnover of water (influx and efflux) and Field Metabolic Rates (FMR—determined using doubly-labelled water) of *Amphibolurus nuchalis* at Shark Bay. A significant decline in both parameters is evident in late summer, commensurate with a fall in activity (see Fig. 5.7) (from Bradshaw & Nagy, unpublished).

and from these one may predict DMIs in the field from water and sodium turnover data. Dry-matter intakes (in $g(kg.d)^{-1}$), were predicted from turnover data on four field trips, one in spring (October/November 1980) and three in summer (January/February 1982; March 1980; and March 1983) and in no case did the mean DMI predicted from water turnover data differ significantly from that derived independently from the turnover of sodium. The data were therefore averaged and are presented also in Fig. 5.12. As would be expected, DMIs reflect the overall changes in water and sodium turnover, with a slight but non-significant fall in January/February followed by a highly significant reduction in March.

Death in late summer of adult *Amphibolurus nuchalis* at Shark Bay is thus associated with reduced rates of metabolism, activity, and food and water intake and these changes are mirrored by declining blood glucose concentrations and increasing circulating levels of adrenal corticosteroids. The *ultimate* cause of death appears to be chronic dehydration, but the reduced DMIs suggest that starvation is also a factor, although the $CI_S$ confirms this only in the case of females. No quantitative data are available on insect abundance at Shark Bay in spring and summer but qualitative observations suggest that there is a marked reduction in both insect diversity and abundance

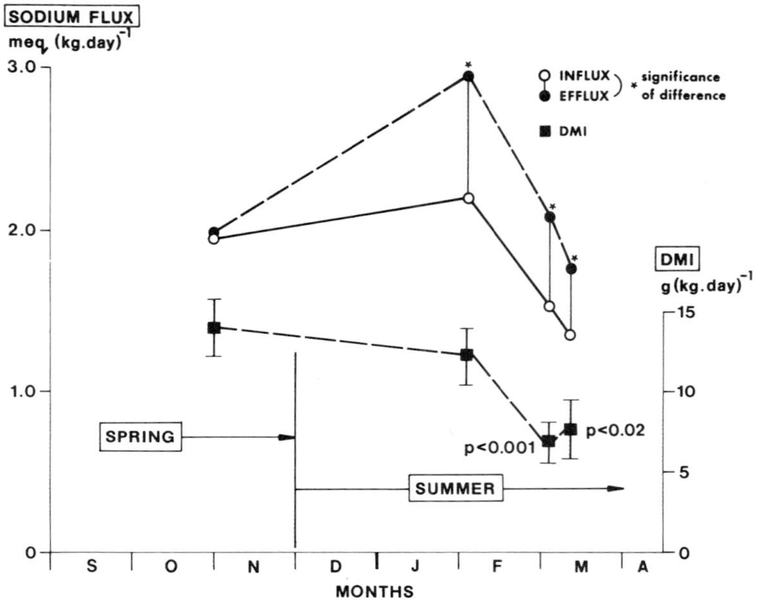


**Fig. 5.12** Seasonal variations in rate of sodium turnover ($^{22}$Na) and estimated dry-matter intake (DMI) of *Amphibolurus nuchalis* at Shark Bay in Western Australia (from Bradshaw & Nagy, unpublished).

with the progression of summer. This would exacerbate the lizards' difficulties in maintaining water and electrolyte balance. Although it is tempting to thus view this species' demise as being the consequence of a harsh, arid environment, exceeding the genetic capacities of this lizard to maintain physiological homeostasis, this ignores the fact that juveniles emerge from the egg phase in late February, and not only survive but grow throughout this period when the adults are dying (Bradshaw, 1981). How is it possible therefore for juveniles to survive under conditions that are inimical to the adults?

In an attempt to answer this question, comparable measurements were made on newly-emerged juveniles (body weights 2–10 g) in April 1981. These results are of some interest. The rates of $CO_2$ production (FMR) measured from the turnover of $^3HH^{18}O$ were very high and averaged 0.675 ± 0.18 $mL(g.h)^{-1}$, which is almost six times greater than that of adults measured at a comparable time of the year. To some extent this difference must reflect higher weight-specific rates of metabolism, which one would expect in smaller individuals, but applying the equation of Bennett & Dawson (1976), relating rate of oxygen consumption to body weight in lizards at 37°C, ($y = 0.424g^{-0.18}$ mL $O_2(g.h)^{-1}$), gives a figure of 0.297 mL $O_2(g.h)^{-1}$, or 0.223 mL $CO_2(g.h.)^{-1}$, assuming an RQ of 0.75. FMR is thus still three times higher in these juveniles than would be predicted on the basis of their body weight, and this increment must reflect increased activity, as the equation is for lizards measured at rest. These figures corroborate personal observations that juveniles are much more active than adult *Amphibolurus nuchalis* at this time of the year, emerging earlier and retiring later in the day.

Rates of water turnover for the juveniles were also considerably higher than those of adults (influx = 32.9 mL$(kg.d)^{-1}$ and efflux = 56.4 mL$(kg.d)^{-1}$) but, as with adults, the fluxes were not balanced. The juveniles were in negative water balance and lost, on average, 2.8% of their body weight per day. To what extent this weight loss resulted from the fact that extremely small individuals (in some cases weighing only 2 g) were being captured, injected, bled and recaptured over a period of seven days, is not known, but one would expect the effect of disturbance to have been greater with juveniles than with adults. Individual growth records are not available for juvenile *Amphibolurus nuchalis* at Shark Bay over this period, but some growth does occur between February and August (Bradshaw, 1981), suggesting that food is assimilated during late summer, prior to the winter period of dormancy. Comparing Condition Indices for the juveniles with those few adults captured in April, 1981, is also instructive and Table LVII shows that the juveniles have a higher $CI_{BW}$ and $CI_{TBW}$, but do not differ in their $CI_S$. This suggests that the juveniles are in better overall condition because of their higher body water content (mean = 83.1 ± 1.58%) and their small body size would again be a factor, as juvenile vertebrates characteristically have higher weight-specific water contents than adults.

Rates of sodium turnover were also much higher than recorded for adults at Shark Bay in late summer (influx = 3.54 ± 0.84 mmol $(kg.d)^{-1}$; efflux = 5.78 ± 0.94 mmol $(kg.d)^{-1}$, $P < 0.001$) and the juveniles were also in negative sodium balance. Water and sodium turnover data can again be used to estimate rates of DMI in the field, although these figures should be viewed with a certain degree of caution as the original laboratory feeding trial did not include juveniles, and their diet in the field is unknown, and almost certainly different from that of the adults. In fact, the DMI estimated from rates of water turnover was significantly different from that predicted from rates of sodium turnover (8.15 ± 1.72 g$(kg.d)^{-1}$ *versus* 22.35 ± 5.73 g$(kg.d)^{-1}$, $P < 0.05$), and further work would be needed to establish which of these two estimates is the more reliable. One other measurement which suggests that feeding rates may be higher in juveniles than in adults, however, is plasma glucose concentration: this averaged 14.94 mmol/L in a sample of five juveniles large enough in size to permit the measurement, which compares with 13.91 mmol/L in the four adults caught on the trip. This difference is highly significant statistically with $P \ll 0.001$.

When this long-term study of the environmental endocrinology of *Amphibolurus nuchalis* was commenced at Shark Bay, it appeared that this annual species died during the long summer from the effects of prolonged drought, in spite of its impressive physiological adaptation to arid conditions, manifested in reduced rates of metabolism and evaporative water loss (Bradshaw, 1970). In fact the situation is more complex than this, and although the *ultimate* cause of death does appear to be dehydration, the *proximate* cause seems to be the stress associated with the process of reproduction and the attenuating effect that this has on the individual's subsequent ability to withstand the delibitating effects of drought and excessively high environmental temperatures. If this were not the case it would be difficult to see how juveniles, with their larger surface/mass ratio and greater vulnerability to the desiccating effects of the environment, could survive at a time when the much larger adults are dying. Subsequent work with *Amphibolurus nuchalis* at Port Hedland, approximately 2000 km north of Perth, Western Australia, where it is found living sympatrically with another species from this genus, *Amphibolurus caudicinctus*, reveals that *nuchalis* die there during summer as at Shark Bay, but that *caudicinctus*, which do not breed until March/April survive the summer drought with little apparent difficulty, despite the fact that they seem physiologically less well-adapted to survive long periods of drought (see Bradshaw, 1970). Further work is needed, and is in progress, on the effects of the *process of reproduction* in males and females on pituitary–adrenal function in an effort to account for the breakdown in physiological homeostasis that is evident in postreproductive specimens of both species. Mortality of males after mating has been described in a number of species of dasyuroid marsupials (see Lee *et al.*, 1977), and ascribed to enhanced adrenocortical activity. It would be of great interest to know whether a similar mechanism is operative in reptiles.

## II. *UROMASTIX ACANTHINURUS* IN THE SAHARA DESERT

This large (0.3–0.7 kg) agamid herbivore, which is generally thought of as one of the most 'desert-adapted' of reptiles, is known in North Africa either as 'Le Fouette-Queue' (Whiptail) in French or 'Dob' in Arabic (see Plate XVI). It has already been discussed in some detail from the point of view of water and electrolyte balance in Chapter 2. *Uromastix acanthinurus* occurs in some of the driest regions of the Sahara (see Plate XVII), always in association with some vegetation. Over 60 species of plants have been identified in its diet (Dubuis *et al.*, 1971). A relatively large and stable population was studied in the region of Béni-Abbès in Algeria by Grenot (1976) and by Lemire (1983). The oasis of Béni-Abbès is a particularly propitious place for such a study as it occurs at the

**Plate XVI** The Fouette-Queue or 'Dob', *Uromastix acanthinurus*, photographed at the entrance to its burrow in the Monts d'Ougarta in Algeria, near Béni-Abbès by Dr C. J. Grenot.

**Plate XVII** Partially-disinterred burrow system of *Uromastix acanthinurus* in southern Morocco, close to the village of Tata on the edge of the Sahara (note size of trenching tool).

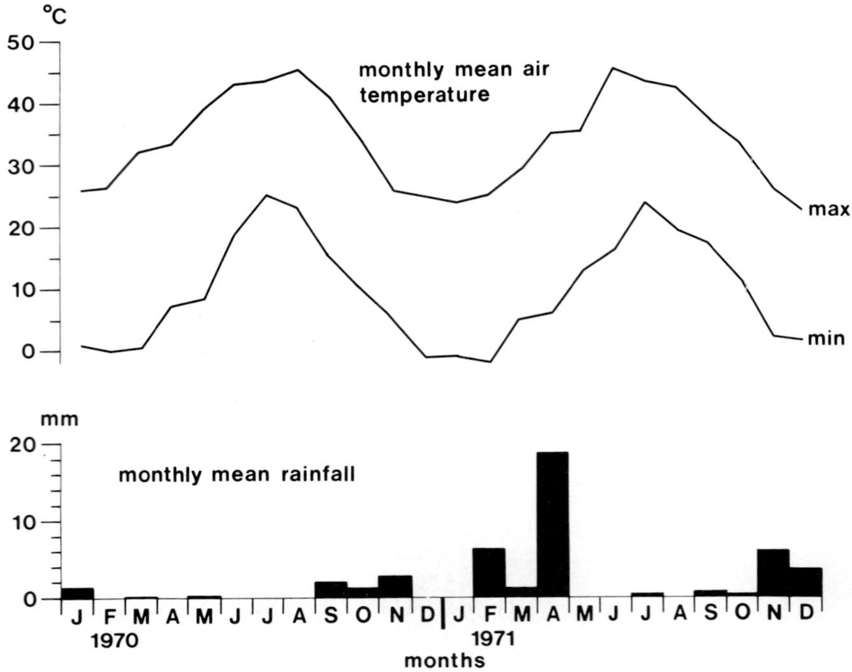


**Fig. 5.13** Variations in climatic factors (monthly mean maxima and minima for air temperatures and rainfall in mm) during 1970 and 1971 in the Ougarta Mountains in Algeria where a population of *Uromastix acanthinurus* was studied (from Lemire, Grenot & Vernet, 1982).

confluence of the four major Saharan biotopes: *djebel* (Monts d'Ougarta), *reg* (Hamada du Guir), *erg* (Grand Erg Occidental), and *oued* (L'oued Saoura). A complete ecophysiological study of this species is still lacking, because of the difficulties associated with studying it in the remote region where it occurs in any abundance. Typical densities are of the order of 0.1–0.2 individuals/ha, but may reach 3–6/ha in very favourable habitats, such as the Monts d'Ougarta (Grenot & Vernet, 1973).

Nothing is known of the lifespan of this animal in its natural habitat, but it is certainly not an annual species. Thus it is of interest to contrast its responses to unfavourable climatic conditions with those of *Amphibolurus nuchalis*. As was seen in Chapter 2, the water content of the vegetation preferred by *Uromastix acanthinurus* declines substantially towards the end of summer (Fig. 2.26). This is associated with a substantial loss of body condition, dehydration and hypernatraemia (see Fig. 2.24 and 2.25). Lemire *et al.* (1982) reported data

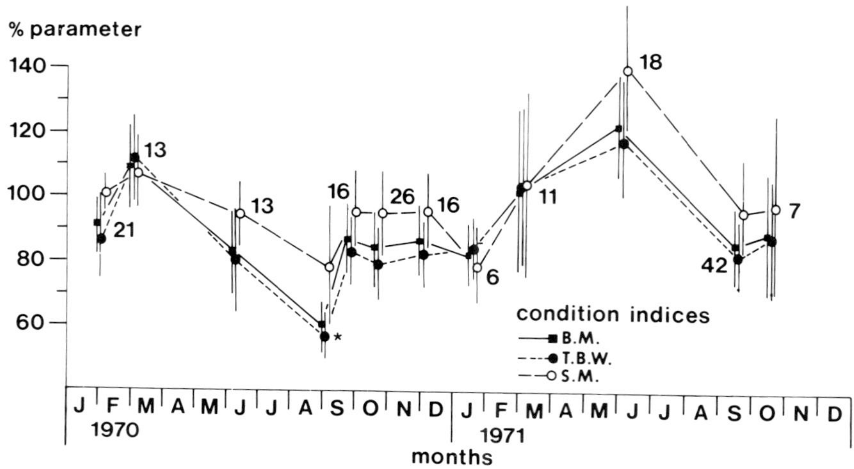


**Fig. 5.14** Variations of Condition Indices (BM = body mass; TBW = total body water; SM = solid matter) of a population of *Uromastix acanthinurus* in the Ougarta Mountains in central Algeria during 1970 and 1971. Data plotted as Means ± SE with the number of individuals for each sample shown in parentheses (from Lemire, Grenot & Vernet, 1982).

from the population of *Uromastix acanthinurus* in the Monts d'Ougarta from 1971–1972 and changes in climatic factors and Condition Indices are shown in Figs. 5.13 and 5.14. The $CI_{BW}$ and $CI_{TBW}$ both decline in late summer (August–September) but, as with *Amphibolurus nuchalis*, the fall in $CI_S$ is not as marked, nor statistically significant. In 1971 conditions were much more favourable than in 1970, with almost 20 mm of rain falling in April. This was followed by a marked improvement in the condition of the animals caught that year and Condition Indices all exceeded 100% in May–June. The $CI_S$ averaged almost 140% due to the presence of eggs in all eight females captured whereas in the previous year only one of six females caught was gravid at the same time of year. During the summer of 1970, which was particularly severe, Lemire *et al.* (1982) were able to discern two physiological groups amongst the individuals that were active: those that were in good condition, fully hydrated, and showing relatively normal plasma sodium concentrations; and a second group which evidenced considerable weight loss due to dehydration and pronounced hypernatraemia (see Table LIX). This is very reminiscent of what Bradshaw (1970) found with slow- and fast-growing *Amphibolurus ornatus* in late summer at Bakers Hill, but no data are available on individual growth rates in *Uromastix acanthinurus*. It can only be speculation whether this pronounced individual variation has a similar basis.

**Table LIX:** Water and electrolyte balance of *Uromastix acanthinurus*, population of Ougarta Mountains (Rhenouma) during summer (June 9–10, 1970) (from Lemire *et al.*, 1982).

| | Hydrated (n = 8) | | Dehydrated (n = 6) |
|---|---|---|---|
| B.M./S.V.L. regression | y = 3.18x − 1.74 (r = 0.93) | | |
| Body mass index (%) | 90.1 ± 9 | | 72.8 ± 11.4 |
| | | P < 0.010 | |
| Total body water index (%) | 90.1 ± 9 | | 64.8 ± 11.8 |
| | | P < 0.005 | |
| Solid matter index (%) | 94.9 ± 9.8 | | 87.6 ± 16.6 |
| | | NS | |
| $Na^+$ (mmol/L)[a] | 159.3 ± 5.6 | | 207.9 ± 13.2 |
| | | P < 0.005 | |
| $K^+$ (mmol/L)[a] | 3.2 ± 0.5 | | 3.4 ± 0.8 |
| | | NS | |
| Total body water (mL/100g)[a] | 75.3 ± 0.9 | | 70.3 ± 0.5 |
| | | P < 0.005 | |

[a] n = 3 for hydrated, n = 4 for dehydrated animals.
Mean values ± S.D., B.M./S.V.L. body mass *vs* snout-vent length.

Some insight into the response of this species to drought conditions can be gleaned from the few measurements that have been made of rates of water turnover in the field, and under semi-natural conditions at Béni-Abbès (Lemire *et al.*, 1979; Lemire, 1983). Thirty three individuals were captured and released in May 1978 at Igli on the Hamada du Guir; 12 were later recaptured and rates of water turnover estimated. These animals were in balance, with rates of influx not differing significantly from efflux rates. Mean influx values for males, females and juveniles were 9.7, 13.0 and 9.4 mL$(kg.d)^{-1}$ respectively. These rates of turnover correspond to figures of 1.21%, 1.38% and 1.27% of the total body water content exchanged per day and are the lowest recorded for any reptile (see Nagy, 1982). Lemire (1983) also reports data from animals studied in September of the previous year and these data, compared with those from May 1978 in Table LX, differ significantly. Rates of water turnover are much higher in these individuals (influx = 75.1 mL$(kg.d)^{-1}$) and the animals are in negative water balance (efflux = 101.3 mL$(kg.d)^{-1}$, $P < 0.005$), and experienced acute dehydration during the period of seven days they were at liberty. The reasons for this are that not only were conditions much warmer at Igli in September, 1977 than in May, 1978, but also the animals were perturbed in some way by the experiment. All abandoned their habitual burrows and dug new burrows in the vicinity, losing considerable amounts of body water in the process. Data from animals maintained in enclosures, exposed to natural conditions, and fed various diets (lucerne and native plants)

**Table LX:** Rates of water turnover of free-ranging *Uromastix acanthinurus* at Igli (Hamada de Guir, Algeria) in September 1977 and May 1978 (from Lemire, 1983).

| | | | | | | | Water turnover | | | | |
|---|---|---|---|---|---|---|---|---|---|---|---|
| | | Days | Body Wt | | TBW | | mL(kg.day)$^{-1}$ | | % TBW/day | | ΔBW |
| Period | N | active | Initial | Final | Initial | Final | Influx | Efflux | Influx | Efflux | (%/day) |
| | | | | | % | | | | | | |
| September 1977 | 5 | 7.0 | 413.2 ±235.8 | 389.5 ±232.6 | 78.3 ±2.2 | 54.2 ±2.5 | 75.1 ±2.0 | 101.3 ±2.0 | 9.6 ±0.58 | 12.7 ±0.60 | −1.0 ± 0.43 |
| | | | | | P < 0.005 | | P < 0.005 | | P < 0.005 | | |
| May 1978 | 12 | 15.3 ±2.6 | 297.0 ±169.9 | 279.7 ±160.0 | 80.7 ±3.2 | 79.5 ±4.5 | 11.4 ±2.4 | 12.8 ±5.6 | 1.35 ±0.36 | 1.67 ±0.72 | −0.38 ± 0.09 |
| | | | | | NS | | NS | | NS | | |

are summarised diagrammatically in Fig. 5.15 and it is evident that, despite the provision of supplementary vegetation, the animals were still not in balance and had relatively high rates of evaporative water loss when compared with *Uromastix* living under natural conditions at Igli in May, 1978. Also shown in Fig. 5.15 are data from animals not feeding and inactive within their burrow. These show that this species is able to balance unavoidable losses of water from metabolic water production under these conditions.

The data available for *Uromastix acanthinurus* are unfortunately too few and fragmentary to allow more than speculation on the causes of death of this lizard in its natural habitat. Grenot (1976) gives no indication of lifespan but states that sexual maturity is achieved at four to five years of age and found that 80% of the young died in their first two years of life, either as a result of low temperatures in winter, or from predation. Grenot & Vernet (1972) found evidence of young *Uromastix acanthinurus* in the stomachs of two Saharan snakes, the Horned-Viper *Cerastes cerastes* and *Psammophis schokari*, as well as in the varanid, *Varanus griseus*. Doubtless the many raptors common in the Sahara, as well as carnivorous mammals such as *Vulpes ruppelli, Felix lybica* and the hedgehog, *Paraechilnus aethiopicus*, are also effective predators on young *Uromastix*. Once adulthood is achieved, however, they would appear to have few predators, other than humans, and their lifespan is probably quite long, depending upon available food resources. Grenot (1976) records biomass figures from 2–29 kg/ha for *Uromastix acanthinurus* in Algeria, which may be quite substantial in the Sahara. The home range of each animal is from 1–5 ha, and this gives a considerable area over which the individual may graze in search of suitable vegetation. It seems clear that this animal's fortunes are inextricably linked to the quality, and especially the water content, of the vegetation. Lemire (1983) clearly established the link between rainfall, growth of the vegetation, enhancement of condition and, ultimately, successful reproduction of *Uromastix acanthinurus* in regions where annual rainfall is always less than 30 mm. This species is thus exposed to habitats approximately 10 times more arid than that occupied by *Amphibolurus nuchalis* at Shark Bay. Although both show evidence of chronic dehydration in late summer, *Uromastix acanthinurus* is able to survive such periods due to its larger size and reduction of activity. Whereas *Amphibolurus nuchalis* remain active throughout summer, and gradually die, *Uromastix acanthinurus* considerably reduce their activity and spend more time in their burrows, thus reducing both their energy expenditure and rate of water loss. Figures from individuals in their burrows show that this species is capable of balancing insensible water losses with metabolic water production and, under these conditions, death would come ultimately from starvation due to exhaustion of lipid and protein reserves. Their capacity to withstand hypernatraemia would also appear to assist their long-term survival. Levels up to 230 mmol/L have been recorded in the field after prolonged drought, as with the Australian agamid, *Amphibolurus ornatus*. Mortality may

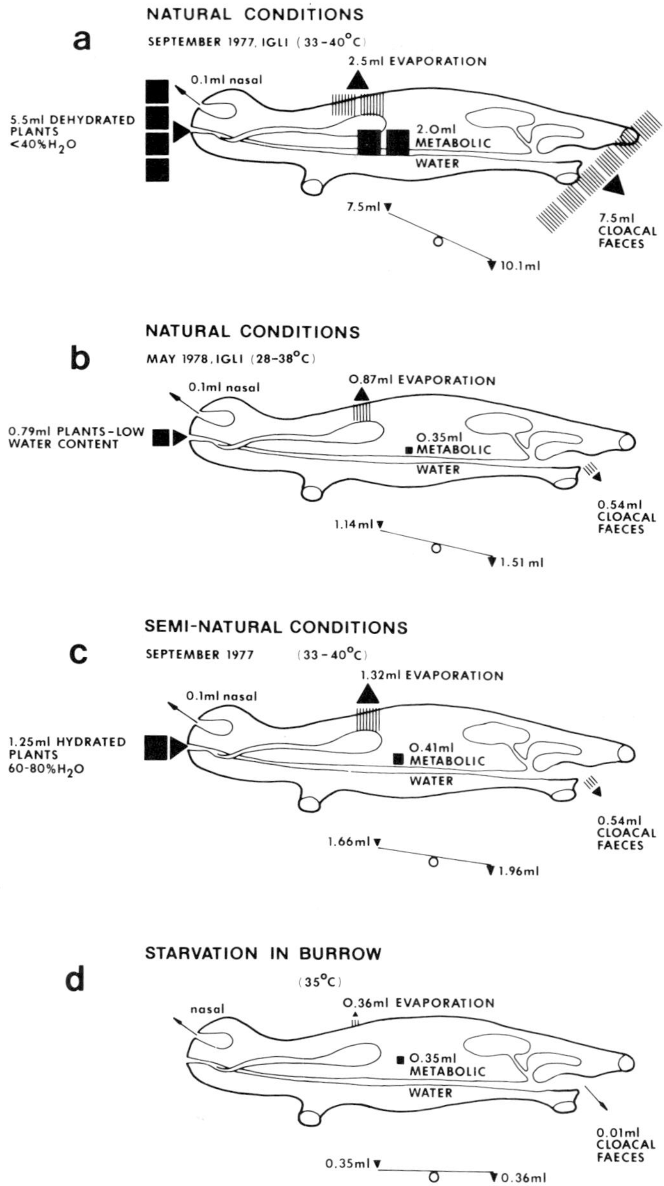


**Fig. 5.15** A schematic representation of the hydric balance of the Saharan agamid lizard *Uromastix acanthinurus* under both natural and simulated conditions in the field in North Africa (modified from Lemire, 1983).

be high in some areas during the winter period of hibernation but it seems reasonable to speculate that the majority of adults would die slowly from starvation following a series of very dry years, during each of which their body condition would decline significantly. Because of inadequate rainfall, the vegetation would be deficient in moisture and nutritive value, so that the lizards would be unable to restore in spring the condition lost during the previous year. It is now known how long such a series of bad seasons would need to persist before a local population of *Uromastix acanthinurus* would be eliminated, but it certainly can survive in regions where rain does not fall for periods of up to 10 years. This emphasises the extraordinary capacity of this species to endure prolonged drought.

## III. *SAUROMALUS OBESUS* IN THE MOJAVE DESERT

The North American chuckwalla, *Sauromalus obesus* is in many ways the ecological equivalent of *Uromastix acanthinurus* in the New World: a large (100–400 g) relatively long-lived herbivore which is common in the Great Basin, Mojave, Colorado and Sonora deserts of North America. Berry (1974) found that chuckwallas at Lone Butte in California grow very slowly, reaching sexual maturity after five to six years. He estimated that some individuals live for as long as 20 years. The ability of this species to reduce body temperature by evaporative cooling has been commented upon by Case (1972). Crawford & Kampe (1971) found that 66% of the metabolic heat production was dissipated at 43.5°C by evaporation of body water. Norris & Dawson (1964) described unique lateral lymph sacs in *Sauromalus obesus* which they suggested serve to expand the extracellular fluid volume above that of other lizards and store water which may be subsequently reclaimed during periods of drought. Smits (1985) has recently extended these early observations on a closely related species, *Sauromalus hispidus*, and shown that the fluid found within the sacs is indeed true lymph and that expansion of the extracellular fluid volume is associated with salt loading, sodium and potassium chloride being equally effective.

Nagy (1972, 1973) studied the ecophysiology of a population of chuckwallas on Black Mountain in the Mojave Desert, and changes in water and electrolyte balance have already been discussed in Chapter 2. Annual plants were fairly abundant in the central Mojave Desert in the spring of 1970, despite a lower than average rainfall the previous year, and perennial plants were lush when this study commenced. By May, however, the water content of the vegetation had declined dramatically, as seen in Fig. 5.16, and the nitrogen content of *Franseria dumosa* leaves, which form an important part of the diet of the chuckwalla, also fell to about 1.5% dry weight. Potassium concentrations increased to about 700 mmol/L (see Fig. 5.17). Whereas animals were abroad

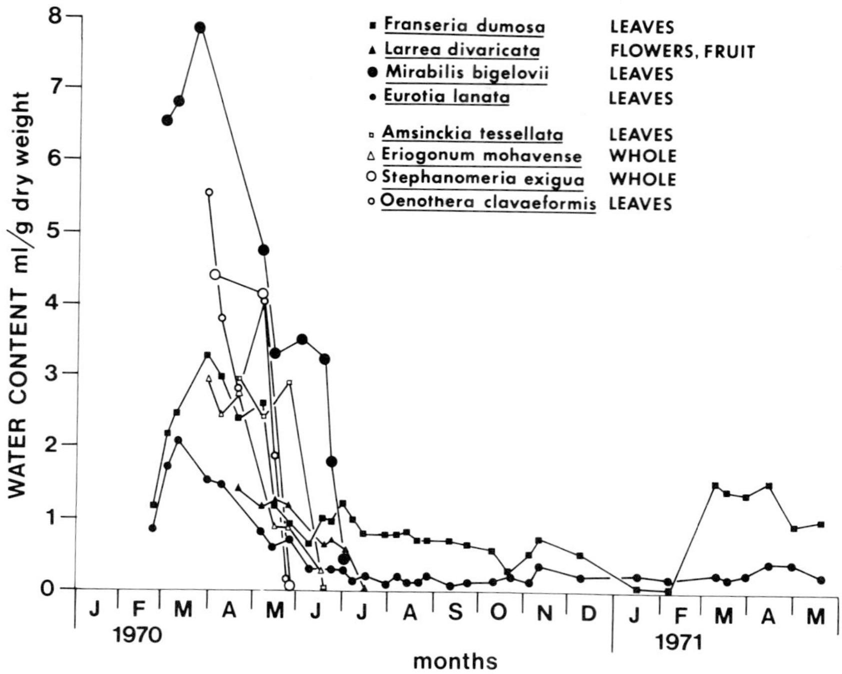


**Fig. 5.16** Seasonal variation in water content of representative plants eaten by the chuckwalla (*Sauromalus obesus*) in the Mojave desert. Closed symbols represent perennials; open symbols indicate annual plants (from Nagy, 1973).

for approximately eight hours each day in April, activity fell markedly as summer progressed and by June total time active per day was much reduced. Only about one third of the individuals were abroad on any one day. After July, feeding ceased entirely, and all activity was restricted to basking for an hour or so at sunset. Fig. 2.23 shows that this change in behaviour was associated with a marked decline in body weight.

Nagy & Shoemaker (1975) used $^{3}HH^{18}O$ to measure rates of $CO_2$ production in free-ranging chuckwallas, and hence field metabolic rates (FMR). These are shown in Fig. 5.18 for the same period. The marked depression of FMR in summer to levels of about 0.05 mL $CO_2(g.h)^{-1}$ is evident and this figure is in fact lower than that predicted for resting animals of the same body weight from the equation of Bennet & Dawson (1976) (0.168 mL $O_2(g.h)^{-1}$). The reasons for the loss of condition of the chuckwallas have been analysed in some detail by Nagy (1975a) in relation to energy and nitrogen balance. The energy budget erected by Nagy (1975a) suggests that this species

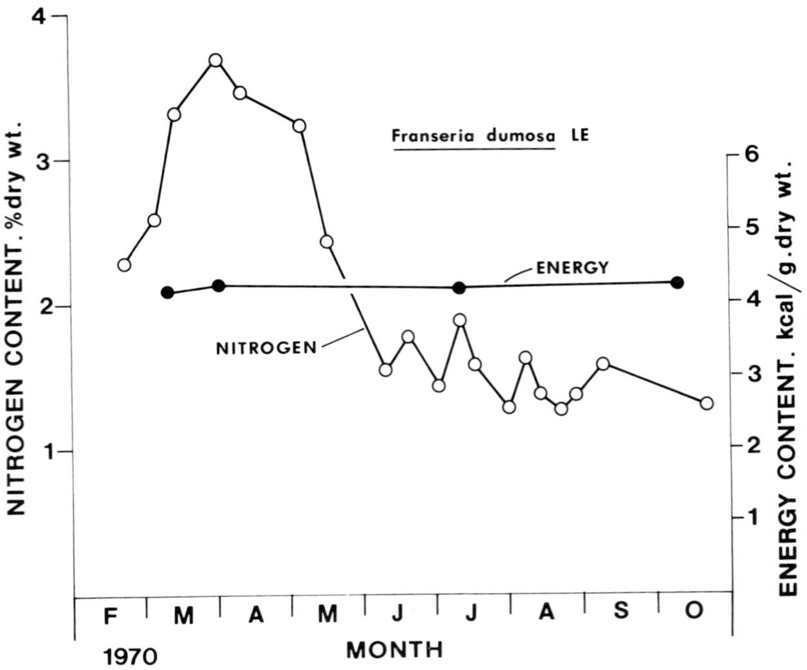


**Fig. 5.17** Nitrogen and energy contents of leaves of *Franseria* (= *Ambrosia*) *dumosa*, eaten by chuckwallas, at various times of the year (from Nagy & Shoemaker, 1975).

requires 21.42 kJ$(100\ g.d)^{-1}$ and that this may be obtained by ingesting 1.24 g dry matter $(100\ g.d)^{-1}$. The lizards were eating about the same amount of food in late May as in April; the energy content of this did not change throughout the year (see Fig. 5.17). As the fat content of these animals remained constant or increased slightly in early summer (Nagy, 1972), these data suggest that the chuckwallas were in positive energy balance during that period. A shortage of energy thus does not appear to be the reason for the loss of weight which occurred in May and Nagy (1975a) suggests that it is because the diet becomes *too dry* to provide sufficient water for the maintenance of fluid balance. Fig. 5.19 shows a theoretical energy and material budget for *Sauromalus obesus* in the field under steady-state conditions and the relatively large fraction of water lost via the faeces and through evaporation is evident. As the plants eaten by chuckwallas desiccate progressively in summer, water intake forcibly declines, but rates of water loss remain about the same. Dry-matter intake (DMI) remained the same, or increased slightly, and because chuckwallas are unable to

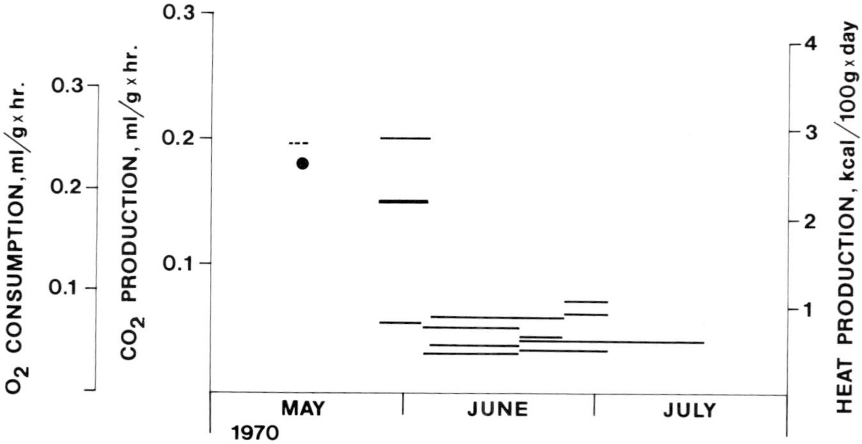


**Fig. 5.18** Rates of $CO_2$ production in free-living chuckwallas (*Sauromalus obesus*) during spring and summer. Dashed line = $O_2$ consumption calculated from water influx rates of lizards muzzled for 3 days in the field; black circle = heat production calculated from energy budget; solid lines = rates of $CO_2$ production measured with doubly-labelled water with the length of the line indicating the duration of the measuring period (from Nagy & Shoemaker, 1975).

digest cellulose (Nagy, 1977), about 40% of the ingested dry matter is voided as faeces. These faeces are always approximately 60% water and this constitutes an important avenue of water loss, so long as the animals continue to eat. As the diet become progressively drier, more and more water must be taken from the body water pool to balance losses. Dehydration becomes chronic and the lizards lose weight, even though they are in positive energy balance. In fact Nagy (1977) found that, when the diet is dry (i.e. less than 40% water), the more an animal eats, the more rapidly it dehydrates. The behavioural response of the chuckwalla to this unfavourable situation is the same as that of *Amphibolurus ornatus* in Australia; to cease eating and reduce activity to an absolute minimum. Even so, Nagy (1972) found that the chuckwallas lost about 40% of body weight from the summer of 1970 until spring 1971. It is clear that they would not be able to survive more than a single adverse year, and a succession of two very dry years would probably result in the extinction of the population. Berry (1974) makes the point that the chuckwalla is only active for five to six months of the year and is a 'spring animal' which endures the other unfavourable periods of the year (summer and winter) by reducing metabolic expenditure.

Nagy (1975b) carried out feeding trials with *Sauromalus obesus* in the laboratory to determine the minimum nitrogen requirements of this species,

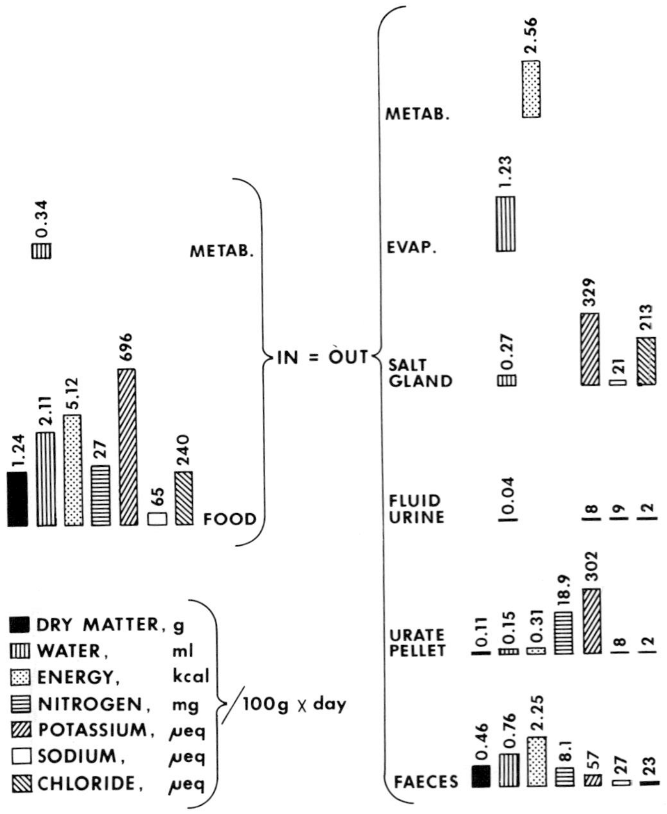


**Fig. 5.19** Energy and material budgets of chuckwallas (*Sauromalus obesus*) maintaining constant body weights in the field and assuming steady-state conditions (from Nagy, 1975a).

and also the relationship between nitrogen assimilation rates and electrolyte intake. The minimum digestible nitrogen requirement was found to be 175 mg N$(kg.d)^{-1}$ or 109 mg N$(kg^{0.75}.d)^{-1}$ for a diet containing normal field levels of water and potassium. Nitrogen was stored when the animals were fed a succulent diet, such as occurs in spring in the field. The relationship between the rate of urinary nitrogen excretion and nitrogen assimilation on low (LK) and high (HK) potassium diets is shown in Fig. 5.20. It is evident that nitrogen excretion rates are elevated with high potassium intake. The combination of a low protein–high potassium diet, such as is available to the chuckwallas in late summer, leads to elevated rates of nitrogen excretion and catabolism of muscle when the protein intake is below minimum daily requirements. Nagy (1975b),

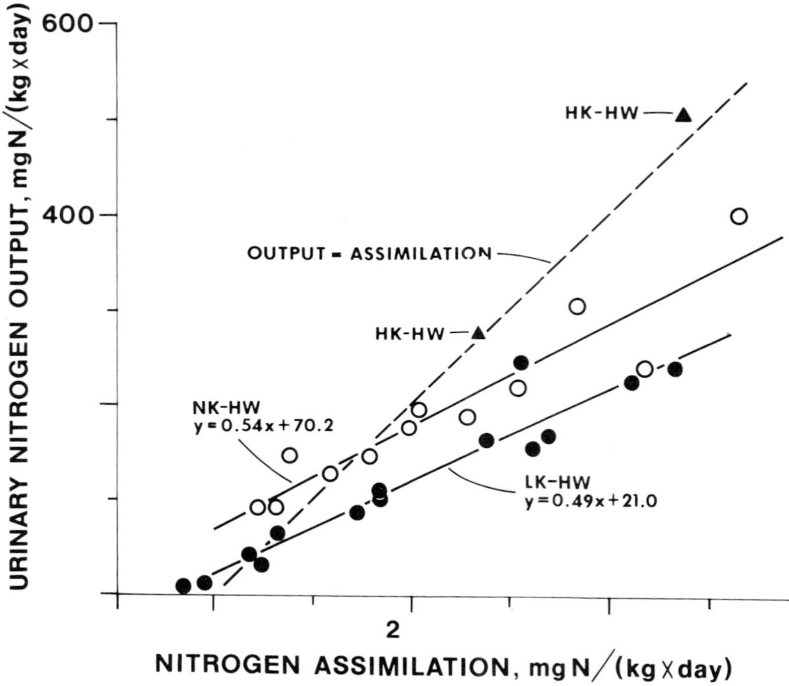


**Fig. 5.20** Relationships between nitrogen excretion and assimilation for chuckwallas (*Sauromalus obesus*) fed a high-potassium, high-water diet (HK–HW, solid triangles); a normal-potassium, high-water diet (NK–HW, open circles) and a low-potassium, high-water diet (LK–HW closed circles) (from Nagy, 1975b).

however, noted, that nitrogen excretion rates decline significantly when nitrogen assimilation falls below the minimum requirement. This would decrease the rate of loss of body nitrogen when feeding on poor-quality vegetation. Excess dietary potassium is also well-handled by chuckwallas, increased amounts being eliminated from the body by the activity of the nasal salt gland and also in the form of insoluble urates precipitated in the cloaca.

From the above one may see there are certain broad similarities between the Sahara agamid *Uromastix acanthinurus* and the North-American iguanid *Sauromalus obesus*. Both are long-lived herbivores with a considerable capacity to survive in desert regions through a combination of behavioural and physiological attributes which, whilst not being unique to these species, certainly account for their success in arid regions. Like the Fouette-Queue, chuckwallas have their enemies. Berry (1974) lists four species of birds, five species of mammals and eight snakes which he considered potential predators of chuckwallas. Coyotes (*Canis latrans*) are known to prey on chuckwallas but the extent of this predation pressure is unknown. Because of their relatively large size, however, predation is likely to be more effective on juvenile chuckwallas than on adults and it seems probably that, as with the Fouette-Queue, the mortality of adults is more a result of unfavourable conditions sustained beyond the individuals' capacity to maintain physiological homeostatis. In this context, *Uromastix acanthinurus* occurs in regions far more arid than those frequented by *Sauromalus obesus*, and its resistance to prolonged drought would appear to be greater, but further work on this Saharan species is sorely needed as knowledge of its ecophysiology is still very limited and, in the absence of this, one may only speculate on the causes and incidence of mortality.

# 6
# Conclusion

It has been the aim of this book to account for the conspicuous success of reptiles as a group in the deserts of the world, as judged by their obvious taxonomic diversity and numerical abundance. Reptiles of one form or another are always present in arid and semi-arid environments, often in great abundance, and in the hyper-arid deserts they are frequently the only representatives of the Vertebrata to be found. The conventional wisdom has it that desert-living species of reptiles are 'highly-adapted' to the exigencies of their chosen environment and that this adaptedness is reflected in various modifications of their morphology, behaviour, physiology, and perhaps also their population dynamics, so that individual survival is enhanced and, ultimately, population persistence secured through the effectiveness of these adaptations. This 'paradigm' in the sense of Kuhn (1970) has underpinned much of the effort of herpetologists during the last 25 years. The literature abounds with studies purporting to identify a host of 'adaptations' which characterise desert reptiles and distinguish them from those reptiles which do not inhabit desert environments. The major difficulty with this approach, as made clear in the Introduction, is that, as more and more data have accumulated on reptiles from a variety of habitats it has become less and less evident that desert reptiles are uniquely adapted to their immediate environment. This is why I have tried to test this implicit assumption through a critical analysis of what is known of the ecophysiology of reptiles from three of the major desert regions of the world.

This analysis shows fairly clearly, I believe, that it is difficult to identify unique features which, at one and the same time, distinguish desert reptiles from close relatives restricted to mesic or humid environments, and also account for their manifest ability to survive and reproduce in environments which are amongst the most extreme found on earth. Population persistence in the arid zone requires that individuals survive for a period of time sufficient to secure reproduction and eventual recruitment, and such survival obviously depends upon the felicitous exploitation of highly variable and limited resources for growth and maturation, as well as the capacity to endure extended periods which are potentially stressful for the individual. At one time I would have pointed to reduced rates of metabolism and evaporative water loss as obvious 'adaptations' characteristic of desert lizards but recent work, detailed in Chapter 2, has shown that herpetologists have grossly underestimated the plasticity of reptilian physiology and the extent to which prior acclimatisation may account for major changes in physiological capacities. A good example of this comes from early work on the thermal relations of Western Australian lizards where it was found that desert-living species had higher lethal temperatures than closely related species from mesic habitats. Subsequent studies showed, however, that suitable acclimation to high environmental temperatures in the laboratory is all that is needed to elevate the lethal temperature of the non-desert species to a comparable level and this is obviously not 'adaptation' in the gentic sense of the word.

I have attempted to overcome this difficulty by considering genetic adaptations as:

> heritable modifications of the phenotype arising through natural selection in a particular habitat which are shared by some, but not all, members of a taxon and which thereby permit those so endowed to occupy a niche or habitat denied other members of the taxon.

This definition does not seem greatly at variance with the rather confused and somewhat indecisive literature on the concept of biological adaptation (see for example, Medawar, 1951; Underwood, 1954; Stern, 1970; Burian, 1983), but its consistent application does have some interesting consequences for attributes which are normally thought of as classical adaptations. The wing of a bird for example can clearly be considered an 'adaptation' in the above sense, but only in relation to other birds and vertebrates which do not fly, as the adaptive advantage of the wing *per se* amongst flying birds is obscure. This highlights the problem with definitions which attempt to link adaptations with persistent directional selection and reproductive superiority of the bearers (e.g. Stern, 1970) as the variation between the fertility of different birds due to variation in wing type is probably quite trivial and overwhelmed in real situations by that due to a host of other ecological factors. In the same way, the pentadactyl limb may be considered an adaptation which characterises the vertebrates when comparing them with other Phyla but, within the Vertebrata, it is the modifications of this limb which are adaptive and not the fact that it conforms to the pentadactyl design. Evolutionary novelties which may arise from time to time within a taxon may thus be considered an 'adaptations' whilst they are differentially distributed throughout the taxon. This will presumably be the same period during which the bearers of these heritable phenotypic modifications will enjoy some relative reproductive advantage over their fellows, because of their exploitation of a new niche. But once all the members of the taxon possess this modification, it can no longer be considered as an adaptation in the same sense, within that particular taxon, be it a species, genus, family or phylum, although it would still be quite proper to consider it as adaptive when comparing the performance of this group of animals with that of other taxa. Once a particular genetic modification becomes the common property of the taxon, our use of the word 'adaptation' essentially changes to the vernacular form. The organisms are now seen to be well-designed to contend with the exigencies of their particular environment through the incorporation of these genetic novelties, but their 'fitness' need not imply persistent directional selection, only normalising or stabilising selection which is necessary to maintain these modifications within the population if it is to persist in that particular environment. The term 'adaptation' is thus being employed in a way analogous to its use by neurophysiologists to describe the way in which neural systems adapt perfectly after a transient response: i.e., the behaviour return precisely to the pre-stimulus level, even though the stimulus remains at the new higher level and homeostasis is thus maintained (see Koshland *et al.*, 1982).

Viewed in this way it is possible to see why there has been confusion over the nature of the desert adaptations in reptiles and why their identification has proven so intractable. There are many factors which promote survival and reproduction of reptiles inhabiting desert environments, but, as we have seen in the preceding chapters, these are not unique prerogatives of desert species and are equally shared by species inhabiting other habitats. There are only two ways of resolving this apparent dilemma: either by continuing to search for more and more subtle 'specific adaptations' to account for the phenomenon (i.e., pursue the 'adaptationist programme' in the sense of Gould & Lewontin, 1979 *ad absurdo*, and some authors suggest that such adaptations will be found through a study of the Old World rather than the New World herpetofaunae—e.g. Mayhew, 1968; Louw & Holm, 1972), or alternatively admit that our definition of what constitutes an 'adaptation' is in need of revision. If we take this second alternative, rather than the first, it is possible to see that reptiles survive and reproduce particularly effectively in desert environments because of a number of fundamental characteristics, which appear common to at least all the terrestrial members of the group, and which enable them to thrive where other animals are at a distinct disadvantage. Foremost amongst these are the following:

(i) their low rates of metabolism, compared with birds and mammals, which result in extremely low rates of resource utilisation and turnover and lead to considerable economy in the handling of water;

(ii) their dependence upon external heat sources for normal activity (ectothermy) and their ability to thermoregulate by behavioural means;
(iii) their ability to excrete the nitrogenous waste products of protein digestion in the form of relatively insoluble uric acid, thereby conserving water that would otherwise be osmotically obligated to a substance such as urea;
(iv) their ability to withstand, for long periods of time, marked perturbations of the *milieu intérieur* which would prove fatal for birds and mammals; and
(v) their widespread utilisation of behavioural avoidance as an effective tactic promoting survival during periods when continued activity would be counterproductive.

Taken together, these features virtually characterise squamate reptiles as we know them, and this is of course the same group of reptiles which is so widespread in desert regions of the world, as well as in other temperate habitats. Their signal success in desert environments is thus not due to any specific adaptations, but rather results from the fact that the reptilian *'bauplan'* uniquely fits them for life in such an austere habitat. Pough (1980) has discussed in some detail the advantages of ectothermy for tetrapods and Nagy (1983) has amply demonstrated how lizards, with their low energy requirements, may exploit habitats which are essentially too poor in resources to sustain endotherms such as birds and mammals. In this context it is interesting to note that nomadism is a tactic frequently utilised by birds and mammals in deserts, either for survival or to take advantage of short-lived resources (see Davies, 1984), but it is one that is almost never resorted to by reptiles.

It is thus possible to reconcile a fruitless search for specific adaptations to desert life in reptiles with their manifestly successful radiation in such habitats, and see this instead as a direct consequence of their relatively modest mode of existence which imposes far fewer demands on the environment than that of endotherms such as birds and mammals. However, squamate reptiles as a group are better-adapted than say, amphibians, for life in deserts because they possess some attributes, listed above, not shared by amphibians (e.g. ectothermy, uric acid excretion as well as internal fertilisation, and the possession of a cleidoic egg) and these are the heritable modifications of their phenotype which enable them to survive and reproduce in such habitats. The only paradox remaining is why repeated periods of exposure to environments that we judge as 'stressful' have not resulted in the gradual accumulation of specific adaptations which would enhance the quality of life for these animals. Stress has been taken here to be manifested by a significant deviation of the *milieu intérieur*, associated with elevated levels of regulatory hormones, but when examining the ability of desert reptiles to maintain thermal and osmoregulatory homeostasis in Chapter 2, it was difficult to identify such periods of stress. This is partly because of the limitations of the data in the case of water and electrolyte balance (too few hormonal studies have been carried out in the field) and partly because, in the case of temperature regulation, their behaviour appears adequate for the maintenance of thermal homeostasis. In the case of lizards such as the Australian agamid *Amphibolurus nuchalis* studied at Shark Bay, which die each summer from chronic dehydration after breeding as nine-month-old animals, it would appear that, in an evolutionary sense, the obvious stress endured by the adults is of little consequence because they have already transmitted their genes to the next generation which survives the summer in the egg phase. The presence of lateral lymph sacs in *Sauromalus obesus* and the development of highly efficient cephalic salt-excreting glands in *Uromastix acanthinurus* and *Dipsosaurus dorsalis* may be suggested as examples of 'specific adaptations' which have evolved in response to the persistent imposition of osmoregulatory stress on the adult phase of relatively long-lived species, but far too little is known of the ecophysiology of these animals at this stage to warrant such a conclusion and there may well be other alternative explanations for their evolution. What is abundantly clear is that the majority of desert reptiles survive and reproduce without recourse to such adjuncts because they are ideally suited as animals to exploit the peculiar exigencies of the desert environment which make survival and long-term persistence there so difficult for most other organisms.

# References


Alcala, A. C. (1967). Population biology of the 'flying' lizard, *Draco volans*, on Negros Island, Philippines. *Natur. Appl. Sci. Bull.* **20**, 335–372.

Alimen, H. (1963). Considérations sur la chronologie du Quaternaire saharien. *Bull. Soc. Géol. Fr.* **5**, 627–634.

Andrews, R. M. (1976). Growth rates in island and mainland anoline lizards. *Copeia* 1976, 477–482.

Andrews, R. M. (1979). Reproductive effort of female *Anolis limifrons* (Sauria: Iguanidae). *Copeia* 1979, 620–626.

Andrews, R. M. (1982). Patterns of growth in reptiles. *In* "Biology of the Reptilia" (C. Gans, F. H. Pough, eds), Vol. 13, 273–320. Academic Press, New York.

Arnold, E. N. and Gallagher, M. O. (1977). Reptiles and amphibians from the mountains of northern Oman. *J. Oman Studies Special Report* **1**, 59–79.

Arsland, M., Lobo, J., Zaidi, A., Jalali, S. and Qazi, M. H. (1978). Annual androgen rhythm in the Spiny-Tailed lizard, *Uromastix hardwicki. Gen. Comp. Endocrinol.* **36**, 16–22.

Asplund, K. K. and Lowe, C. H. (1964). Reproductive cycles of the iguanid lizards, *Urosaurus ornatus* and *Uta stansburiana*, in south-eastern Arizonia. *J. Morphol.* **115**, 27–33.

Atsatt, S. R. (1939). Color changes as controlled by temperature and light in the lizards of the desert regions of southern California. *Univ. Calif. Publ. Biol. Sci.* **1**, 237–276.

Aubreville, A. (1949). "Climats, Forêts et Désertification de l'Afrique Tropicale". Société d'Editions Géographiques, Maritimes et Coloniales, Paris.

Avery, R. A. (1982). Field studies of body temperatures and thermoregulation. *In* "Biology of the Reptilia" (C. Gans, F. H. Pough, eds), Vol. 12, 93–166. Academic Press, New York.

Axelrod, D. I. (1950). The evolution of desert vegetation. *Carnegie Inst. of Washington Pub.* No. **590**, 216–306.

Axelrod, D. I. (1958). Evolution of the Madro-Tertiary Geoflora. *Bot. Rev.* **24**, 433–509.

Badir, N. (1958). Seasonal variation of the male urinogenital organs of *Scincus scincus* L. and *Chalcides chalcides* Forsk. *Z. Wiss. Zool.* **160**, 290–351.

Badir, N. (1968). Structure and function of corpus luteum during gestation in the viviparous lizard *Chalcides ocellatus*. *Anat. Anz. Bd.* **122**, 1–10.

Baeyens, D. A. and Rountree, R. L. (1983). A comparative study of evaporative water loss and epidermal permeability in an arboreal snake, *Opheodrys aestivus*, and a semi-aquatic snake, *Nerodia rhombifera*. *Comp. Biochem. Physiol.* **76A**, 301–304.

Bagnara, J. T. and Hadley, M. E. (1972). "Chromatophores and Color Change". Prentice-Hall, New York.

Bagnouls, F. and Gaussen, H. (1953). Saison sèche et indice xérothermique. *Bull. Soc. Hist. Nat., Toulouse*, **88**, 193–239.

Bagnouls, F. and Gaussen, H. (1957). Les climats biologiques et leur classification. *Ann. Géogr.* **66**, 193–220.

Baker, B. I. and Rance, T. A. (1983). Further observations on the distribution and properties of teleost Melanin Concentrating Hormone. *Gen. Comp. Endocrinol.* **50**, 423–431.

Baker, L. A., Weathers, W. W. and White, F. N. (1972). Temperature induced peripheral blood flow changes in lizards. *J. Comp. Physiol.* **80**, 313–323.

Bakken, G. S. and Gates, D. M. (1975). Heat-transfer analysis of animals: some implications for field ecology, physiology, and evolution. *In* "Perspectives of Biophysical Ecology" (D. M. Gates, R. B. Schmerl, eds), 255–290. Springer-Verlag.

Bakker, H. R. and Bradshaw, S. D. (1977). Effect of hypothalamic lesions on water metabolism of the toad *Bufo marinus*. *J. Endocrinol.* **75**, 161–172.

Ballinger, R. E. (1976). Evolution of life history strategies: implications of recruitment in a lizard population following density manipulations. *Southwest Natur.* **21**, 203–208.

Ballinger, R. E. (1977). Reproductive strategies: food availability as a source of proximal variation in a lizard. *Ecology* **58**, 628–635.

Ballinger, R. E. and Schrank, G. D. (1970). Acclimation rate and variability of the critical thermal maximum in the lizard *Phrynosoma cornutum*. *Physiol. Zool.* **43**, 19–22.

Balme, B. E. (1962). Palynological report No. 98: Lake Eyre No. 20 Bore, South Australia. *In* "Investigation of Lake Eyre" (R. K. Johns, N. H. Ludbrook, eds). *Rep. Invest. Dep. Mines S. Aust.* No. **24**, Pts 1 & 2, 89–102.

Balout, L. (1952). Pluviaux interglaciaires et préhistoire saharienne. *Trav. Ins. Rech. Sahar.* **8**, 9–21.

Barbault, R. (1975). Observations écologiques sur la reproduction des lézards tropicaux: les stratégies de ponte en forêt et en savanne. *Bull. Soc. Zool. de France* **100**, 153–168.

Barbault, R. and Halffter, G. (eds) (1981). "Ecology of the Chihuahuan Desert". Publication No. 8, Instituto de Ecologia, A. C., Mexico. (MAB, UNESCO).

Barber, B. J. and Crawford, E. C. (1977). A stochastic dual-limit hypothesis for behavioral thermoregulation in lizards. *Physiol. Zool.* **50**, 53–60.

Barber, B. J. and Crawford, E. C. (1979). Dual threshold control of peripheral thermoregulation in the lizard *Dipsosaurus dorsalis*. *Physiol. Zool.* **52**, 250–263.

Barker Jørgensen, C. (1983). Ecological physiology: background and perspectives. *Comp. Biochem. Physiol.* **75A**, 5–7.

Bartholomai, A. and Howie, A. (1970). Vertebrate fauna from the lower Trias of Australia. *Nature* **225**, 1063.

Bartholomew, G. A. (1964). The roles of physiology and behaviour in the maintenance of homeostasis in the desert environment. *Symp. Soc. exp. Biol.* **18**, 7–29.

Bartholomew, G. A. (1982). Physiological control of body temperature. *In* "Biology of the Reptilia" (C. Gans, F. H. Pough, eds), Vol. 12, 167–211. Academic Press, New York.

Bartholomew, G. A. and Lasiewski, R. C. (1965). Heating and cooling rates, heart rate and simulated diving in the Galápagos marine iguana. *Comp. Biochem. Physiol.* **16**, 573–582.

Bartholomew, G. A. and Tucker, V. A. (1964). Size, body temperature, thermal conductance, oxygen consumption, and heart rate in Australian varanid lizards. *Physiol. Zool.* **37**, 341–354.

Bartholomew, G. A. and Vleck, D. (1979). The relation of oxygen consumption to body size and to heating and cooling in the Galápagos marine iguana, *Amblyrhynchus cristatus*. *J. Comp. Physiol.* **132**, 285–288.

Bartholomew, G. A., and Tucker, V. A. and Lee, A. K. (1965) Oxygen consumption, thermal conductance, and heart rate in the Australian skink *Tiliqua scincoides*. *Copeia* 1966, 169–173.

Bartlett, P. N. and Gates, D. M. (1967). The energy budget of a lizard on a tree trunk. *Ecology* **48**, 315–322.

Baverstock, P. R. (1972). The evolution and maintenance of genetic variability in growth rates in the lizard, *Amphibolurus ornatus*. Ph.D. thesis, University of Western Australia.

Baverstock, P. R. (1975). Effect of variations in rate of growth on physiological parameters in the lizard *Amphibolurus ornatus*. *Comp. Biochem. Physiol.* **51A**, 619–631.

Baverstock, P. R. (1978). The probable basis of the relationship between growth rate and winter mortality in the lizard *Amphibolurus ornatus* (Agamidae). *Oecologia* **37**, 101–107.

Baverstock, P. R. and Bradshaw, S. D. (1975). Variation in rate of growth and adrenal corticosteroidogenesis in field and laboratory populations of the lizard *Amphibolurus ornatus*. *Comp. Biochem. Physiol.* **52A**, 557–565.

Beadle, N. C. W. (1966). Soil phosphate and its role in moulding segments of the Australian flora and vegetation, with special reference to xeromorphy and scleromorphy. *Ecology* **47**, 992–1007.

Beadle, N. C. W. (1981). Origins of the Australian angiosperm flora. *In* "Ecological Biogeography of Australia" (A. Keast, ed.), Vol. 1, 409–426. Junk, den Haage.

Beard, J. S. (1978). The evolution of Australian desert plants. *In* "Evolution of Desert Biota" (D. W. Goodall, ed.), 51–63. University of Texas Press, Austin.

Beard, J. S. (1981). Bioclimates in Southern Australia. *In* "Ecologia Mediterranea" (P. Quézel, ed.), Tome VIII, 89–93. Scientific Affairs Division, NATO.

Bennett, A. F. and Dawson, W. R. (1976). Metabolism. *In* "Biology of the Reptilia" (C. Gans, W. R. Dawson, eds), Vol. 5, 127–223. Academic Press, New York.

Bennett, A. F. and Licht, P. (1975). Evaporative water loss in scaleless snakes. *Comp. Biochem. Physiol.* **52A**, 213–215.

Bennett, A. F. and Nagy, K. A. (1977). Energy expenditure in free-ranging lizards. *Ecology* 58, 697–700.

Bentley, P. J. (1959). Studies on the water and electrolyte metabolism of the lizard *Trachysaurus rugosus* (Gray). *J. Physiol.* **145**, 37–47.

Bentley, P. J. (1962). Studies on the permeability of the large intestine and urinary bladder of the tortoise (*Testudo graeca*) with special reference to effects of neurohypophyseal and adrenocortical hormones. *Gen. Comp. Endocrinol.* **2**, 323–328.

Bentley, P. J. (1971). "Endocrines and Osmoregulation". Springer-Verlag, Heidelberg.

Bentley, P. J. (1974). Action of neurohypophyseal peptides in amphibians, reptiles and birds. *In* "Handbook of Physiology" (S. R. Geiger, ed.), 19–28. American Physiological Society, Washington.

Bentley, P. J. (1976). Osmoregulation. *In* "Biology of the Reptilia" (C. Gans, W. R. Dawson, eds), Vol. 5, 365–412. Academic Press, New York.

Bentley, P. J. and Bradshaw, S. D. (1972). Electrical potential difference across the cloaca and colon of the Australian lizards *Amphibolurus ornatus* and *A. inermis*. *Comp. Biochem. Physiol.* **42A**, 465–471.

Bentley, P. J. and Schmidt-Nielsen, K. (1965). Permeability to water and sodium of the crocodilian, *Caiman sclerops*. *J. Cell. Comp. Physiol.* **66**, 303–310.

Bentley, P. J. and Schmidt-Nielsen, K. (1966). Cutaneous water loss in reptiles. *Science* **151**, 1547–1549.

Berk, M. L. and Heath, J. E. (1975a). Effects of preoptic, hypothalamic and telencephalic lesions on thermoregulation in the lizard, *Dipsosaurus dorsalis*. *J. Thermal Biol.* **1**, 65–78.

Berk, M. L. and Heath, J. E. (1975b). An analysis of behavioral thermoregulation in the lizard *Dipsosaurus dorsalis*. *J. Thermal Biol.* **1**, 15–22.

Berry, K. H. (1974). The ecology and social behaviour of the chuckwalla, *Sauromalus obesus obesus* Baird. *Univ. Calif. Publ. Zool.* **101**, 1–60.

Blackburn, D. G. (1982). Evolutionary origins of viviparity in the Reptilia. *Amphibia-Reptilia* **3**, 185–205.

Blackburn, D. G. (1985). Evolutionary origins of viviparity in the Reptilia. II: Serpentes, Amphisbaenia and Ichthyosauria. *Amphibia-Reptilia* **6**: 259–291.

Blair, W. F. (1960). "The Rusty Lizard: A Population Study". University of Texas Press, Austin.

Blair, W. F., Hulse, A. C. and Mares, M. A. (1976). Origins and affinities of vertebrates of the North American Sonoran Desert and the Monte Desert of northwest Argentina. *J. Biogeog.* **3**, 1–18.

Blount, R. F. (1929). Seasonal cycles of the interstitial cells in the testis of the horned toad (*Phrynosoma solare*). *J. Morphol.* **48**, 317–343.

Bona-Gallo, A., Licht, P., MacKenzie, D. S. and Lofts, B. (1980). Annual cycles in levels of pituitary and plasma gonadotropin, gonadal steroids and thyroid activity in the Chinese Cobra (*Naja naja*). *Gen. Comp. Endocrinol.* **42**, 477–493.

Bons, J. (1967). Recherches sur la biogéographie et la biologie des amphibiens et des reptiles du Maroc. Thèse d'Etat, Université de Montpellier, France.

Bons, J. and Bons, N. (1960). Notes sur la reproduction et le développement de *Chamaeleo chamaeleon* (L.). *Bull. Soc. Sc. Nat. Phys. Maroc* **40**, 161–197.

Bons, J. and Saint Girons, H. (1982). Le cycle sexuel des reptiles au Maroc et ses rapports avec la répartition géographique et le climat. *Bull. Soc. Zool. de France* **107**, 71–86.

Bons, N. (1969). Le cycle sexuel du mâle chez *Acanthodactylus erythrurus lieomaculatus* Dum. et Bibr. (Sauria, Lacertidae). *Bull. Soc. Sc. Nat. Phys. Maroc* **49**, 161–197.

Borst, J. G. G., De Vries, L. A., Van Leeuwen, A. M., Den Ottolander, G. J. H. and Cejka, V. (1960). The maintenance of circulatory stability at the expense of volume and electrolyte stability. *Clin. Chim. Acta* **5**, 887–914.

Borut, A., Horowitz, M. and Castel, A. (1972). Blood volume regulation in the spiny mouse: capillary changes due to dehydration. *Symp. Zool. Soc. Lond.* **31**, 175–189.

Botte, V. and Angelini, F. (1980). Endocrine control of reproduction in reptiles: the refractory period. *In* "Steroids and Their Mechanism of Action in Non-mammalian Vertebrates" (G. Delrio, J. Brachet, eds), 201–212. Raven Press, New York.

Bourne, A. R. and Seamark, R. F. (1975). Seasonal changes in 17$\beta$-hydroxysteroids in the plasma of the male lizard (*Tiliqua rugosa*). *Comp. Biochem. Physiol.* **50B**, 535–536.

Bourne, A. R. and Seamark, R. F. (1978). Seasonal variation in steroid biosynthesis by the testis of the lizard *Tiliqua rugosa. Comp. Biochem. Physiol.* **59B**, 363–367.

Bowler, J. M. (1976). Aridity in Australia: age, origins and expression in aeolian landforms and sediments. *Earth Sciences Review* **12**, 279–310.

Bradshaw, S. D. (1970). Seasonal changes in the water and electrolyte metabolism of *Amphibolurus* lizards in the field. *Comp. Biochem. Physiol.* **36**, 689–718.

Bradshaw, S. D. (1971). Growth and mortality in a field population of *Amphibolurus* lizards exposed to seasonal cold and aridity. *J. Zool.* **165**, 1–25.

Bradshaw, S. D. (1972). The endocrine control of water and electrolyte metabolism in desert reptiles. *Gen. Comp. Endocrinol.* Suppl. **3**, 360–373.

Bradshaw, S. D. (1975). Osmoregulation and pituitary-adrenal function in desert reptiles. *Gen. Comp. Endocrinol.* **25**, 230–248.

Bradshaw, S. D. (1976). Effect of hypothalamic lesions on kidney and adrenal function of the lizard *Amphibolurus ornatus. Gen. Comp. Endocrinol.* **25**, 123.

Bradshaw, S. D. (1978a). Volume regulation in desert reptiles and its control by pituitary and adrenal hormones. *In* "Osmotic and Volume Regulation" (C. Barker Jørgensen, E. Skadhauge, eds), 38–53. Alfred Benzon Symposium XI, Munksgaard.

Bradshaw, S. D. (1978b). Aspects of hormonal control of osmoregulation in desert reptiles. *In* "Comparative Endocrinology" (P. J. Gaillard, H. H. Boer, eds), 213–216. Elsevier/North Holland Biomedical Press, Amsterdam.

Bradshaw, S. D. (1981). Ecophysiology of Australian desert lizards: studies on the genus *Amphibolurus. In* "Ecological Biogeography of Australia" (A. Keast, ed.), Vol. 2, 1394–1434. Junk, den Haage.

Bradshaw, S. D. (1985). Hydro-mineral balance in reptiles and its hormonal control. *In* "Current Trends in Comparative Endocrinology" (B. Lofts, D. K. O. Chan, eds.), Vol. 2, pp. 939–943. Hong Kong University Press.

Bradshaw, S. D. and Grenot, C. J. (1976). Plasma aldosterone levels in two reptilian species, *Uromastix acanthinurus* and *Tiliqua rugosa*, and the effect of several experimental treatments. *J. Comp. Physiol.* **111**, 71–76.

Bradshaw, S. D. and Main, A. R. (1968). Behavioural attitudes and regulation of temperature in *Amphibolurus* lizards. *J. Zool. Lond.* **154**, 193–221.

Bradshaw, S. D. and Rice, G. E. (1981). The effects of pituitary and adrenal hormones on renal and post-renal reabsorption of water and electrolytes in the lizard, *Varanus gouldii* (Gray). *Gen. Comp. Endocrinol.* **44**, 82–93.

Bradshaw, S. D. and Shoemaker, V. H. (1967). Aspects of water and electrolyte changes in a field population of *Amphibolurus* lizards. *Comp. Biochem. Physiol.* **20**, 855–865.

Bradshaw, S. D. and Waring, H. (1969). Comparative studies on the biological activity of melanin-dispersing hormone (MDH). *In* "La Spécificité Zoologique des Hormones Hypophysaires et de leurs Activités" (M. Fontaine, ed.). Coll. CNRS **177**, 135–152.

Bradshaw, S. D., Shoemaker, V. H. and Nagy, K. A. (1972). The rôle of adrenal corticosteroids in the regulation of kidney function in the desert lizard *Dipsosaurus dorsalis*. *Comp. Biochem. Physiol.* **43A**, 621–635.

Bradshaw, S. D., Bradshaw, J. A. and Lachiver, F. (1976). Quelques observations sur l'écophysiologie d'*Agama mutabilis* Merren, dans le sud tunisien. *C.R. Acad. Sci. (Paris)* **282**, 93–96.

Bradshaw, S. D., Gans, C. and Saint Girons, H. (1980). Behavioral thermoregulation in a pygopodid lizard, *Lialis burtonis*. *Copeia* 1980, 738–743.

Bradshaw, S. D., Tom, J. A. and Bunn, S. E. (1984). Corticosteroids and control of nasal salt gland function in the lizard *Tiliqua rugosa*. *Gen. Comp. Endocrinol.* **54**, 308–313.

Bradshaw, S. D., Lemire, M., Vernet, R. and Grenot, C. J. (1984). Aldosterone and the control of secretion by the nasal gland of the North African desert lizard, *Uromastix acanthinurus*. *Gen. Comp. Endocrinol.* **54**, 314–323.

Bradshaw, S. D., den Hollander, J., Mutton, L. A., Prince, G., Boulden, L. R. and Monzu, N. (1970). Nocturnalism in geckos. Honours thesis, Zoology Department, University of Western Australia.

Braysher, M. (1971). The structure and function of the nasal salt gland from the Australian sleepy lizard *Trachydosaurus* (formerly *Tiliqua*) *rugosus*: Family Scincidae. *Physiol. Zool.* **44**, 129–136.

Braysher, M. and Green, B. (1970). Absorption of water and electrolytes from the cloaca of an Australian lizard, *Varanus gouldii* (Gray). *Comp. Biochem. Physiol.* **35**, 607–614.

Brodsky, W. A. and Schilb, T. P. (1960). Electrical and osmotic characteristics of the isolated turtle bladder. *J. Clin. Invest.* **39**, 974–975.

Brodsky, W. A. and Schilb, T. P. (1965). Osmotic properties of isolated turtle bladder. *Amer. J. Physiol.* **208**, 46–57.

Brodsky, W. A. and Schilb, T. P. (1966). Ionic mechanisms for sodium and chloride transport across turtle bladders. *Amer. J. Physiol.* **210**, 987–996.

Brooks, G. R. (1967). Population ecology of the ground skink, *Lygosoma laterale* (Say). *Ecol. Monogr.* **37**, 71–87.

Burbidge, A. A., Kirsch, J. A. W. and Main, A. R. (1974). Relationships within the Chelidae (Testudines: Pleurodira) of Australia and New Guinea. *Copeia* 1974, 392–409.

Burian, R. M. (1983). Adaptation. *In* "Dimensions of Darwinism" (M. Grene, ed.), 287–314. Cambridge University Press, London.

Buscarlet, L. A., Proux, J. and Gerster, R, (1978). Utilisation du double marquage $HT^{18}O$ dans une étude de bilan métabolique chez *Locusta migratoria migratorioides*. *J. Insect Physiol.* **24**, 225–232.

Bustard, H. R. (1967). Activity cycle and thermoregulation in the Australian gecko *Gehyra variegata*. *Copeia* 1967, 753–758.

Butler, D. G. and Knox, W. H. (1970). Adrenalectomy of the painted turtle (*Chrysemys picta belli*): effect on ion regulation and tissue glycogen. *Gen. Comp. Endocrinol.* **14**, 551–556.

Cabanac, M. (1979). Le comportement thermorégulateur. *J. Physiol. Paris* **75**, 115–178.

Cabanac, M., Hammel, H. T. and Hardy, J. D. (1967). *Tiliqua scincoides*: temperature-sensitive units in the lizard brain. *Science* **158**, 1050–1051.

Cagle, F. R. (1946). The growth of the Slider Turtle, *Pseudemys scripta elegans*. *Amer. Midl. Nat.* **36**, 685–729.

Callard, G. V., Petro, Z. and Ryan, K. J. (1977). Identification of aromatase in the reptilian brain. *Endocrinology* **100**, 1214–1218.

Callard, G. V., Petro, Z. and Ryan, K. J. (1978). Conversion of androgen to estrogen and other steroids in the vertebrate brain. *Amer. Zool.* **18**, 511–523.

Callard, I. P. and Callard, G. V. (1978). The adrenal gland in Reptilia. *In* "General, Comparative and Clinical Endocrinology of the Adrenal Cortex" (I. Chester Jones, I. W. Henderson, eds), Vol. 2, 370–418. Academic Press, London.

Callard, I. P. and Doolittle, J. P. (1973). The influence of intrahypothalamic implants of progesterone on ovarian growth and function in the ovoviviparous lizard, *Sceloporus cyanogenys*. *Comp. Biochem. Physiol.* **44A**, 625–630.

Callard, I. P. and Lance, V. (1977). The control of reptilian follicular cycles. *In* "Reproduction and Evolution", 4th Int. Symp. on Comp. Biology of Reproduction (J. H. Calaby, C. H. Tyndale-Biscoe, eds), 199–209. Australian Academy of Science, Canberra.

Callard, I. P., Bayne, C. G. and McConnell, W. F. (1972b). Hormones and reproduction in the female lizard *Sceloporus cyanogenys*. *Gen. Comp. Endocrinol.* **18**, 175–194.

Callard, I. P., Callard, G. V., Lance, V. and Eccles, S. (1976). Seasonal changes in testicular structure and function and the effects of gonadotropins in the freshwater turtle, *Chrysemys picta*. *Gen. Comp. Endocrinol.* **30**, 347–356.

Callard, I. P., Doolittle, J., Banks, W. C. and Chan, S. W. C. (1972a). Recent studies on the control of the reptilian ovarian cycle. *Gen. Comp. Endocrinol.* Suppl. **3**, 65–75.

Calvet, C. (1982). Une diagnose de la végétation méditerranéenne basée sur le critère d'évaporation. *In* "Ecologia Mediterranea" (P. Quézel, ed.), Tome VIII, 49–58. Scientific Affairs Division, NATO.

Camazine, B., Garstka, W., Tokarz, R. and Crews, D. (1980). Effects of castration and androgen replacement on male courtship behavior in the red-sided garter snake (*Thamnophis sirtalis parietalis*). *Horm. Behav.* **14**, 318–327.

Camps, G. (1960). Massinissa ou les débuts de l'histoire. *Libyca.* **8**, 1–320.

Capot-Rey, R. (1951). Une carte de l'indice d'aridité au Sahara francais. *Bull. Assoc. Géogr. Fr.* **216**, 73–76.

Case, T. J. (1972). Thermoregulation and evaporative cooling in the chuckwalla, *Sauromalus obesus. Copeia* 1972, 145–150.

Case, T. J. (1978). On the evolution and adaptive significance of postnatal growth rates in the terrestrial vertebrates. *Quart. Rev. Biol.* **53**, 243–282.

Caughley, G. (1966). Mortality patterns in mammals. *Ecology* **47**, 906–918.

Chan, D. K. O., Callard, I. P. and Chester Jones, I. (1970). Observations on the water and electrolyte composition of the iguanid lizard *Dipsosaurus dorsalis* (Baird & Girard) with special reference to the control by the pituitary gland and the adrenal cortex. *Gen. Comp. Endocrinol.* **15**, 374–387.

Chan, S. W. C., Ziegel, S. and Callard, I. P. (1973). Plasma progesterone in snakes. *Comp. Biochem. Physiol.* **44A**, 631–637.

Chavaillon, J. (1964). Les formations quaternaires du Sahara nord-occidental. *Mém. Cent. Rech. Zones Arides*, CNRS **5**, 1–394.

Chui, K. W., Lofts, B. and Tsui, H. W. (1970). The effects of testosterone on the sloughing cycle and epidermal glands of the female Gecko, *Gekko gecko* L. *Gen. Comp. Endocrinol.* **15**, 12–19.

Claussen, D. L. (1967). Studies of water loss in two species of lizards. *Comp. Biochem. Physiol.* **20**, 115–130.

Cloudsley-Thompson, J. L. (1972). Temperature regulation in desert reptiles. *Symp. Zool. Soc. Lond.* **31**, 39–59.

Clutton-Brock, T. H. (1984). Reproductive effort and terminal investment in iteroparous animals. *Anim. Nat.* **123**, 212–229.

Cody, M. L. and Diamond, L. M. (1975). "Ecology and Evolution of Communities". Belknap Press, Harvard University.

Cogger, H. G. (1961). An investigation of the Australian members of the family Agamidae (Lacertilia) and their phylogenetic relationships. M.Sc. thesis, Sydney University.

Cogger, H. G. (1969). A study of the ecology and biology of the mallee dragon (*Amphibolurus fordi*) and its adaptations to survival in an arid environment. Ph.D. thesis, Macquarie University.

Cogger, H. G. (1974). Thermal relations of the mallee dragon *Amphibolurus fordi* (Lacertilia: Agamidae). *Aust. J. Zool.* **22**, 319–339.

Cogger, H. G. (1975). "Reptiles and Amphibians of Australia". A. H. & A. W. Reed, Sydney.

Cogger, H. G. and Heatwole, H. (1981). The Australian reptiles: origins, biogeography, distribution patterns and island evolution. *In* "Ecological Biogeography of Australia" (A. Keast, ed.), Vol. 2, 1333–1373. Junk, den Haage.

Cohen, A. (1975). Some factors affecting water economy in snakes. *Comp. Biochem. Physiol.* **51A**, 361–368.

Cole, L. C. (1954). The population consequences of life history phenomena. *Quart. Rev. Biol.* **29**, 103–137.

Cole, L. L. (1943). Experiments on toleration of high temperatures in lizards with reference to adaptive coloration. *Ecology* **24**, 94–108.

Colwell, R. K. (1974). Predictability, constancy, and contingency of periodic phenomena. *Ecology* **55**, 1148–1153.

Congdon, J. D., King, W. W. and Nagy, K. A. (1978). Validation of the HTO-18 method for determination of $CO_2$ production of lizards (genus *Sceloporus*). *Copeia* 1978, 360–362.

Congdon, J. D., Dunham, A. E. and Tinkle, D. W. (1982). Energy budgets and life histories of reptiles. *In* "Biology of the Reptilia" (C. Gans, F. H. Pough, eds), Vol. 13, 233–271. Academic Press, New York.

Congdon, J. D., Vitt, L. J., Van Loben Sels, R. and Ohmart, R. D. (1982). The ecological significance of water flux rates in arboreal desert lizards of the genus *Urosaurus*. *Physiol. Zool.* **55**, 317–322.

Conrad, G. (1971). Synthèse de l'évolution continentale post-hercynienne du Sahara algérien. *Bull. Serv. Carte Géol. Algér.* **41**, 143–159.

Cooper, W. E. and Ferguson, G. W. (1972a). Steroids and color change during gravidity in the lizard *Crotaphytus collaris*. *Gen. Comp. Endocrinol.* **18**, 69–72.

Cooper, W. E. and Ferguson, G. W. (1972b). Relative effectiveness of progesterone and testosterone as inductors of orange spotting in female collared lizards. *Herpetologica* **28**, 64–65.

Cothran, M. L. and Hutchison, D. H. (1979). Effect of melatonin on thermal selection by *Crotaphytus collaris* (Squamata: Iguanidae). *Comp. Biochem. Physiol.* **63A**, 461–466.

Courrier, R. (1927). Étude sur le déterminisme des caractères sexuels secondaires de quelques Mammifères à activité testiculaire périodique. *Arch. Biol. Liège* **37**, 173–334.

Courrier, R. (1929). Les modifications saisonnières de l'appareil urogénital chez *Uromastix acanthinurus* Bell. *Arch. Anat. Micr.* **25**, 388–394.

Courty, Y. and Dufaure, J. P. (1979). Levels of testosterone in the plasma and testis of the viviparous lizard (*Lacerta vivipara* Jacquin) during the annual cycle. *Gen. Comp. Endocrinol.* **39**, 336–342.

Cowles, R. B. (1958). Possible origin of dermal temperature regulation. *Evolution* **12**, 347–357.

Cowles, R. G. and Bogert, C. M. (1944). A preliminary study of the thermal requirements of desert reptiles. *Bull. Amer. Mus. Nat. Hist.* **83**, 265–296.

Crawford, E. C. (1972). Brain and body temperatures in a panting lizard. *Science* **177**, 431–433.

Crawford, E. C. and Kampe, G. (1971). Physiological responses of the lizard *Sauromalus obesus* to changes in ambient temperature. *Amer. J. Physiol.* **220**, 1256–1260.

Crenshaw, J. W. (1955). The life history of the southern spiny lizard, *Sceloporus undulatus* Latreille. *Amer. Midl. Nat.* **54**, 257–298.

Crews, D. (1973). Coition-induced inhibition of sexual receptivity in female *Anolis carolinensis*. *Physiol. Behav.* **11**, 463–468.

Crews, D. (1975). Psychobiology of reptilian reproduction. *Science* **189**, 1059–1065.

Crews, D. (1977). The annotated Anole: studies on the control of lizard reproduction. *Amer. Sci.* **65**, 428–434.

Crews, D. (1979a). Endocrine control of reptilian reproductive behaviour. *In* "Endocrine Control of Sexual Behaviour" (C. Beyer, ed.), 167–222. Raven Press, New York.

Crews, D. (1979b). Neuroendocrinology of lizard reproduction. *Biol Reprod.* **20**, 51–73.

Crews, D. (1979c). The hormonal control of behaviour in a lizard. *Sci. Amer.* **241**, 158–165.

Crews, D. (1980). Interrelationships among ecological, behavioral, and neuroendocrine processes in the reproductive cycle of *Anolis carolinensis* and other reptiles. *In* "Advances in the Study of Behavior" (J. S. Rosenblatt, R. A. Hinde, C. G. Beer, M. C. Busnell, eds), **11**, 1–74. Academic Press, New York.

Crews, D. (1983). Regulation of reptilian reproductive behavior. *In* "Advances in Vertebrate Neuroethology" (J.-P. Ewert, R. R. Capranica, D. J. Engle, eds), NATO ASI Series A, Vol. 56, 997–1032.

Crews, D. (1984). Gamete production, sex hormone secretion, and mating behaviour uncoupled. *Horm. Behav.* **18**, 22–28.

Crews, D. and Garstka, W. (1982). The ecological physiology of a garter snake. *Sci. Amer.* 1982, 136–144.

Crews, D. and Licht, P. (1974). Inhibition by corpora atretica of ovarian sensitivity to environmental and hormonal stimulation in the lizard *Anolis carolinensis*. *Endocrinology* **95**, 102–106.

Crews, D., Rosenblatt, J. S. and Lehrman, D. S. (1974). Effect of unseasonal environmental regime, group presence, group composition and males' physiological state on ovarian recrudescence in the lizard, *Anolis carolinensis*. *Endocrinology* **94**, 541–547.

Crews, D., Traina, V., Wetzel, F. D. and Muller, C. (1978). Hormonal control of male reproductive behavior in the lizard *Anolis carolinensis*: role of testosterone, dihydrotestosterone, and estradiol. *Endocrinology* **103**, 1814–1821.

Crisp, M., Cook, L. M. and Hereward, P. V. (1979). Color and heat balance in the lizard *Lacerta dugesii*. *Copeia* 1979, 250–258.

Crisp, T. M. (1964). Seasonal reproductive cycles of the male lizard *Sceloporus cyanogenys*. *Texas Rep. Biol. Med.* **22**, 205–206.

Crook, I. G. (1975). The Tuatara. *In* "Biogeography and Ecology in New Zealand" (G. Kuschel, ed.), 331–352. Junk, den Haage.

Crook, K. A. W. (1981). The break-up of the Australian-Antarctic segment of Gondwanaland. *In* "Ecological Biogeography of Australia" (A. Keast, ed.), Vol. 1, 3–14. Junk, den Haage.

Cuellar, O., Roth, J. J., Fawcett, J. D. and Jones, R. E. (1972). Evidence for sperm sustenance by secretion of the renal sexual segment of male lizards, *Anolis carolinensis*. *Herpetologica* **28**, 53–57.

Dantzler, W. H. (1967). Glomerular and tubular effects of arginine vasotocin in water snakes (*Natrix sipedon*). *Amer. J. Physiol.* **212**, 83–91.

Dantzler, W. H. (1970). Kidney function in desert vertebrates. *In* "Hormones and the Environment" (G. K. Benson, J. G. Phillips, eds). *Mem. Soc. Endocrinol.* No. **18**, 157–189.

Dantzler, W. H. (1976). Renal function (with special emphasis on nitrogen excretion). *In* "Biology of the Reptilia" (C. Gans, W. R. Dawson, eds.), Vol. 5, 447–503. Academic Press, New York.

Dantzler, W. H. (1978). Some renal glomerular and tubular mechanisms involved in osmotic and volume regulation in reptiles and birds. *In* "Osmotic and Volume Regulation", Alfred Benzon Symposium XI (C. B. Jørgensen, E. Skadhauge, eds), 187–201. Munksgaard, Copenhagen.

Dantzler, W. H. and Braun, E. J. (1980). Comparative nephron function in reptiles, birds and mammals. *Amer. J. Physiol.* **239**, R197–R213.

Dantzler, W. H. and Holmes, W. N. (1974). Water and mineral metabolism in Reptilia. *In* "Chemical Zoology" (M. Florkin, B. T. Scheer, eds), Vol. IX, 277–336.

Dantzler, W. H. and Schmidt-Nielsen, B. (1966). Excretion in fresh-water turtle (*Pseudemys scripta*) and desert tortoise (*Gopherus agassizii*). *Amer. J. Physiol.* **210**, 198–210.

Dantzler, W. H., Shaffner, D. P., Chiu, P. J. S. and Sawyer, W. H. (1970). Probenecid inhibition of arginine vasotocin in snakes, frogs, and on frog bladders *in vitro*. *Amer. J. Physiol.* **218**, 929–936.

Daugherty, D. R. and Callard, I. P. (1972). Plasma corticosterone levels in the male iguanid lizard *Sceloporus cyanogenys* under various physiological conditions. *Gen. Comp. Endocrinol.* **19**, 69–79.

Davies, S. J. J. F. (1984). Nomadism as a response to desert conditions in Australia. *J. Arid Envts.* **7**, 183–195.

Dawson, W. R. (1975). On the physiological significance of the preferred body temperature of reptiles. *In* "Perspectives of Biophysical Ecology", *Ecological Studies* Vol. 12 (D. M. Gates, R. B. Schmerl, eds), 443–473. Springer-Verlag.

Dawson, W. R. and Templeton, J. R. (1963). Physiological responses to temperature in the lizard *Crotaphytus collaris*. *Physiol. Zool.* **36**, 219–236.

Deevey, E. S. (1947). Life tables for natural populations of animals. *Quart. Rev. Biol.* **22**, 283–314.

den Hollander, J. (1970). Nocturnalism in geckos from the aspect of water relations. Honours thesis, Zoology Department, University of Western Australia.

De Witt, C. B. (1967a). Precision of thermoregulation and its relation to environmental factors in the desert iguana, *Dipsosaurus dorsalis*. *Physiol. Zool.* **40**, 49–66.

De Witt, C. B. (1967b). Behavioral thermoregulation in the desert iguana. *Science* **158**, 809–810.

De Witt, C. B. and Friedman, R. M. (1979). Significance of skewness in ectotherm thermoregulation. *Amer. Zool.* **19**, 195–209.

Dial, B. E. (1978). The thermal ecology of two sympatric nocturnal *Coleonyx* (Lacertilia: Gekkonidae). *Herpetologica* **34**, 194–201.

Di Castri, F. (1981). Mediterranean shrublands of the world. *In* "Ecosystems of the World: II. Mediterranean-Type Shrublands" (F. Di Castri, D. W. Goodall, R. L. Specht, eds), 1–52. Elsevier, Amsterdam.

Dmi'el, R. (1967). Studies on reproduction, growth, and feeding in the snake *Spalerosophis cliffordi* (Colubridae). *Copeia* 1967, 332–346.

Dmi'el, R. (1972). Effect of activity and temperature on metabolism and water loss in snakes. *Amer. J. Physiol.* **223**, 510–516.

Dmi'el, R. and Borut, A. (1972). Thermal behavior, heat exchange, and metabolism in the desert snake *Spalerosophis cliffordi. Physiol. Zool.* **45**, 78–94.

Dodt, E. (1973). The parietal eye (pineal and parietal organs) of lower vertebrates. *In* "Handbook of Sensory Physiology" (R. Jung, ed.), Vol. 7, 113–140. Springer-Verlag, Heidelberg.

Dubief, J. (1950). Évaporation et coefficients climatiques au Sahara. *Trav. Inst. Rech. Sahara* **6**, 13–44.

Dubief, J. (1952). Le vent et le déplacement du sable au Sahara. *Trav. Inst. Rech. Sahara* **8**, 123–126.

Dubief, J. (1971). Die Sahara, eine Klima-Wüste. *In* "Die Sahara und ihre Randgebiete" (I. H. Schiffers, ed.), 273–348. München.

Dubuis, A., Faurel, L., Grenot, C. and Vernet, R. (1971). Sur le régime alimentaire du lézard saharien *Uromastix acanthinurus* Bell. *C.R. Acad. Sci. (Paris)* **273**, 500–503.

Dufaure, J. P. and Chambon, M. (1978). Uptake of [$^3$H] Testosterone in several organs of the male viviparous lizard (*Lacerta vivipara* Jacquin) and selective retention by the epididymis. *Gen. Comp. Endocrinol.* **36**, 23–29.

Duggan, R. T. and Lofts, B. (1978). Adaptation to fresh water in the sea snake *Hydrophis cyanocinctus*: tissue electrolytes and peripheral corticosteroids. *Gen. Comp. Endocrinol.* **36**, 510–520.

Duggan, R. T. and Lofts, B. (1979). The pituitary-adrenal axis in the sea snake, *Hydrophis cyanocinctus* Daudin. *Gen. Comp. Endocrinol.* **38**, 374–383.

Dunham, A. E. (1978). Food availability as a proximate factor influencing individual growth rates in the iguanid lizard *Sceloporus merriami. Ecology* **59**, 770–778.

Dunham, A. E. (1981). Populations in a fluctuating environment: the comparative population ecology of the iguanid lizards *Sceloporus merriami* and *Urosaurus ornatus. Misc. Publ. Mus. Zool. Univ. Mich.* **158**, 1–62.

Dunham, A. E. (1982). Demographic and life-history variation among populations of the iguanid lizard *Urosaurus ornatus*: implications for the study of life-history phenomena in lizards. *Herpetologica* **38**, 208–221.

Dunson, W. A. (1968). Salt gland secretion. *In* Uti pelagic sea snake *Pelarmis. Amer. I. Physiol.* **215**, 1512–1517.

Dunson, W. A. (1969). Reptilian salt glands. *In* "Exocrine Glands" (S. Y. Botelho, F. P. Brooks, W. B. Shelley, eds), 83–103. University of Pennsylvania Press, Philadelphia.

Dunson, W. A. (1970). Some aspects of electrolyte and water balance in three estuarine reptiles, the diamondback terrapin, American and "salt water" crocodiles. *Comp. Biochem. Physiol.* **32**, 161–174.

Dunson, W. A. (1974). Salt gland excretion in a mangrove monitor lizard. *Comp. Biochem. Physiol.* **474**, 1245–1255.

Dunson, W. A. (1976). Salt glands in reptiles. *In* "Biology of the Reptilia" (C. Gans, W. R. Dawson, eds), Vol. 5, 413—445. Academic Press, New York.

Dunson, W. A. (1979). Control mechanisms in reptiles. *In* "Mechanisms of Osmoregulation in Animals" (R. Gilles, ed.), 273—322. John Wiley & Sons, Chichester.

Dunson, W. A. and Dunson, M. K. (1974). Interspecific differences in fluid concentration and secretion rate of sea snake salt glands. *Amer. J. Physiol.* **227**, 430–438.

Dunson, W. A. and Taub, A. M. (1967). Extrarenal salt excretion in sea snakes (*Laticauda*). *Amer. J. Physiol.* **213**, 975–982.

Durand-Dastès, F. (1968). Aride (domaine). Les Climats. *In* "Encyclopaedia Universalis" 2, 1968, 370–373, Carte 1.

Duval, D., Guillette, L. J. and Jones, R. E. (1982). Environmental control of reptilian reproductive cycles. *In* "Biology of the Reptilia", Vol. 13 (C. Gans, F. H. Pough, eds), 201–231. Academic Press, New York.,

Edmonds, C. J. and Pilcher, D. (1972). Sodium transport mechanisms of the large intestine. *In* "Transport Across the Intestine" (W. L. Burland, P. D. Samuels, eds), 43–57. Churchill-Livingstone, Edinburgh.

Elick, G. E. and Selander, J. A. (1972). Comparative water loss in relation to habitat selection in small colubrid snakes. *Amer. Midl. Nat.* **88**, 429–439.

Emberger, L. (1930). La végétation de la région méditerranéenne. Essai d'une classification de groupements végétaux. *Rev. Gen. Bot.* **42**, 641–662, 705–721.

Emberger, L. (1933). Nouvelle contribution à l'étude de la classification des groupements végétaux. *Rev. Gen. Bot.* **45**, 473–486.

Emberger, L. (1954). Une classification biogéographique des climats. *Rec. Trav. Lab. Bot. Géol. Zool. Univ. Montpellier, Sér. Bot.* **7**, 3–43.

Engbretson, G. A. and Hutchison, V. H. (1976). Parietalectomy and thermal selection in the lizard *Sceloporus magister. J. Exp. Zool.* **198**, 29–38.

Ensor, D. M. (1978). "Comparative Endocrinology of Prolactin". Chapman and Hall, London.

Estes, R. (1983). The fossil record and early distribution of lizards. *In* "Advances in Herpetology and Evolutionary Biology" (A. C. J. Rhodin, K. Miyata, eds), 365–398. Harvard University, Cambridge, Mass.

Ferguson, G. W. (1976). Color change and reproductive cycling in female collared lizards (*Crotaphytus collaris*). *Copeia* 1976, 491–494.

Ferguson, G. W. and Brockman, T. (1980). Geographic differences of growth rate of *Sceloporus* lizards (Sauria: Iguanidae). *Copeia* 1980, 259–264.

Ferguson, G. W., Bohlen, C. H. and Woolley, H. P. (1980). *Sceloporus undulatus*: comparative life history and regulation of a Kansas population. *Ecology* **61**, 313–322.

Fergusson, B., Bradshaw, S. D. and Cannon, J. R. (1985). Hormonal control of femoral gland secretion in the lizard *Amphibolurus ornatus. Gen. Comp. Endocrinol.* **57**, 371–376.

Firth, B. T. and Heatwole, H. (1976). Panting thresholds in lizards: the role of the pineal complex in panting responses in an agamid, *Amphibolurus muricatus. Gen. Comp. Endocrinol.* **29**, 388–401.

Firth, B. T. and Turner, J. S. (1982). Sensory, neural, and hormonal aspects of thermoregulation. *In* "Biology of the Reptilia", Vol. 12 (C. Gans, F. H. Pough, eds), 213–274. Academic Press, New York.

Fisher, R. A. (1930). "The Genetical Theory of Natural Selection". Clarendon Press, Oxford.

Fitch, H. S. (1956). An ecological study of the collared lizard (*Crotaphytus collaris*). *Univ. Kans. Publ. Mus. Nat. Hist.* **8**, 213–274.

Fitch, H. S. (1970). Reproductive cycles of lizards and snakes. *Misc. Publ. Univ. Kans. Mus. Nat. Hist.* **52**, 1–247.

Fitzpatrick, E. A. (1964). Seasonal distribution of rainfall in Australia analysed by Fourier methods. *Arch. Met. Geoph. Bioklim. Ser. B.* **13**, 270–286.

Folk, G. E. (1974). "Textbook of Environmental Physiology", 2nd ed. Lea & Febiger, Philadelphia.

Frakes, L. A. and Kemp, E. M. (1972). Influence of continental positions on early Tertiary climates. *Nature* **240**, 97–100.

Francaz, J-M., Dudemaine, M., Vernet, R. and Grenot, C. J. (1976). Étude de l'évolution de la température interne et du rythme cardiaque chez le lézard saharien: *Varanus griseus* Daud. par radiotélémetrie. *C.R. Acad. Sci. (Paris)* **282**, 1199–1201.

Froudist, J. (1970). Thermoregulation in geckoes with respect to its possible influence on nocturnal habit. Honours thesis, Zoology Department, University of Western Australia.

Fukada, H. (1961). Biological studies on the snakes. VIII. On the growth formulae of snakes and their application to other reptiles. *Bull. Kyoto Gakugei Univ.* **B 17**, 16–40.

Gabe, M. and Saint Girons, H. (1976). Contribution à la morphologie comparée des fosses nasales et de leurs annexes chez les Lépidosauriens. *Mém. Mus. Nat. Hist. Nat.*, **A98**, 1–87.

Galli-Gallardo, S. M., Oguro, C. and Pang, P. K. T. (1979). Renal responses of the Chilean toad, *Calyptocephalella caudiverbera*, and the mudpuppy, *Necturus maculosus*, to mestocin. *Gen. Comp. Endocrinol.* **37**, 134–136.

Galloway, R. W. and Kemp, E. M. (1981). Late Cainozoic environments in Australia. *In* "Ecological Biogeography of Australia" (A. Keast, ed.), Vol. 1, 53–80. Junk, den Haage.

Gans, C. and Pough F. H. (1982). Physiological ecology: its debt to reptilian studies, its value to students of reptiles. *In* "Biology of the Reptilia", Vol. 12 (C. Gans, F. H. Pough, eds), 1–13. Academic Press, New York.

Gans, C., Krakauer, T. and Paganelli, C. V. (1968). Water loss in snakes: interspecific and intraspecific variability. *Comp. Biochem. Physiol.* **27**, 747–761.

Gapp, D. A., Ho, S. M. and Callard, I. P. (1979). Plasma levels of vitellogenin in *Chrysemys picta* during the annual gonadal cycle: measurement by specific radioimmunoassay. *Endocrinology* **104**, 784–790.

Garrick, L. D. (1974). Reproductive influences on behavioural thermoregulation in the lizard *Sceloporus cyanogenys*. *Physiol. Behav.* **12**, 85–91.

Garrick, L. D. (1979). Lizard thermoregulation: operant responses for heat at different thermal intensities. *Copeia* 1979, 258–266.

Garstka, W. and Crews, D. (1981). Female sexual pheromone in the skin and circulation of a garter snake. *Science* **214**, 681–683.

Garstka, W., Camazine, B. and Crews, D. (1982). Interactions of behavior and physiology during the annual reproductive cycle of the red-sided gartersnake *Thamnophis sirtalis parietalis. Herpetologica* **38**, 104–123.

Garstka W., Schwende, F., Novotny, M. and Crews, D. (1983). Biochemical aspects of snake mating pheromone. (Unpubl.—cited in Crews, 1983).

Gauthier, R. (1967). Écologie et éthologie des reptiles du Sahara Nord-occidental (Région de Béni-Abbès). *Annales du Musée Royale de l'Afrique Centrale*, Tervuren, Belgique, Série 8, **155**, 1–83.

Gibbons, J. W. (1967). Variation in growth rates in three populations of the Painted Turtle, *Chrysemys picta. Herpetologica* **23**, 296–303.

Gibbons, J. W. (1976). Aging phenomena in reptiles. *In* "Special Review of Experimental Aging Research" (M. F. Elias, B. E. Eleftheriou, P. K. Elias, eds), 453–475. EAR, Inc., Bar Harbor, Maine.

Gilles-Baillien, M. (1969). Seasonal changes in the permeability of the isolated vesical epithelium of *Testudo hermanni hermanni* Gmelin. *Biochim. Biophys. Acta* **193**, 129–136.

Gilmore, C. W. (1928). Fossil lizards of North America. *Mem. Nat. Acad. Sci.* **22**, 1–201.

Gist, D. G. and DeRoos, R. (1966). Corticoids of the alligator adrenal gland and the effects of ACTH and progesterone on their production *in vitro. Gen. Comp. Endocrinol.* **7**, 304–313.

Glaser, R. (1958). Increase in locomotor activity following shielding of the parietal eye in night lizards. *Science* **128**, 1577–1578.

Goldberg, S. R. (1971). Reproductive cycle of the ovoviviparous iguanid lizard *Sceloporus jarrovi* Cope. *Herpetologica* **27**, 123–131.

Gonzalez, C., Shamoo, Y. E. and Brodsky, W. A. (1967a). Electrical nature of active chloride transport across short-circuited turtle bladders. *Amer. J. Physiol.* **212**, 641–650.

Gonzalez, C., Shamoo, Y. E., Wyssbrod, H. R., Solinger, R. E. and Brodsky, W. A. (1967b). Electrical nature of transport across the isolated turtle bladder. *Amer. J. Physiol.* **213**, 333–340.

Gould, S. J. and Lewontin, R. C. (1979). The spandrels of San Marco and the Panglossian paradigm: a critique of the adaptationist programme. *Proc. Roy. Soc. Lond.* **B205**, 581–598.

Green, B. (1972). Aspects of renal function in the lizard *Varanus gouldii. Comp. Biochem. Physiol.* **43A**, 747–756.

Greer, A. E. (1977). The systematics and evolutionary relationships of the scincid lizards *Lygosoma. J. Nat. Hist.* **11**, 515–540.

Greer, A. E. (1980). Critical thermal maximum temperatures in Australian scincid lizards: their ecological and evolutionary significance. *Aust. J. Zool.* **28**, 91–102.

Grenot, C. J. (1968a). Sur l'excrétion nasale de sels chez le lézard saharien *Uromastix acanthinurus. C.R. Acad. Sci. (Paris)* **226**, 1871–1874.

Grenot, C. J. (1968b). Étude comparative de la résistance à la chaleur d'*Uromastix acanthinurus* et de *Varanus griseus. Terre et la Vie* **4**, 390–409.

Grenot, C. J. (1974). Physical and vegetational aspects of the Sahara desert. *In* "Desert Biology", Vol. 2 (G. W. Brown, ed.), 103–164. Academic Press, New York.

Grenot, C. J. (1976). Écophysiologie du lézard saharien *Uromastix acanthinurus* Bell 1825 (Agamidae, herbivore). *Publ. Lab. Zool. E.N.S.* **7**, 1–323.

Grenot, C. J. and Loirat, F. (1973). L'activité et le comportement thermorégulateur du lézard saharien *Uromastix acanthinurus* Bell. *Terre et la Vie* **27**, 435–455.

Grenot, C. J. and Serrano, V. (1982). Distribution spatiale et structure des communautés de petits vertébrés du désert de Chihuahua. *C.R. Soc. Biogéogr.* **58**, 159–191.

Grenot, C. J. and Vernet, R. (1972a). Place des reptiles dans l'écosystème du désert pierreux, au Sahara occidental. *Bull. Nat. Orléanais* Sér. **3**, No. 5, 23.

Grenot, C. J. and Vernet, R. (1972b). Les reptiles dans l'écosystème au Sahara occidental. *C.R. Soc. Biogéogr.* **433**, 96–112.

Grenot, C. J. and Vernet, R. (1973). Sur une population d'*Uromastix acanthinurus* Bell isolée au milieu du Grand Erg Occidental (Sahara algérien). *C. R. Acad. Sci. (Paris)* **276**, 1349–1352.

Griffiths, I. (1962). Skeletal lamellae as an index of age in heterothermous tetrapods. *Ann. Mag. Nat. Hist.* **13**, 4, 449–465.

Guillette, L. J. (1981). On the occurrence of oviparous and viviparous forms of the Mexican lizard *Sceloporus aeneus*. *Herpetologica* **37**, 11–15.

Guillette, L. J. (1982). The evolution of viviparity and placentation in the high elevation, Mexican lizard *Sceloporus aeneus*. *Herpetologica* **38**, 94–103.

Guillette, L. J. and Casas-Andreu, G. (1980). Fall reproductive activity in the high altitude Mexican lizard, *Sceloporus grammicus microlepidotus*. *J. Herpetol.* **14**, 143–147.

Guillette, L. J., Spielvogel, S. and Moore, F. L. (1981). Luteal development, placentation, and plasma progesterone concentration in the viviparous lizard *Sceloporus jarrovi*. *Gen. Comp. Endocrinol.* **43**, 20–29.

Guillette, L. J., Jones, R. E., Fitzgerald, K. T. and Smith, H. M. (1980). Evolution of viviparity in the lizard genus *Sceloporus*. *Herpetologica* **36**, 201–215.

Guillette, L. J., Lavia, L. A., Walker, N. J. and Roberts, D. K. (1984). Luteolysis induced by prostaglandin $F_{2a}$ in the lizard *Anolis carolinensis*. *Gen. Comp. Endocrinol.* **56**, 271–277.

Harris, V. A. (1964). "The Life of the Rainbow Lizard". Hutchinson, London.

Hart, R. P. (1980). Ecology of the salmonella in a wild marsupial, the quokka *Setonix brachyurus*. Ph.D. thesis, University of Western Australia.

Heath, J. E. (1965). Temperature regulation and diurnal activity in Horned Lizards. *Univ. Calif. Publ. Zool.* **64**, 97–136.

Heath, J. E. (1970). Behavioral regulation of body temperature in poikilotherms. *Physiologist* **13**, 399–410.

Heatwole, H. (1970). Thermal ecology of the desert dragon *Amphibolurus inermis*. *Ecol. Monog.* **40**, 425–457.

Heatwole, H. (1976). "Reptile Ecology". University of Queensland Press, St Lucia.

Heatwole, H. and Butler, H. (1981). Structure of an assemblage of lizards on Barrow Island, Western Australia. *Aust. J. Herpet.* **1**, 37–44.

Heatwole, H. and Firth, B. T. (1982). Voluntary maximum temperature of the Jackie Lizard, *Amphibolurus muricatus*. *Copeia* 1982, 824–829.

Heatwole, H. and Johnson, C. R. (1979). Thermoregulation in the red-bellied black snake, *Pseudechis porphyriacus* (Elapidae). *Zool. J. Linn. Soc.* **65**, 83–101.

Heatwole, H. and Veron, J. E. N. (1977). Vital limit and evaporative water loss in lizards (Reptilia, Lacertilia): a critique and new data. *J. Herpetol.* **11**, 341–348.

Heatwole, H., Firth, B. T. and Staddart, H. (1975). Influence of season, photoperiod and thermal acclimation on the panting threshold of *Amphibolurus muricatus*. *J. Exp. Zool.* **191**, 183–192.

Heatwole, H., Firth, B. T. and Webb, G. J. W. (1973). Panting thresholds of lizards. I. Some methodological and internal influences on the panting threshold of an agamid, *Amphibolurus muricatus*. *Comp. Biochem. Physiol.* **46A**, 799–826.

Henzell, R. P. (1965). An analysis of competition between lizards in the field: water relations. Honours thesis, Zoology Department, University of Western Australia.

Henzell, R. P. (1972). Adaptation to aridity in lizards of the *Egernia whitei* species-group. Ph.D. thesis, University of Adelaide.

Herlant, M. (1933). Recherches histologiques et expérimentales sur les variations cycliques du testicule et des caractères sexuels secondaires des Reptiles. *Arch. Biol. Liège* **44**, 347–468.

Highfill, D. R. and Mead, R. A. (1975a). Sources and levels of progesterone during pregnancy in the garter snake, *Thamnophis elegans*. *Gen. Comp. Endocrinol.* **27**, 389–400.

Highfill, D. R. and Mead, R. A. (1975b). Function of corpora lutea of pregnancy in the viviparous snake, *Thamnophis elegans*. *Gen. Comp. Endocrinol.* **27**, 401–407.

Hirshfield, M. F. and Tinkle, D. W. (1975). Natural selection and the evolution of reproductive effort. *Proc. Nat. Acad. Sci. U.S.A.* **72**, 2227–2231.

Hoffstetter, R. (1961). Squamates. *In* "Le Gisement de Vertébrés miocènes de Béni-Mellal (Maroc)". *Notes et Mém. Serv. Géol. Maroc* **155**, 95–101.

Holleman, D. F., White, R. G. and Luick, J. R. (1982). Application of the isotopic method for measuring total body water, body composition and body water turnover. *In* "Use of Tritiated Water in Studies of Production and Adaptation in Ruminants", 9–32. Institute of Atomic Energy Agency, Vienna.

Holmes, W. N. and McBean, R. L. (1964). Some aspects of electrolyte excretion in the green turtle, *Chelonia mydas mydas*. *J. Exp. Biol.* **41**, 81–90.

Holmes, W. N. and Pearce, R. B. (1979). Hormones and osmoregulation in the vertebrates. *In* "Mechanisms of Osmoregulation in Animals" (R. Gilles, ed.), 413–533. John Wiley & Sons, Chichester.

Horowitz, M. and Borut, A. (1970). Effect of acute dehydration on body fluid compartments in three rodent species, *Rattus norvegicus, Acomys cahirinus* and *Meriones crassus*. *Comp. Biochem. Physiol.* **35**, 283–290.

Horowitz, M. and Borut, A. (1975). Blood volume regulation in dehydrated rodents: plasma protein turnover and sedimentation coefficients. *Comp. Biochem. Physiol.* **51A**, 827–831.

Horton, D. R. (1972). Evolution in the genus *Egernia* (Lacertilia: Scincidae). *J. Herpetology* **6**, 101–109.

Huang, D. P., Vinson, G. P. and Phillips, J. G. (1969). The metabolism of pregnenolone and progesterone by cobra adrenal tissue *in vitro* and the effect of ACTH on product yield-time curves. *Gen. Comp. Endocrinol.* **12**, 637–643.

Huey, R. B. (1982). Temperature, physiology, and the ecology of reptiles. *In* "Biology of the Reptilia", Vol. 12 (C. Gans, F. H. Pough, eds), 25–91. Academic Press, New York.

Huey, R. B. and Pianka, E. R. (1977a). Patterns of niche overlap among broadly sympatric versus narrowly sympatric Kalahari lizards (Scincidae: Mabuya). *Ecology* **58**, 119–128.

Huey, R. B. and Pianka, E. R. (1977b). Seasonal variation in thermoregulatory behavior and body temperature of diurnal Kalahari lizards. *Ecology* **58**, 1066–1075.

Huey, R. B. and Slatkin, M. (1976). Cost and benefits of lizard thermoregulation. *Quart. Rev. Biol.* **52**, 363–384.

Hulse, A. C. (1976). Growth and morphometrics of *Kinosternon sonoriense* (Reptilia, Testudines, Kinosternidae). *J. Herpetol.* **10**, 341–348.

Hutchinson, G. E. (1957). Concluding remarks. *Cold Spring Harbor Symposia on Quantitative Biology* **22**, 415–427.

Hutchison, V. H. and Kosh, R. J. (1974). Thermoregulatory function of the parietal eye in the lizard *Anolis carolinensis. Oecologia* **16**, 173–177.

Hutchinson, V. H., Dowling, H. G. and Vinegar, H. (1966). Thermoregulation in a brooding female python, *Python molurus bivittatus. Science* **151**, 694–696.

Huxley, J. (1942). "Evolution: the modern synthesis". George Allen & Unwin, London.

Jackson, D. G. Jr. (1970). A biometrical study of growth in *Pseudemys concinna suwanniensis. Copeia* 1970, 528–534.

Jacob, F. (1977). Evolution and tinkering. *Science* **196**, 1161–1166.

James, F. C. and Porter, W. P. (1979). Behavior-microclimate relationships in the African rainbow lizard, *Agama agama. Copeia* 1979, 585–593.

Johnson, A. W. (1968). The evolution of desert vegetation in western North America. *In* "Desert Biology" (G. W. Brown, ed.), Vol. 1, 101–140. Academic Press, New York.

Joleaud, L. (1938). Histoire de la formation d'un désert: Paléogéographie du Sahara. *In* "La Vie dans la Région Désertique nord-tropicale de l'Ancien Monde". *Mém. Soc. Biogéogr.* 6, 21–45.

Jones, R. E. (1975). Endocrine control of clutch size in reptiles. IV. Estrogen-induced hyperemia and growth of ovarian follicles in the lizard *Anolis carolinensis. Gen. Comp. Endocrinol.* **25**, 211–222.

Jones, R. E. and Guillette, L. J. (1982). Hormonal control of oviposition and parturition in lizards. *Herpetologica* **38**, 80–93.

Jones, R. E., Summer, C. H. and Lopez, K. H. (1983). Adrenergic inhibition of uterine contractions and oviposition in the lizard *Anolis carolinensis. Gen. Comp. Endocrinol.* **51**, 77–83.

Jones, R. E., Sedgley, N. N., Gerrard, A. M. and Roth, J. J. (1974). Endocrine control of clutch size in reptiles. III. $\Delta^5$-3$\beta$-hydroxy-steroid dehydrogenase activity in different-sized ovarian follicles of *Anolis carolinensis. Gen. Comp. Endocrinol.* **22**, 448–453.

Joss, J. M. P. (1983). Ovarian steroid production in an oviparous lizard. XIIth Confce Europ. Comp. Endocrinologists. Sheffield, August 1983 (abstr.).

Joss, J. M. P. (1985). Ovarian steroid production in oviparous lizards of the genus *Lampropholis* (Scincidae). In "Biology of Australian Frogs and Reptiles" (G. Grigg, R. Shine, H. Ehmann, eds) 319–326. Surrey Beatty & Sons, Chipping Norton.

Junqueira, L. C. U., Malnic, G. and Monge, C. (1966). Reabsorptive function of the ophidian cloaca and large intestine. *Physiol. Zool.* **39**, 151–159.

Keast, A. (1959). The reptiles of Australia. *Monographiae Biologicae* **8**, 115–135.

Kehl, R. (1944). Étude de quelques problèmes d'endocrinologie génitale chez certains reptiles du Sud Algérien. *Rev. Canad. Biol. Montréal* **3**, 131–219.

Kehl, R. and Combescot, C. (1955). Reproduction in the Reptilia. *Mem. Soc. Endocr.* **4**, 57–74.

Kemp, E. M. (1981). Tertiary palaeogeography and the evolution of Australian climate. *In* "Ecological Biogeography of Australia" (A. Keast, ed.), Vol. 1, 33–49.

Kennaway, D. J., Frith, R. G., Phillipou, G., Matthews, C. D. and Seamark, R. F. (1977). A specific radioimmunoassay for melatonin in biological tissue and fluids and its validation by gas chromatography-mass spectrophotometry. *Endocrinology* **101**, 119–127.

Kennett, J. P. and Shakelton, N. J. (1976). Oxygen isotopic evidence for the development of the psychrosphere 38 Myr ago. *Nature* **160**, 513–515.

Kershaw, A. P. (1981). Quaternary vegetation and environments. *In* "Ecological Biogeography of Australia" (A. Keast, ed.), Vol. 1, 83–101.

Kiester, A. R. (1971). Species density of North American amphibians and reptiles. *Syst. Zool.* **29**, 127–137.

King, D. (1980). The thermal biology of free-living sand goannas (*Varanus gouldii*) in southern Australia. *Copeia* 1980, 755–767.

King, M. (1977). Reproduction in the Australian gecko *Phyllodactylus marmoratus* (Gray). *Herpetologica* **33**, 7–13.

King, M. and King, D. (1975). Chromosomal evolution in the lizard genus *Varanus* (Reptilia). *Aust. J. Biol. Sci.* **28**, 89–108.

Kitchener, D. J., Chapman, A., Dell, J. and Muir, B. G. (1980). Lizard assemblage and reserve size and structure in the Western Australian wheatbelt—some implications for conservation. *Biol. Conserv.* **17**, 25–62.

Kluge, A. G. (1967a). Systematics, phylogeny and zoogeography of the lizard genus *Diplodactylus* Gray (Gekkonidae). *Aust. J. Zool.* **15**, 1007–1108.

Kluge, A. G. (1967b). Higher taxonomic categories of gekkonid lizards and their evolution. *Bull. Amer. Mus. Nat. Hist.* **135**, 1–59.

Kluger, M. J., Tarr, R. S. and Heath, J. E. (1973). Posterior hypothalamic lesions and disturbances in behavioral thermoregulation. *Physiol. Zool.* **46**, 79–84.

Knowles, H. C. (1956). Hypernatraemia. *Metabolism* **5**, 508–518.

Kobayashi, D., Mautz, W. J. and Nagy, K. A. (1983). Evaporative water loss: humidity acclimation in *Anolis carolinensis* lizards. *Copeia* 1983, 701–704.

Koike, T. I., Pryor, L. R., Neldon, H. L. and Venable, R. S. (1977). Effect of water deprivation on plasma radio-immunoassayable arginine vasotocin in conscious chickens (*Gallus domesticus*). *Gen. Comp. Endocrinol.* **33**, 359–364.

Köppen, W. and Geiger, R. (1936). "Handbuch der Klimatologie". Bornträger, Berlin, Band 1, Teil C, pp C, 42–43.

Koshland, D. E., Goldbeter, A. and Stock, J. B. (1982). Amplification and adaptation in regulatory and sensory systems. *Science* **217**, 220–225.

Krekorian, C. O. (1984). Life history of the desert iguana, *Dipsosaurus dorsalis*. *Herpetologica* **40**, 415–424.

Kuhn, T. S. (1970). The structure of scientific revolutions. *In* "International Encyclopaedia of Unified Science". Vol. 2. University of Chicago Press, Chicago.

Kutcher, C. (1967). Plasma volume change during water deprivation in gerbils, hamsters, guinea pigs and rats. *Comp. Biochem. Physiol.* **25**, 929–936.

Lambert, S. and Reid, W. H. (1981). Biogeography of the Colorado Herpetofauna. *Amer. Midl. Nat.* **106**, 145–156.

Lang, J. W. (1979). Thermophilic response of the American alligator and the American crocodile to feeding. *Copeia* 1979, 48–59.

Langlois, J. P. (1902). La régulation thermique chez les poikilothermes. *J. Physiol. Path. Génerale* **4**, 249–256.

LaPointe, J. L. (1969). Effect of ovarian steroids and neurohypophysial hormones on the oviduct of the viviparous lizard, *Klauberina riversiana*. *J. Endocr.* **43**, 197–205.

LaPointe, J. L. (1977). Comparative physiology of neurohypophysial hormone action on the vertebrate oviduct-uterus. *Amer. Zool.* **17**, 763–773.

Lauder, G. (1982). Historical biology and the problem of design. *J. Theor. Biol.* **97**, 57–67.

Lavocat, R. (1961). Conclusions générales. *In* "Le Gisement de Vertébrés miocènes de Béni-Mellal (Maroc)". *Notes et Mem. Serv. Géol. Maroc* **155**, 109–115.

Le Brie, S. J. and Elizondo, R. S. (1969). Saline loading and aldosterone in water snakes *Natrix cyclopion*. *Amer. J. Physiol.* **217**, 426–430.

Lee, A. K., Bradley, A. J. and Braithwaite, R. W. (1977). Corticosteroid levels and male mortality in *Antechinus stuartii*. *In* "The Biology of Marsupials" (B. Stonehouse, D. Gilmore, eds), 209–220. Macmillan, London.

LeFevre, M. E. (1973). Effects of aldosterone on the isolated substrate-depleted turtle bladder. *Amer. J. Physiol.* **225**, 1252–1256.

Le Houérou, H. N. (1959). Recherches écologiques et floristiques sur la végétation de la Tunisie méridionale. *Mém. Hors Ser. Inst. Rech. Sahariennes*, Algers, Algeria.

Le Houérou, H. N. (1979). North Africa. *In* "Arid-Land Ecosystems: Structure Functioning and Management" (D. W. Goodall, R. A. Perry, K. M. W. Howes, eds), Vol. 1, 83–107. IBP, Cambridge University Press, London.

Le Houérou, H. N. (1982). The arid bioclimates in the Mediterranean isoclimatic zone. *In* "Ecologia Mediterranea", Tome VIII (P. Quézel, ed.), 103–114. Scientific Affairs Division, NATO.

Lemire, M. (1983). Contribution à l'étude des structures nasales des Sauriens. Structure et fonction de la glande "à sels" des lézards déserticoles. Thèse d'État de l'Université de Pierre et Marie Curie, Paris.

Lemire, M. and Vernet, R. (1982). Effects of different loads on the nasal gland secretion of the lizard *Uromastix acanthinurus* (Agamidae) and relations with its function. *Comp. Biochem. Physiol.* **72A**, 65–72.

Lemire, M. and Vernet, R. (1983). Acclimatation of nasal salt secretion in the lizard *Uromastix acanthinurus* (Agamidae). Effects of repetitive salt loads. *J. Arid Environ.* **6**, 145–153.

Lemire, M., Deloince, R. and Grenot, C. J. (1970). Étude des cavités nasales du Lézard Fouette-Queue, *Uromastix acanthinurus* Bell. *C.R. Acad. Sci. (Paris)* **270**, 817–820.

Lemire, M., Deloince, R. and Grenot, C. J. (1972). Étude histoenzymologique de la glande "à sels" du lézard saharien *Uromastix acanthinurus* Bell. *C.R. Acad. Sci. (Paris)* **274**, 3414–3416.

Lemire, M., Grenot, C. J. and Vernet, R. (1979). La balance hydrique d'*Uromastix acanthinurus* Bell (Sauria, Agamidae) au Sahara dans des conditions semi-naturelles. *C.R. Acad. Sci. (Paris)* **288**, 359–362.

Lemire, M., Grenot, C. J. and Vernet, R. (1982). Water and electrolyte balance of free-living Saharan lizards, *Uromastix acanthinurus* (Agamidae). *J. Comp. Physiol.* **146**, 81–93.

Lemire, M., Vernet, R. and Grenot, C. J. (1980). Electrolyte excretion by the nasal gland of an herbivorous Saharan lizard, *Uromastix acanthinurus* (Agamidae). Effects of single NaCl and KCl loads. *J. Arid Environ.* **3**, 325–330.

Lerner, I. M. (1958). "The Genetic Basis of Selection". John Wiley, New York.

Licht, P. (1967a). Environmental control of annular testicular cycles in the lizard *Anolis carolinensis*. II. Seasonal variations in the effects of photoperiod and temperature on testicular recrudescence. *J. Exp. Zool.* **166**, 243–253.

Licht, P. (1967b). Interaction of prolactin and gonadotropins on appetite, growth, and tail regeneration in the lizard, *Anolis carolinensis*. *Gen. Comp. Endocrinol.* **9**, 49–63.

Licht, P. (1972). Problems in experimentation on timing mechanisms for annual physiological cycles in reptiles. *In* "Hibernation and Hypothermia, Perspectives and Challenges" (F. E. South, J. P. Hannon, J. R. Willis, E. T. Pengelley, N. R. Alpert, eds), 681–711. Elsevier, Amsterdam.

Licht, P. (1973). Influence of temperature and photoperiod on the annual ovarian cycle in the lizard, *Anolis carolinensis*. *Copeia* 1973, 465–472.

Licht, P. (1974). Endocrinology of Reptilia. The pituitary system. *In* "Chemical Zoology" (M. Florkin, B. T. Scheer, eds), Vol. 9, 399–448. Academic Press, New York.

Licht, P. (1979). Reproductive endocrinology of reptiles and amphibians: gonadotropins. *Ann. Rev. Physiol.* **41**, 337–351.

Licht, P. (1982). Endocrine patterns in the reproductive cycles of turtles. *Herpetologica* **38**, 51–61.

Licht, P. (1984). Reptiles. *In* "Marshall's Physiology of Reproduction" (G. E. Lamming, ed.), Vol. 1, 206–282. Churchill Livingstone, Edinburgh.

Lifson, N. and McClintock, R. (1966). Theory and use of the turnover rates of body water for measuring energy and material balance. *J. Theoret. Biol.* **12**, 46–74.

Lillywhite, H. B. and Maderson, P. F. A. (1982). Skin structure and permeability. *In* "Biology of the Reptilia" (C. Gans, F. H. Pough, eds), Vol. 12C, 397–442. Academic Press, New York.

Lofts, B. (1977). Patterns of spermatogenesis and steroidogenesis in male reptiles. *In* "Reproduction and Evolution" (J. H. Calaby, C. H. Tyndale-Biscoe, eds), 127–136. 4th Int. Symp. on Comp. Biology of Reproduction. Australian Academy of Science, Canberra.

Lofts, B. and Chui, K. W. (1968). Androgens and reptilian spermatogenesis. *Arch. Anat. Histol. Embryol.* **51**, 409–418.

Lofts, B. and Tsui, H. W. (1977). Histological and histochemical changes in the gonads and epididymides of the male soft-shelled turtle, *Trionyx sinensis*. *J. Zool.* **181**, 57–68.

Logan, R. F. (1968). Causes, climates and distribution of deserts. *In* "Desert Biology" (G. W. Brown, ed.), Vol. 1, 21–50. Academic Press, New York.

Louw, G. N. and Holm, E. (1972). Physiological, morphological and behavioural adaptations of the ultrapsammophilous Namib Desert lizard, *Aporosaura anchietae* (Bocage). *Madoqua* Ser. II, **Vol. 1**, 67–85.

Louw, G. N., Young, B. A. and Bligh, J. (1976). Effect and thyroxine and noradrenaline on thermoregulation, cardiac rate and oxygen consumption in the monitor lizard *Varanus albigularis albigularis*. *J. Thermal. Biol.* **1**, 189–193.

Low, B. S. (1978). Environmental uncertainty and the parental strategies of marsupials and placentals. *Amer. Nat.* **112**, 197–213.

MAB (Anon.) (1979). Map of the world distribution of arid regions. *MAB Technical Notes* 7, 54. UNESCO, Paris.

Macchi, I. A. (1967). Regulation of reptilian and avian adrenocortical secretion and effect of ACTH on the *in vitro* conversion of progesterone to corticoids. Proc. II Int. Congr. on Hormonal Steroids, Milan 1966 (L. Martini, F. Fraschini, M. Motta, eds), *Exp. Med. Int. Congr. Ser.* No. **132**, 1094–1103.

Macchi, I. A. and Phillips, J. G. (1966). *In vitro* effect of adrenocorticotropin on corticoid secretion in the turtle, snake and bullfrog. *Gen. Comp. Endocrinol.* **6**, 170–182.

MacCulloch, R. D. and Secoy, D. M. (1983). Demography, growth and food of Western Painted Turtles, *Chrysemys picta belli* (Gray), from southern Saskatchewan. *Can. J. Zool.* **61**, 1499–1509.

Macfarlane, W. V., Morris, R. J. H. and Howard, B. (1963). Turnover and distribution of water in desert camels, sheep, cattle and kangaroos. *Nature* **197**, 270–271.

Mackay, R. S. (1964). Galápagos tortoise and marine iguana deep body temperatures measured by radiotelemetry. *Nature* **204**, 355–358.

MacMahon, J. A. (1978). North American deserts: their floral and faunal components. *In* "Evolution of Desert Biota" (D. W. Goodall, ed.), 21–82. University of Texas Press, Austin.

MacMahon, J. A. (1981). Introduction. *In* "Arid-Land Ecosystems: Structure, Functioning and Management" (D. W. Goodall, R. A. Perry, K. M. W. Howes, eds), Vol. 2, 263–269. IBP, Cambridge University Press, London.

Maderson, P. F. A., Mayhew, W. W. and Sprague, G. (1970). Observations on the epidermis of desert-living iguanids. *J. Morph.* **130**, 25–36.

Main, A. R. (1978). Adaptations of Australian vertebrates to desert conditions. *In* "Evolution of Desert Biota" (D. W. Goodall, ed.), 101–131. University of Texas Press, Austin.

Main, A. R. (1983). Macropod ecophysiology: a possible integration with ecological and evolutionary studies. *In* "Research on Rottnest Island" (S. D. Bradshaw, ed.). *J. Roy. Soc. W.A.* **66**, 5–9.

Main, A. R. and Bakker, H. R. (1981). Adaptation of macropod marsupials to aridity. *In* "Ecological Biogeography of Australia" (A. Keast, ed.), Vol. 3, 1489–1519. Junk, den Haage.

Main, A. R., Lee, A. K. and Littlejohn, M. J. (1958). Evolution in three genera of Australian frogs. *Evolution* **12**, 224–233.

Mainoya, J. R., Bern, H. A. and Regan, J. W. (1974). Influence of ovine prolactin on transport of fluid and sodium chloride by the mammalian intestine and gall bladder. *J. Endocr.* **63**, 311–317.

Martin, H. A. (1981). The Tertiary flora. *In* "Ecological Biogeography of Australia" (A. Keast, ed.), Vol. 1, 393–406. Junk, den Haage.

Martonne, E. de (1927). "Traité de Géographie Phytique", Tome I. A. Colin, Paris.

Mautz, W. J. (1979). The metabolism of reclusive lizards, the Xantusiidae. *Copeia* 1979, 577–584.

Mautz, W. J. (1982). Patterns of evaporative water loss. *In* "Biology of the Reptilia" (C. Gans, F. H. Pough, eds), Vol. 12C, 443–481. Academic Press, New York.

Mayhew, W. W. (1963). Temperature preferences of *Sceloporus orcutti*. *Herpetologica* **18**, 217–233.

Mayhew, W. W. (1968). Biology of desert amphibians and reptiles. *In* "Desert Biology" (G. W. Brown, ed.), Vol. 1, 195–356. Academic Press, New York.

Mayhew, W. W. (1971). Reproduction in the desert lizard *Dipsosaurus dorsalis*. *Herpetologica* **27**, 57–77.

Mayhew, W. W. and Wright, S. J. (1970). Seasonal changes in testicular histology of three species of the lizard genus *Uma*. *J. Morphol.* **130**, 163–186.

McGinnis, S. M. (1966). *Sceloporus occidentalis*: preferred body temperature of the Western Fence Lizard. *Science* **152**, 1090–1091.

McGinnis, S. M. and Dickson, L. L. (1967). Thermoregulation in the desert iguana *Dipsosaurus dorsalis*. *Science* **156**, 1757–1759.

McGinnis, S. M. and Voigt, W. G. (1971). Thermoregulation in the desert tortoise *Gopherus agassizii*. *Comp. Biochem. Physiol.* **40A**, 119–126.

McGowran, B. (1978). Early Tertiary foraminiferal biostratigraphy in southern Australia: progress report. *In* "The Crespin Volume: Essays in Honour of Irene Crespin". *Bur. Min. Res. Aust. Bull.* **192**, 83–95.

McNicol, D. and Crews, D. (1979). Estrogen/progesterone synergy in the control of female sexual receptivity in the lizard, *Anolis carolinensis*. *Gen. Comp. Endocrinol.* **38**, 68–74.

McPherson, R. J., Boots, L. R., MacGregor, R. III and Marion, K. R. (1982). Plasma steroids associated with seasonal reproductive changes in a multi-clutched freshwater turtle, *Sternotherus odoratus*. *Gen. Comp. Endocrinol.* **48**, 440–451.

Medawar, P. B. (1951). Problems of adaptation. *New Biology* **11**, 10–26.

Medica, P. A., Bury, R. B. and Turner, F. B. (1975). Growth of the Desert Tortoise (*Gopherus agassizii*) in Nevada. *Copeia* 1975, 639–643.

Medica, P. A., Turner, F. B. and Smith, D. D. (1973). Hormonal induction of color change in female leopard lizards *Crotaphytus wislizenii*. *Copeia* 1973, 658–661.

Meigs, P. (1952). World distribution of arid and semi-arid homoclimates. *In* "Reviews of Research on Arid Zone Gydrology", *Arid Zone Research* 1, 203–210. UNESCO, Paris.

Milewski, A. V. (1981). A comparison of reptile communities in relation to soil fertility in the mediterranean and adjacent arid parts of Australia and southern Africa. *J. Biogeog.* **8**, 493–503.

Milewski, A. V. (1983). A comparison of ecosystems in mediterranean Australia and southern Africa: nutrient-poor sites at the Barrens and the Caledon Coast. *Ann. Rev. Ecol. Syst.* **14**, 57–76.

Miller, M. R. (1948). The seasonal histological changes occurring in the ovary, corpus luteum and testis in the viviparous lizard, *Xantusia vigilis*. *Univ. Calif. Publ. Zool.* **47**, 197–224.

Miller, M. R. (1952). The normal histology and experimental alteration of the adrenal of the viviparous lizard, *Xantusia vigilis*. *Anat. Rec.* **113**, 309–323.

Minnich, J. E. (1970). Water and electrolyte balance of the desert iguana, *Dipsosaurus dorsalis*, in its natural habitat. *Comp. Biochem. Physiol.* **35**, 921–933.

Minnich, J. E. (1972). Excretion of urate salts by reptiles. *Comp. Biochem. Physiol.* **41A**, 535–549.

Minnich, J. E. (1979). Reptiles. *In* "Comparative Physiology of Osmoregulation in Animals" (G. M. O. Maloiy, ed.), Vol. 1, 391–641. Academic Press, London.

Minnich, J. E. (1982). The use of water. *In* "Biology of the Reptilia" (C. Gans, F. H. Pough, eds), Vol. 12C, 325–395. Academic Press, New York.

Minnich, J. E. and Piehl, P. A. (1972). Spherical precipitates in the urine of reptiles. *Comp. Biochem. Physiol.* **41A**, 551–554.

Minnich, J. E. and Shoemaker, V. H. (1970). Diet, behavior and water turnover in the desert iguana, *Dipsosaurus dorsalis*. *Amer. Midl. Nat.* **84**, 496–509.

Minnich, J. E. and Shoemaker, V. H. (1972). Water and electrolyte turnover in a field population of the lizard *Uma scoparia*. *Copeia* 1972, 650–659.

Monod, Th. (1973). "Les Déserts". Editions Horizons de France, Paris.

Moore, R. G. (1978). Seasonal and daily activity patterns and thermoregulation in the southwestern speckled rattlesnake (*Crotalus mitchelli pyrrhus*) and the Colorado desert sidewinder (*Crotalus cerastes laterorepens*). *Copeia* 1978, 439–442.

Moratka, D. J. (1977). "A Biogeographical Analysis of the Chihuahuan Desert Through its Herpetofauna". Junk, Den Haage.

Morgareidge, K. R. and White, F. N. (1969). Cutaneous vascular changes during heating and cooling in the Galápagos marine iguana. *Nature* **223**, 587–591.

Moritz, C. (1983). Parthenogenesis in the endemic Australian lizard *Heteronotia binoei* (Gekkonidae). *Science* **220**, 735–737.

Morton, S. R. (1979). Diversity of desert-dwelling mammals: a comparison of Australia and North America. *J. Mamm.* **60**, 253–264.

Munsick, R. A. (1966). Chromatographic and pharmacologic characterization of the neurohypophyseal hormones of an amphibian and a reptile. *Endocrinology* **78**, 591–599.

Murphy, G. I. (1968). Pattern in life history and the environment. *Amer. Nat.* **102**, 391–403.

Murrish, D. E. and Schmidt-Nielsen, K. (1970). Water transport in the cloaca of lizards: active or passive? *Science* **170**, 324–326.

Mutton, L. A. (1970). The response of geckos to sunlight. Honours thesis, Zoology Department, University of Western Australia.

Myhre, K. and Hammel, H. T. (1969). Behavioral regulation in internal temperature in the lizard *Tiliqua scincoides*. *Amer. J. Physiol.* **217**, 1490–1495.

Naftolin, F., Ryan, K. J., Davies, I. J., Reddy, V. V., Flores, F., Petro, Z., Kuhn, M., White, R. J., Takaoka, Y. and Wolin, L. (1975). The formation of estrogens by central neuroendocrine tissues. *Recent Prog. Horm. Res.* **31**, 295–319.

Nagy, K. A. (1972). Water and electrolyte budgets of a free-living desert lizard, *Sauromalus obesus*. *J. Comp. Physiol.* **79**, 39–62.

Nagy, K. A. (1973). Behaviour, diet and reproduction in a desert lizard, *Sauromalus obesus*. *Copeia* 1973, 93–102.

Nagy, K. A. (1975a). Water and energy budgets of free-living animals: measurement using isotopically-labeled water. *In* "Environmental Physiology of Desert Organisms" (N. F. Hadley, ed.), 227–245. Dowden Hutchinson and Ross Inc., Stroudsburg, Pennsylvania.

Nagy, K. A. (1975b). Nitrogen requirement and its relation to dietary water and potassium content in the lizard *Sauromalus obesus*. *J. Comp. Physiol.* **104**, 49–58.

Nagy, K. A. (1977). Cellulose digestion and nutrient assimilation in *Sauromalus obesus*, a plant-eating lizard. *Copeia* 1977, 355–362.

Nagy, K. A. (1980). $CO_2$ production in animals: analysis of potential errors in the doubly-labeled water method. *Amer. J. Physiol.* **238**, R466–R473.

Nagy, K. A. (1982). Field studies of water relations. *In* "Biology of the Reptilia" (C. Gans, F. H. Pough, eds), Vol. 12C, 483–501. Academic Press, London.

Nagy, K. A. (1983). Ecological energetics. *In* "Lizard Ecology: Studies of a Model Organism" (R. B. Huey, E. R. Pianka, T. W. Schoener, eds), 24–54, 429–433. Harvard University Press, Cambridge, Mass.

Nagy, K. A. and Costa, D. P. (1980). Water flux in animals: analysis of potential errors in the tritiated water method. *Amer. J. Physiol.* **238**, R454–R465.

Nagy, K. A. and Shoemaker, V. H. (1975). Energy and nitrogen budgets of the free-living desert lizard *Sauromalus obesus*. *Physiol. Zool.* **43**, 252–262.

Nahal, I. (1981). The mediterranean climate from a biological viewpoint. *In* "Ecosystems of the World—II: Mediterranean-type Shrublands" (F. Di Castri, D. W. Goodall, R. L. Specht, eds), 63–86. Elsevier, Amsterdam.

Noble, G. K. and Bradley, H. T. (1933). The mating behavior of lizards: its bearing on the theory of sexual selection. *Annals N.Y. Acad. Sci.* **XXXV**, 25–100.

Noble, G. K. and Greenberg, B. (1941). Effects of seasons, castration and crystalline sex hormones upon the urinogenital system and sexual behavior of the lizard (*Anolis carolinensis*). I. The adult female. *J. Exp. Zool.* **88**, 451–479.

Norris, K. S. (1967). Color adaptation in desert reptiles and its thermal relationship. *In* "Lizard Ecology" (W. W. Milstead, ed.), 162–299. University of Missouri Press, Columbia.

Norris, K. S. and Dawson, W. R. (1964). Observations on the water economy and electrolyte excretion of Chuckwallas (Lacertilia, *Sauromalus*). *Copeia* 1964, 638–646.

Norris, K. S. and Lowe, C. H. (1964). An analysis of background color-matching in amphibians and reptiles. *Ecology* **45**, 565–580.

Nussbaum, R. A. and Diller, L. V. (1976). The life history of the Side-Blotched Lizard, *Uta stansburiana* Baird and Girard, in North-Central Oregon. *Northwest Science* **50**, 243–260.

Ozenda, P. (1958). "Flore du Sahara septentrional et central". CNRS, Paris.

Packard, G. C. (1966). The influence of ambient temperature and aridity on modes of reproduction and excretion of amniote vertebrates. *Amer. Nat.* **100**, 667–682.

Packard, G. C., Tracy, C. R. and Roth, J. J. (1977). The physiological ecology of reptilian eggs and embryos, and the evolution of viviparity within the Class Reptilia. *Biol. Rev.* **52**, 71–105.

Pang, P. K. T. and Sawyer, W. H. (1978). Renal and vascular responses of the bullfrog (*Rana catesbeiana*) to mesotocin. *Amer. J. Physiol.* **235**, F151–F155.

Parker, W. S. (1972). Notes on *Dipsosaurus dorsalis* in Arizona. *Herpetologica* **28**, 226–229.

Parks, J. R. (1970). Growth curves and the physiology of growth. I. Animals. *Amer. J. Physiol.* **219**, 833–843.

Parmenter, C. J. and Heatwole, H. (1975). Panting thresholds of lizards. IV. The effects of dehydration on the panting threshold of *Amphibolurus barbatus* and *Amphibolurus muricatus*. *J. Exp. Zool.* **191**, 327–332.

Parmenter, R. G. (1980). Effects of food availability and water temperature on the feeding ecology of pond sliders. *Copeia* 1980, 503–514.

Peaker, M. (1978). Excretion of potassium from the orbital region in *Testudo carbonaria*: a salt gland in a terrestrial tortoise? *J. Zool.* **184**, 421–442.

Peaker, M. and Linzell, L. J. (1975). "Salt Glands in Birds and Reptiles". Cambridge University Press, London.

Pearl, R. and Miner, J. R. (1935). Experimental studies on the duration of life. XIV. The comparative mortality of certain lower organisms. *Quart. Rev. Biol.* **10**, 60–79.

Pearson, O. P. (1977). The effect of substrate and skin color on the thermoregulation of a lizard. *Comp. Biochem. Physiol.* **58A**, 353–358.

Peters, R. H. (1976). Tautology in evolution and ecology. *Amer. Nat.* **110**, 1–12.

Phillips, J. G., Chester Jones, I. and Bellamy, D. (1962). Biosynthesis of adrenocortical hormones by adrenal glands of lizards and snakes. *J. Endocrinol.* **25**, 233–237.

Pianka, E. R. (1966). Convexity, desert lizards and spatial heterogeneity. *Ecology* **47**, 1055–1059.

Pianka, E. R. (1967). Lizard species diversity. *Ecology* **48**, 333–351.

Pianka, E. R. (1969). Habitat specificity, speciation, and species density in Australian desert lizards. *Ecology* **50**, 498–502.

Pianka, E. R. (1970a). On 'r' and 'K' selection. *Amer. Nat.* **104**, 592–597.

Pianka, E. R. (1970b). Comparative autecology of the lizard *Cnemidophorus tigris* in different parts of its geographic range. *Ecology* **51**, 703–720.

Pianka, E. R. (1971). Lizard species density in the Kalahari desert. *Ecology* **52**, 1024–1029.

Pianka, E. R. (1972). Zoogeography and speciation of Australian desert lizards: an ecological perspective. *Copeia* 1972, 127–145.

Pianka, E. R. (1973). The structure of lizard communities. *Ann. Rev. Ecol. Syst.* **4**, 53–74.

Pianka, E. R. (1974). Niche overlap and diffuse competition. *Proc. Nat. Acad. Sci. U.S.A.* **71**, 2141–2145.

Pianka, E. R. (1975). Niche relations of desert lizards. *In* "Ecology and Evolution of Communities" (M. Cody, J. Diamond, eds.), 292–314. Harvard University Press, Cambridge, Mass.

Pianka, E. R. (1976). Natural selection of optimal reproductive tactics. *Amer. Zool.* **16**, 775–784.

Pianka, E. R. (1978). Reptilian species diversity. *In* "Biology of the Reptilia" (D. W. Tinkle, C. Gans, eds), Vol. 7, 1–34. Academic Press, New York.

Pianka, E. R. (1981). Diversity and adaptive radiations of Australian desert lizards. *In* "Ecological Biogeography of Australia" (A. Keast, ed.), Vol. 2, 1377–1392. Junk, den Haage.

Pianka, E. R. and Parker, W. S. (1975). Age-specific reproductive tactics. *Amer. Nat.* **109**, 453–464.

Pianka, E. R. and Schall, J. J. (1981). Species densities of Australian vertebrates. *In* "Ecological Biogeography of Australia" (A. Keast, ed.), Vol. 3, 1677–1694. Junk, den Haage.

Pielou, E. C. (1981). The usefulness of ecological models: a stock-taking. *Quart. Rev. Biol.* **56**, 17–31.

Popper, K. (1959). "The Logic of Scientific Discovery". Hutchinson, London.

Porter, W. P. and Gates, D. M. (1969). Thermodynamic equilibria of animals with environment. *Ecol. Monogr.* **39**, 227–244.

Porter, W. P. and James, F. C. (1979). Behavioral implications of mechanistic ecology. II: The African Rainbow Lizard, *Agama agama. Copeia* 1979, 594–619.

Porter, W. P., Mitchell, J. W., Beckman, W. A. and De Witt, C. B. (1973). Behavioral implications of mechanistic ecology. *Oecologia* **13**, 1–54.

Pough, F. H. (1980). The advantages of ectothermy for tetrapods. *Amer. Nat.* **115**, 92–112.

Powell, C. McA. and Johnson, B. D. (1980). Constraints on the Cenozoic position of Sundaland. *Tectonophys.* **63**, 91–109.

Powell, C. McA., Johnson, B. D. and Veevers, J. J. (1981). The early cretaceous break-up of Eastern Gondwanaland, the separation of Australia and India, and their interaction with Southeast Asia. *In* "Ecological Biogeography of Australia" (A. Keast, ed.), Vol. 1, 17–29. Junk, den Haage.

Prange, H. D. and Schmidt-Nielsen, K. (1969). Evaporative water loss in snakes. *Comp. Biochem. Physiol.* **28**, 973–975.

Prasad, M. R. N. and Reddy, P. R. K. (1972). Physiology of the sexual segment of the kidney in reptiles. *Gen. Comp. Endocrinol.* Suppl. **3**, 649–662.

Presch, W. (1980). Evolutionary history of the South American Microteiid lizards (Teiidae: Gymnophthalminae). *Copeia* 1980, 36–56.

Presch, W. (1982). The Family-Subfamily taxon in lizards: effects of plate tectonics on biogeographic patterns. *In* "Advances in Cladistics", Vol. 2 (N. J. Platnick, V. A. Funk, eds), 191–198. Columbia University Press, New York.

Quay, W. B. (1965). Retinal and pineal hydroxyindole-O-methyl transferase. *Life Sci.* **4**, 983–991.

Quay, W. B. (1969). Factors in the measurement of pineal acetylserotonin methyltransferase activity in the lizard, *Sceloporus occidentalis*. *Gen. Comp. Endocrinol.* **17**, 220–226.

Quézel, P. (1965). "La végétation du Sahara du Tchad à la Mauritanie". Gustav Fischer Verlag, Stuttgart.

Quézel, P. and Barbero, M. (1982). Definition and characterization of Mediterranean-type ecosystems. *In* "Ecologia Mediterranea", Tome VIII (P. Quézel, ed.), 15–29. Scientific Affairs Division, NATO.

Quilty, P. G. (1977). Cenozoic sedimentation cycles in Western Australia. *Geology* **5**, 336–340.

Ralph, C. L. (1984). Pineal bodies and thermoregulation. *In* "The Pineal Gland" (R. J. Reiter, ed.), 193–219. Raven Press, New York.

Ralph, C. L., Firth, B. T., Gern, W. A. and Owens, D. W. (1979). The pineal complex and thermoregulation. *Biol Rev.* **54**, 41–72.

Raven, P. H. and Axelrod, D. I. (1975). History of the flora and fauna of Latin America. *Amer. Sci.* **63**, 520–529.

Reagan, D. P. (1974). Habitat selection in the three-toed box turtle, *Terrapene carolina triunguis*. *Copeia* 1974, 512–527.

Regal, P. J. (1966). Thermophilic responses following feeding in certain reptiles. *Copeia* 1966, 588–590.

Regal, P. J. (1967). Voluntary hypothermia in reptiles. *Science* **155**, 1551–1553.

Renous, S. (1980). Développement de l'aspect historique de la biogéographie par la superposition de deux thèses: proposition d'une hypothèse phylogénetique batie selon les principes Hennigiens et théorie de la dérivé des continents. *C.R. Soc. Biogéogr.* **57**, 81–102.

Rice, G. E. (1982). Plasma arginine vasotocin concentrations in the lizard *Varanus gouldii* (Gray) following water loading, salt loading and dehydration. *Gen. Comp. Endocrinol.* **47**, 1–6.

Rice, G. E. and Bradshaw, S. D. (1980). Changes in dermal reflectance and vascularity and their effects on thermoregulation in *Amphibolurus nuchalis* (Reptilia: Agamidae). *J. Comp. Physiol.* **135**, 139–146.

Rice, G. E., Bradshaw, S. D. and Prendergast, F. J. (1982). The effects of bilateral adrenalectomy on renal function in the lizard *Varanus gouldii* (Gray). *Gen. Comp. Endocrinol.* 47, 182–189.

Riquier, J. and Rossetti, Ch. (1976). Considérations méthodologiques sur l'établissement d'une carte des risques de désertification. Rapport d'une consultation technique. FAO, Rome.

Roberts, J. S. and Schmidt-Nielsen, B. (1966). Renal ultrastructure and excretion of salt and water by three terrestrial lizards. *Amer. J. Physiol.* **211**, 476–486.

Roberts, L. A. (1968). Water loss in the desert lizard, *Uta stansburiana*. *Comp. Biochem. Physiol.* **27**, 583–589.

Rosenbloom, A. A. and Fisher, D. A. (1974). Radioimmunoassay of arginine vasotocin. *Endocrinology* **95**, 1726–1732.

Ross, P. and Crews, D. (1977). Influence of the seminal plug on mating behaviour in the garter snake. *Nature* **267**, 344–345.

Roth, J. J. and Ralph, C. L. (1976). Body temperature of the lizard (*Anolis carolinensis*): effect of parietalectomy. *J. Exp. Zool.* **198**, 17–28.

Roth, J. J. and Ralph, C. L. (1977). Thermal and photic preferences in intact and parietalectomized *Anolis carolinensis*. *Behav. Biol.* **19**, 341–348.

Rundel, P. W. (1979). Ecological impact of fires on mineral and sediment pools and fluxes. *In* "Fire and Fuel Management in Mediterranean-Climate Ecosystems: Research Priorities and Programmes", 17–21. MAB Technical Note 11, UNESCO, Paris.

Saint Girons, H. (1952). Étude Biogéographique des serpents Paléarctiques. *C.R. Som. Séances Soc. Biogéogr.* **250**, 50–67.

Saint Girons, H. (1963). Spermatogenèse et évolution cyclique des caractères sexuels secondaires chez les Squamata. *Ann. Sci. Nat. Zool.* (Paris) 12$^{e}$ sér. **5**, 461–478.

Saint Girons, H. (1965). Les critères d'age chez les reptiles et leurs applications à l'étude de la structure des populations sauvages. *La Terre et la Vie* **4**, 341–360.

Saint Girons, H. (1967). Le cycle sexuel et les corrélations hypophysogénitales chez *Agama bibroni* Duméril au Maroc. *Bull. Biol. Fr. Belg.* **101**, 321–344.

Saint Girons, H. (1975). Critères d'age, structure et dynamique des populations de reptiles. *In* "Problèmes d'Écologie: La Démographie des Populations de Vertébrés" (M. Lamotte, F. Bourlière, eds), 233–252. Masson et Cie, Paris.

Saint Girons, H. (1980). Thermoregulation in reptiles with special reference to the tuatara and its ecophysiology. *Tuatara* **24**, 59–80.

Saint Girons, H. (1982a). Influence des climats de type méditerranéen sur l'écophysiologie et la répartition des reptiles. *In* "Ecologia Mediterranea" (P. Quézel, ed.), Tome VIII, 245–252. Scientific Affairs Division, NATO.

Saint Girons, H. (1982b). Reproductive cycles of male snakes and their relationships with climate and female reproductive cycles. *Herpetologica* **38**, 5–16.

Saint Girons, H. (1982b). Reproductive cycles of male snakes and their relationships with climate and female reproductive cycles. *Herpetologica* **38**, 5–16.

Saint Girons, H. (1984). Les cycles sexuels des lézards males et leurs rapports avec le climat et les cycles reproducteurs des femelles. *Ann. Sci. Nat. Zool.* (Paris) 13$^{e}$ sér. **6**, 221–243.

Saint Girons, H. (1985). Comparative data on lepidosaurian reproduction and some time-tables. *In* "Biology of the Reptilia", Vol. 15 (C. Gans, ed.). 35–58, John Wiley & Sons, New York.

Saint Girons, H. and Saint Girons, M. C. (1956). Cycle d'activité et thermorégulation chez les reptiles (lézards et serpentes). *Vie et Milieu* **7**, 133–226.

Saint Girons, H. and Bradshaw, S. D. (1981). Preliminary observations of behavioural thermoregulation in an elapid snake, the dugite, *Pseudonaja affinis* Gunther. *J. Roy. Soc. W.A.* **4**, 13–16.

Saint Girons, H., Lemire, M. and Bradshaw, S. D. (1977). Structure de la glande nasale externe de *Tiliqua rugosa* (Reptilia, Scincidae) et rapports avec sa fonction. *Zoomorphologie* **88**, 277–288.

Saint Girons, H., Rice, G. E. and Bradshaw, S. D. (1981). Histologie comparée et ultrastructure de la glande nasale externe de quelques Varanidae (Reptilia: Lacertilia). *Ann. Sci. Nat. Zool.* (Paris) 13$^e$ Série **3**, 15–21.

Sandor, T. (1972). Corticosteroids in Amphibia, Reptilia and Aves. *In* "Steroids in Non-Mammalian Vertebrates" (D. R. Idler, ed.), 253–327. Academic Press, New York.

Sandor, T. and Mehdi, A. Z. (1980). Corticosteroids and their role in the extrarenal electrolyte secreting organs of non-mammalian vertebrates. *In* "Steroids and their Mechanism of Action in Non-Mammalian Vertebrates" (G. Delrio, J. Brachet, eds.), 33–49. Raven Press, New York.

Sauvages, Ch. (1963). Étages bioclimatiques. *In* "Atlas du Maroc", *Com. Nat. Géog. Maroc*, Rabat, pp. 44 (annexes).

Savage, J. M. (1960). Evolution of a peninsular herpetofauna. *Syst. Zool.* **9**, 184–212.

Savage, J. M. (1966). The origins and history of the Central American Herpetofauna. *Copeia* 1966, 719–766.

Sawyer, W. H. and Pang, P. K. T. (1975). Endocrine adaptation to osmotic requirements of the environment. Endocrine factors in osmoregulation by lung fishes and amphibians. *Gen. Comp. Endocrinol.* **25**, 224–229.

Sawyer, W. H. and Pang, P. K. T. (1979). Responses by vertebrates to neurohypophyseal principles. *In* "Hormones and Evolution" (E. J. Barrington, ed.), Vol. 2, 493–523. Academic Press, London.

Sawyer, W. H. and Sawyer, M. K. (1952). Adaptive responses to neurohypophysial fractions in vertebrates. *Physiol. Zool.* **25**, 84–98.

Sawyer, W. H., Munsick, R. A. and Van Dyke, H. B. (1959). Pharmacological evidence for the presence of arginine vasotocin and oxytocin in neurohypophyseal extracts from cold-blooded vertebrates. *Nature* **184**, 1464–1465.

Schaefer, W. H. (1933). Hypophysectomy and thyroidectomy of snakes. *Proc. Soc. Exp. Biol. Med.* **30**, 1363–1365.

Schall, J. J. and Pianka, E. R. (1978). Geographical trends in numbers of species. *Science* **201**, 679–686.

Schmidt-Nielsen, B. (1975). Comparative physiology of cellular ion and volume regulation. *J. Exp. Zool.* **194**, 207–220.

Schmidt-Nielsen, B. and Skadhauge, E. (1967). Function of the excretory system of the crocodile (*Crocodylus acutus*). *Amer. J. Physiol.* **212**, 973–980.

Schmidt-Nielsen, K. and Dawson, W. R. (1964). Terrestrial animals in dry heat: desert reptiles. *In* "Handbook of Physiology", Section 4, Adaptation to the Environment (D. B. Dill, ed.), 467–480. American Physiological Society, Bethesda, Maryland.

Schmidt-Nielsen, K. and Fänge, R. (1958). Salt glands in marine reptiles. *Nature* **182**, 783–785.

Schmidt-Nielsen, K., Borut, A., Lee, P. and Crawford, E. (1963). Nasal salt excretion and the possible function of the cloaca in water conservation. *Science* **142**, 1300–1301.

Schoener, T. W. and Schoener, A. (1978). Estimating and interpreting body-size growth in some *Anolis* lizards. *Copeia* 1978, 390–405.

Shield, J. W. (1972). Acclimation and energy metabolism of the dingo, *Canis dingo*, and the coyote, *Canis latrans*. *J. Zool.* **168**, 483–501.

Shine, R. (1980). "Costs" of reproduction in reptiles. *Oecologia* **46**, 92–100.

Shine, R. (1983a). Reptilian viviparity in cold climates: testing the assumptions of an evolutionary hypothesis. *Oecologia* **57**, 397–405.

Shine, R. (1983b). Reptilian reproductive modes: the oviparity-viviparity continuum. *Herpetologica* **39**, 1–8.

Shine, R. (1985). The evolution of viviparity in reptiles: an ecological analysis. *In* "Biology of the Reptilia" (C. Gans, F. Billett, eds), Vol. 15 (Development B), 606–684. John Wiley & Sons, New York.

Shine, R. and Berry, J. F. (1978). Climatic correlates of live-bearing in squamate reptiles. *Oecologia* **33**, 261–268.

Shine, R. and Bull, J. J. (1979). The evolution of live-bearing in lizards and snakes. *Amer. Nat.* **113**, 905–923.

Shoemaker, V. H. and Nagy, K. A. (1977). Osmoregulation in amphibians and reptiles. *Ann. Rev. Physiol.* **39**, 449–471.

Shoemaker, V. H., Licht, P. and Dawson, W. R. (1967). Thermal dependence of water and electrolyte excretion in two species of lizards. *Comp. Biochem. Physiol.* **23**, 255–262.

Shoemaker, V. H., Nagy, K. A. and Bradshaw, S. D. (1972). Studies on the control of electrolyte excretion by the nasal gland of the lizard *Dipsosaurus dorsalis*. *Comp. Biochem. Physiol.* **42A**, 749–757.

Skadhauge, E. (1967). *In vivo* perfusion studies of the cloacal water and electrolyte resorption in the fowl (*Gallus domesticus*). *Comp. Biochem. Physiol.* **23**, 483–501.

Skadhauge, E. (1969). The mechanism of salt and water absorption in the intestine of the eel (*Anguilla anguilla*) adapted to waters of various salinities. *J. Physiol.* **204**, 135–158.

Skadhauge, E. (1978). Hydromineral regulation in lower vertebrates. *In* "Comparative Endocrinology" (P. J. Gaillard, H. H. Boer, eds), 197–207. Elsevier/North Holland Biomedical Press, Amsterdam.

Skadhauge, E. and Duvdevani, I. (1977). Cloacal absorption of NaCl and water in the lizard *Agama stellio*. *Comp. Biochem. Physiol.* **56A**, 275–279.

Slobodkin, K. B. and Rapoport, A. (1974). An optimal strategy of evolution. *Quart. Rev. Biol.* **49**, 181–200.

Smith, D. C. (1977). Interspecific competition and the demography of two lizards. Ph.D. thesis, University of Michigan, Ann Arbor.

Smith, E. N. (1975). Oxygen consumption, ventilation, and oxygen pulse of the American alligator during heating and cooling. *Physiol. Zool.* **48**, 326–337.

Smith, E. N. (1979). Behavioral and physiological thermoregulation of crocodilians. *Amer. Zool.* **19**, 239–247.

Smith, H. T. U. (1968). Geologic and geomorphic aspects of deserts. *In* "Desert Biology" (G. W. Brown, ed.), Vol. 1, 51–100. Academic Press, New York.

Smits, A. W. (1985). Accessory lymph sacs and body fluid partitioning in the lizard, *Sauromalus hispidus*. *J. Exp. Biol.* **121**, 165–177.

Smyth, M. and Smith, M. J. (1968). Obligatory sperm storage in the skink *Hemiergis peronii*. *Science* **161**, 575–576.

Sokolov, V. E., Sukhov, V. P. and Chernyshov, Y. M. (1975). A radiotelemetric study of diurnal fluctuations of body temperature in the desert monitor (*Varanus griseus*). *Zool. Zh.* **54**, 1347–1356.

Specht, R. L. (1981). Ecophysiological principles determining the biogeography of major vegetation formations in Australia. *In* "Ecological Biogeography of Australia" (A. Keast, ed.), Vol. 1, 310–333. Junk, den Haage.

Spellerberg, I. F. (1972). Temperature tolerances of southeast Australian reptiles examined in relation to reptile thermoregulatory behaviour and distribution. *Oecologia* **9**, 23–46.

Spellerberg, I. F. (1974). Influence of photo-period and light intensity on lizard voluntary temperatures. *Brit. J. Herp.* **5**, 412–420.

Spotila, J. R., Lommen, P. W., Bakken, G. S. and Gates, D. M. (1973). A mathematical model for body temperature of large reptiles: implications for dinosaur ecology. *Amer. Nat.* **107**, 391–404.

Spray, D. C. and Belkin, D. B. (1972). Heart rate-cloacal temperature hysteresis in *Iguana* is a result of thermal lag. *Nature* **239**, 337–338.

Stamps, J. and Tanaka, S. (1981). The influence of food and water on growth rates in a tropical lizard (*Anolis aeneus*). *Ecology* **62**, 33–40.

Stearns, S. C. (1976). Life history tactics: a review of the ideas. *Quart. Rev. Biol.* **51**, 1–47.

Stearns, S. C. (1977). The evolution of life history traits: a critique of the theory and a review of the data. *Ann. Rev. Ecol. Syst.* **8**, 145–172.

Stearns, S. C. (1980). A new view of life-history evolution. *Oikos* **35**, 266–281.

Stearns, S. C. (1982). The role of development in the evolution of life histories. *In* "Evolution and Development" (J. T. Bonner, ed.), 237–258. Dahlem Konferenzen, Springer-Verlag, Berlin.

Stearns, S. C. (1984). The effects of size and phylogeny on patterns of covariation in the life history traits of lizards and snakes. *Amer. Nat.* **123**, 56–72.

Stearns, S. C. (1985). Reply to Vitt and Seigel. *Amer. Nat.* **125**: 616.

Stebbins, R. C. and Barwick, R. E. (1968). Radiotelemetric study of thermoregulation in a lace monitor. *Copeia* 1968, 541–547.

Stebbins, R. C. and Cohen, N. W. (1973). The effect of parietalectomy on the thyroid and gonads in free-living western fence lizards *Sceloporus occidentalis*. *Copeia* 1973, 662–668.

Stebbins, R. C. and Eakin, R. M. (1958). The role of the 'third eye' in reptilian behaviour. *Amer. Mus. Novit.* **1870**, 1–40.

Stebbins, R. C. and Wilhoft, D. C. (1966). Influence of the parietal eye on activity in lizards. *In* "The Galápagos" (R. I. Bowman, ed.), 258–268. Proc. Galápagos Int. Sci. Proj., University of California Press, Los Angeles.

Stern, J. T. (1970). The meaning of 'adaptation' and its relation to the phenomenon of natural selection. *In* "Evolutionary Biology" (Th. Dobzhansky, M. K. Hecht, W. L. Steere, eds), Vol. 4, 39–66. Appleton-Century-Crofts, New York.

Stewart, G. R. (1965). Thermal ecology of the garter snakes *Thamnophis sirtalis concinnus* (Hallowell) and *Thamnophis ordinoides* (Barid & Girard). *Herpetologica* **21**, 81–102.

Stolte, H., Schmidt-Nielsen, B. and Davis, L. (1977). Single nephron function in the kidney of the lizard, *Sceloporus cyanogenys*. *Zool. Jahrb. Physiol. Bd.* **81**, 219–244.

Storr, G. M. (1964). Some aspects of the geography of Australian reptiles. *Senck. Biol.* **45**, 577–589.

Storr, G. M. (1982). Revision of the Bearded Dragons (Lacertilia: Agamidae) of Western Australia with notes on the dismemberment of the genus *Amphibolurus*. *Res. West. Aust. Mus.* **10**, 199–214.

Streit, B. (1982). Water turnover rates and half times in animals studied by use of labelled and non-labelled water. *Comp. Biochem. Physiol.* **72A**, 445–454.

Taplin, L. E. and Grigg, G. C. (1981). Salt glands in the tongue of the estuarine crocodile *Crocodylus porosus*. *Science* **212**, 1045–1047.

Templeton, J. R. (1960). Respiration and water loss at higher temperatures in the desert iguana, *Dipsosaurus dorsalis*. *Physiol. Zool.* **33**, 136–145.

Templeton, J. R. (1964). Nasal salt excretion in terrestrial lizards. *Comp. Biochem. Physiol.* **11**, 223–229.

Templeton, J. R. (1966). Responses of the lizard nasal salt gland to chronic hypersalemia. *Comp. Biochem. Physiol.* **18**, 563–572.

Templeton, J. R. (1970). Reptiles. *In* "Comparative Physiology of Thermoregulation", Vol. 1: Invertebrates and Non-Mammalian Vertebrates (G. C. Whittow, ed.), 167–221. Academic Press, New York.

Templeton, J. R. (1972). Salt and water balance in desert lizards. *Symp. Zool. Soc. Lond.* **31**, 61–77.

Templeton, J. R., Murrish, D. E., Elizondo, R. S. and Le Brie, S. J. (1969). Adrenal-renal function in water snakes *Natrix cyclopion*. *Amer. J. Physiol.* **217**, 419–425.

Templeton, J. R., Murrish, D. E., Randall, E. M. and Mugaas, J. N. (1968). The effect of aldosterone and adrenalectomy on nasal salt excretion of the desert iguana, *Dipsosaurus dorsalis*. *Amer. Zool.* **8**, 818–819. (abstract).

Templeton, J. R., Murrish, D. E., Randall, E. M. and Mugaas, J. N. (1972a). Salt and water balance in the desert iguana, *Dipsosaurus dorsalis*. I. The effect of dehydration, rehydration and full hydration. *Z. Vergl. Physiologie* **76**, 245–254.

Templeton, J. R., Murrish, D. E., Randall, E. M. and Mugaas, J. N. (1972b). Salt and water balance in the desert iguana, *Dipsosaurus dorsalis*. II. The effect of aldosterone and adrenalectomy. *Z. Vergl. Physiologie* **76**, 255–269.

Terpin, K. M., Spotilia, J. R. and Folley, R. E. (1979). Thermoregulatory adaptations and heat energy budget analysis of the American alligator *Alligator mississippiensis*. *Physiol. Zool.* **52**, 296–312.

Thomas, D. H. and Skadhauge, E. (1979). Chronic aldosterone therapy and the control of transepithelial transport of ions and water by the colon and coprodaeum of the domestic fowl (*Gallus domesticus*) *in vivo*. *J. Endocrinol.* **83**, 239–250.

Thomas, D. H., Skadhauge, E. and Read, M. V. (1979). Acute effects of aldosterone on water and electrolyte transport in the colon and coprodaeum of the domestic fowl (*Gallus domesticus*) *in vivo*. *J. Endocrinol.* **83**, 229–237.

Thornthwaite, C. W. (1948). An approach towards a rational classification of climate. *Géogr. Rev.* **38**, 55–94.

Thorpe, G. S. and Kontogiannis, J. E. (1977). Evaporative water loss in isolated populations of the coastal side-blotched lizard, *Uta stansburiana hesperis*. *Comp. Biochem. Physiol.* **57**, 133–137.

Tihen, J. A. (1964). Tertiary changes in the herpetofauna of temperate North America. *Senck. Biol.* **45**, 265–279.

Tinkle, D. W. (1967). The life and demography of the side-blotched lizard *Uta stansburiana*. *Misc. Publ. Mus. Zool. Univ. Mich.* **132**, 1–182.

Tinkle, D. W. (1969). The concept of reproductive effort and its relation to the evolution of life histories of lizards. *Amer. Nat.* **103**, 501–516.

Tinkle, D. W. (1972a). The dynamics of a Utah population of *Sceloporus undulatus*. *Herpetologica* **28**, 351–359.

Tinkle, D. W. (1972b). The role of environment in the evolution of life history differences within and between lizard species. *In* "A Symposium on Ecosystematics" (R. T. Allen, F. C. James, eds), *Univ. Kansas Occas. Papers* **4**, 77–100.

Tinkle, D. W. and Ballinger, R. E. (1972). *Sceloporus undulatus*: a study of the intraspecific comparative demography of a lizard. *Ecology* **53**, 570–584.

Tinkle, D. W. and Dunham, A. E. (1983). Demography of the tree lizard, *Urosaurus ornatus*, in central Arizona. *Copeia* 1983, 585–598.

Tinkle, D. W. and Gibbons, J. W. (1977). The distribution and evolution of viviparity in reptiles. *Misc. Publ. Mus. Zool. Univ. Mich.* **154**, 1–55.

Tinkle, D. W. and Hadley, N. F. (1975). Lizard reproductive effort: caloric estimates and comments on its evolution. *Ecology* **56**, 427–434.

Tinkle, D. W., Wilbur, H. M. and Tilley, S. G. (1970). Evolutionary strategies in lizard reproduction. *Evolution* **24**, 55–74.

Tokarz, R. R. and Crews, D. (1981). Effects of prostaglandins on sexual receptivity in the female lizard *Anolis carolinensis*. *Endocrinology* **109**, 451–458.

Tokarz, R. R., Crews, D. and McEwen, B. S. (1981). Estrogen-sensitive progestin binding sites in the brain of the lizard, *Anolis carolinensis*. *Brain Res.* **220**, 95–105.

Tracy, C. R. (1982). Biophysical modeling in reptilian physiology and ecology. *In* "Biology of the Reptilia", Vol. 12 (C. Gans, F. H. Pough, eds), 275–321. Academic Press.

Turc, L. (1951). Nouvelle formule pour le calcul bilan de l'eau en fonction des valeurs moyennes annuelles des précipitations et de la température. *C.R. Acad. Sci.* (*Paris*) **230**, 633–635.

Turc, L. (1961). Évaluation des besoins en eau d'irrigation évapotranspiration potentielle. *Ann. Agron.* **12**, 13–49.

Turner, F. B. (1977). The dynamics of populations of squamates, crocodilians and rhynchocephalians. *In* "Biology of the Reptilia" (C. Gans, D. W. Tinkle, eds), Vol. 7, 157–264. Academic Press, New York.

Turner, F. B., Hoddenbach, G. A., Medica, P. A. and Lannom, J. R. (1970). The demography of the lizard, *Uta stansburiana* Baird and Girard, in southern Nevada. *J. Anim. Ecol.* **39**, 505–519.

Underwood, G. (1954). Categories of adaptation. *Evolution* **8**, 365–377.

Underwood, G. (1957). On lizards of the family Pygopodidae. A contribution to the morphology and phylogeny of the Squamata. *J. Morph.* **100**, 207–268.

Van Louvering, J. A., Berggren, W. A., Drake, R. E., Aguione, E. and Curtis, G. H. (1976). The terminal Miocene event. *Marine Micropal.* **1**, 263–286.

Vannier, G. (1983). The importance of ecophysiology for both biotic and abiotic studies of the soil. *In* "New Trends in Soil Biology", VIIIth Int. Colloq. of Soil Zoology, Aug/Sept 1982, Louvain-la-Neuve, Belgium (in press).

Veith, W. J. (1974). Reproductive biology of *Chamaeleo pumilus pumilus* with special reference to the role of the corpus luteum and progesterone. *Zoologica Africana* **9**, 161–183.

Vernet, R. (1977). Recherches sur l'écologie de *Varanus griseus* Daudin (Reptilia, Sauria, Varanidae) dans les écosystèmes sableux du Sahara nord-occidental (Algérie). Thèse de Doctorat de Troisième Cycle, Université Pierre et Marie Curie, Paris.

Vernet, R. and Grenot, C. J. (1972). Étude du milieu et structure trophique du peuplement reptilien dans le Grand Erg Occidental (Sahara algérien). *C. R. Soc. Biogéogr.* **433**, 112–123.

Vinegar, M. B. (1975). Demography of the striped plateau lizard, *Sceloporus virgatus*. *Ecology* **56**, 172–182.

Vinson, G. P., Whitehouse, B. J., Goddard, C. and Sibley, C. P. (1979). Comparative and evolutionary aspects of aldosterone secretion and zona glomerulosa function. *J. Endocrinol.* **81**, 5P–24P.

Vitt, L. J. and Seigel, R. A. (1985). Life history traits of lizards and snakes. Amer. Nat. **125**, 480–484.

Vitt, L. J., Van Lobel Sels, R. C. and Ohmart, R. D. (1978). Lizard reproduction: annual variations and environmental correlates in the iguanid lizard *Urosaurus graciosus*. *Herpetologica* **34**, 241–253.

Warburg, M. R. (1965). The influence of ambient temperature and humidity on the body temperature and water loss from two Australian lizards, *Tiliqua rugosa* (Gray) (Scincidae) and *Amphibolurus barbatus* Cuvier (Agamidae). *Aust. J. Zool.* **13**, 331–350.

Waring, H. (1963). "Color Change Mechanisms of Cold-Blooded Vertebrates". Academic Press, New York.

Weathers, W. W. (1971). Some cardiovascular aspects of temperature regulation in the lizard *Dipsosaurus dorsalis*. *Comp. Biochem. Physiol.* **40A**, 503–515.

Webb, G. J. W., and Heatwole, H. (1971). Patterns of heat distribution within the bodies of some Australian pythons. *Copeia* 1971, 209–220.

Webb, G. J. W., Johnson, C. R. and Firth, B. T. (1972). Head-body temperature differences in lizards. *Physiol. Zool.* **45**, 130–142.

Weekes, H. C. (1935). A review of placentation among reptiles with particular regard to the function and evolution of the placenta. *Proc. Zool. Soc. Lond.* 1935, 625–645.

Weil, M. R. and Aldridge, R. D. (1981). Seasonal androgenesis in the male water snake, *Nerodia sipedon*. *Gen. Comp. Endocrinol.* **44**, 44–53.

Weintraub, J. D. (1969). Size relationships of the granite spiny lizard, *Sceloporus orcutti*. *Herpetologica* **25**, 25–29.

Wells, P. V. (1977). Post-glacial origin of the present Chihuahuan desert less than 11,500 years ago. *In* "Trans. Symp. on the Biological Resources of the Chihuahuan Desert Region" (R. H. Wauer, D. H. Riskind, eds), 67–83. U.S. Dept. Int. Nat. Park Service Trans. & Proc. Series, No. 3.

Werner, Y. L. (1982). Herpetofaunal survey of the Sinai Peninsula (1967–77) with emphasis on the Saharan sand community. *In* "Herpetological Communities". A Symposium of the Society for the Study of Amphibians and Reptiles and the Herpetologists' League. *Wildl. Res. Rep.* **13**, 153–161. U.S. Fish and Wildlife Service.

Werner, Y. L. and Goldblatt, A. (1978). Body temperature in a basking gekkonid lizard, *Ptyodactylus hasselquistii* (Reptilia, Lacertilia, Gekkonidae). *J. Herpetol.* **12**, 408–411.

Werner, Y. L. and Whitaker, A. H. (1978). Observations and comments on the body temperature of some New Zealand reptiles. *N.Z. J. Zool.* **5**, 375–393.

White, F. N. (1976). Circulation. *In* "Biology of the Reptilia", Vol. 5 (C. Gans, W. R. Dawson, eds), 275–334. Academic Press, New York.

Whitfield, C. L. and Livezey, R. L. (1973). Thermoregulatory patterns in lizards. *Physiol. Zool.* **46**, 285–296.

Wilbur, H. M., Tinkle, D. W. and Collins, J. P. (1974). Environmental certainty, trophic level, and resource availability in life history evolution. *Amer. Nat.* **108**, 805–817.

Wilhoft, D. C. (1963). Gonadal histology and seasonal changes in the tropical Australian skink *Leiolopisma rhomboidalis*. *J. Morphol.* **113**, 185–204.

Wilhoft, D. C. (1966a). The metabolic response to thyroxine of lizards maintained in a thermal gradient. *Gen. Comp. Endocrinol.* **7**, 445–451.

Wilhoft, D. C. (1966b). The effect of thyroidectomy on preferred body temperature of a lizard. *Amer. Zool.* **6**, 601.

Williams, M. B. (1970). Deducing the consequences of evolution: a mathematical model. *J. Theoretical Biol.* **29**, 343–385.

Williams, O. B. (1979). Ecosystems of Australia. *In* "Arid-Land Ecosystems: Structure, Functioning and Managment" (D. W. Goodall, R. A. Perry, K. M. W. Howes, eds), Vol. 1, 145–212. IBP, Cambridge University Press, London.

Wooley, P. (1959). The effect of posterior lobe pituitary extracts on blood pressure in several vertebrate classes. *J. Exp. Biol.* **36**, 453–458.

Wright, A. and Chester Jones, I. (1957). The adrenal gland in lizards and snakes. *J. Endocrinol.* **15**, 83–99.

Yaron, Z. (1972). Endocrine aspects of gestation in viviparous reptiles. *Gen. Comp. Endocrinol.* Suppl. **3**, 663–674.

Yaron, Z. (1985). Reptilian placentation and gestation: Structure, function and endocrine control. *In* "Biology of the Reptilia" (C. Cans, F. Billett, eds.), Vol. 15 (Development B), 527-603, John Wiley & Sons, New York.

Zucker, A. (1980). Procedural and anatomical considerations of the determination of cutaneous water loss in squamate reptiles. *Copeia* 1980, 425–439.

Zweifel, R. G. and Lowe, C. H. (1966). The ecology of a population of *Xantusia vigilis*, the desert night lizard. *Amer. Mus. Novit.* **2247**, 1–57.

# Index

**E**



**F**



**G**

6 7 8 9 0 1 2 3 4 5
A B C D E F G H I J